The Management of Insects in Recreation and Tourism

Insects such as cockroaches, mosquitoes and bed-bugs are usually
not highly sought amongst travellers or recreationists, yet each year,
collectors, butterfly enthusiasts, dragonfly-hunters and apiarists
collect, visit, document and raise insects for recreational purposes.

Illustrating a range of human-insect encounters from an
interdisciplinary perspective, this book provides the first insight into
the booming industry of insect recreation.

Case studies and examples demonstrate the appeal of insects,
ranging from the captivating beauty of butterflies to the curious
fascination of tiger beetles, and challenge the notion that animals
lacking anthropomorphic features hold little or no interest for
humans.

Throughout the book, the emphasis is on the innovators, the
educators, the dedicated researchers and activists who, through
collaboration across fields ranging from entomology to sociology and
anthropology, have brought insects from the recreational fringes to
the forefront of many conservation and leisure initiatives.

RAYNALD HARVEY LEMELIN is an Associate Professor in the School
of Outdoor Recreation, Parks and Tourism at Lakehead University,
Canada. His research interests focus around human–animal
dynamics, originally in the context of polar bear viewing and more
recently in examining the human dimensions of insect conservation.

The Management of Insects in Recreation and Tourism

Edited by

RAYNALD HARVEY LEMELIN
Lakehead University, Canada

CAMBRIDGE UNIVERSITY PRESS
Cambridge, New York, Melbourne, Madrid, Cape Town,
Singapore, São Paulo, Delhi, Mexico City

Cambridge University Press
The Edinburgh Building, Cambridge CB2 8RU, UK

Published in the United States of America by Cambridge University Press, New York

www.cambridge.org
Information on this title: www.cambridge.org/9781107012882

© Cambridge University Press 2013

First published 2013

Printed and Bound in the United Kingdom by the MPG Books Group

A catalogue record for this publication is available from the British Library

Library of Congress Cataloguing in Publication data
  The management of insects in recreation and tourism / edited by Raynald
  Harvey Lemelin.
    pages   cm
  Includes bibliographical references and index.
  ISBN 978-1-107-01288-2
  1. Insects.   2. Human-animal relationships.   3. Wildlife-related recreation.
  4. Ecotourism.   I. Lemelin, Raynald Harvey, editor of compilation.
  QL463.M23   2012
  595.7–dc23   2012018837

ISBN 978-1-107-01288-2 Hardback

Two little boys sit identically entranced by the insects they have captured. Suddenly a stunning yellow butterfly comes floating by and Raymond, the sickly little one, asks his brother Camile if he could capture the butterfly without hurting it. After a somewhat hectic chase, Camile finally captures the butterfly and brings the creature back to his brother. 'La jouissance de son frère en voyant le papillion/his brother's jubilation upon seeing the butterfly makes the effort all worth it,' thinks Camile. After gazing at it for several minutes, Raymond releases the creature back to its airy domain. Perhaps one day, he muses, 'je serais libre comme ce papillion/I will be free, like this butterfly.'

The two little boys in this story are my uncles. Although identical twins, Raymond was the smaller sicklier of the two. This disability was offset by Camile's strength and dedication to his brother. Ironically, years later, it was Raymond who would protect not only his brother, but his entire family and his nation by enlisting in World War II. Landing on the beaches of Junot, Normandy, France, in 1944, my uncle helped to free the land of our ancestors. Months later he was wounded in Belgium and sent home.

The story of little boys and little girls pursuing insects was a common one to rural French-Canada. Yet only two generations later, it would seem that this jouissance for nature has been lost. Disconnected from our roots and our land, we have become indifferent or, worse, intolerant of insects and other creatures. Growing up in an urban environment, I would think nothing of using pesticides and various other deterrents to annihilate insects. Fortunately, my wife, also from a farming background, began to practice organic gardening and created a butterfly garden in our backyard. From her readings, she learnt about the importance of insects and co-existence. It was also during this time that I attended the Great Lakes Odonata Meeting in Fort

Frances, Ontario. By running around in various fields and jumping into ponds, I was brought face-to-face with the beauty of these magnificent animals. During these outings I also had the opportunity to learn about Odonata from dedicated and passionate instructors. Soon thereafter, I would have an epiphany and would refocus my research and dedicate myself to raising our awareness of insects and protecting these amazing creatures. While I would not call myself an entomophile, we did raise 14 monarch butterflies in our house; they have all been released. So perhaps, like my uncles, I can also now take the time to gaze upon the butterfly, to be amazed by the dragonfly, to be awed by the firefly.

This book is dedicated to my uncles Raymond and Camile, to my aunt Georgette, to my wife Elaine, to my son Gabriel, and to all my family and those individuals who care so much about insects and all the animals. May you always take the time to gaze and to be amazed. Ona ki:wa he (may you always be well on your journey).

Contents

Contributors

Ms Alaine Auger
School of Outdoor Recreation, Parks and Tourism
Lakehead University
Thunder Bay, ON, Canada

Ms Kathleen Brager
University of Alberta
Canada

Professor Christopher M. Buddle
Department of Natural Resource Science
McGill University
Ste-Anne-de-Bellevue, QC, Canada

Professor Jaret Daniels
Department of Entomology and Nematology
and
Assistant Director of Exhibits and Public Programs
Florida Museum of Natural History
University of Florida
Gainesville, FL, USA

Professor Thomas A. Delamere
Department of Recreation and Tourism Management
Vancouver Island University
Nanaimo, BC, Canada

Dr Adam Dodd
Department of Culture Studies and Oriental Languages
University of Oslo
Oslo, Norway

Ms Crystal M. Ernst
Department of Natural Resource Science
McGill University
Ste-Anne-de-Bellevue, QC, Canada

Professor Brian D. Farrell
Cambridge, MA, USA
Curator of Entomology
Museum of Comparative Zoology
Harvard University

Professor Adrian Franklin
School of Sociology and Social Work
University of Tasmania
Hobart, Australia

Professor Donna J. Giberson
Department of Biology
University of Prince Edward Island
Charlottetown, PE, Canada

Professor C. Michael Hall
Department of Management
University of Canterbury
Christchurch, New Zealand

Professor Yupa Hanboonsong
Faculty of Agriculture
Khon Kaen University
Khon Kaen, Thailand

Professor Glen T. Hvenegaard
University of Alberta, Augustana Campus
Camrose, AB, Canada

Ms Kelsey Johansen
School of Outdoor Recreation, Parks and Tourism
Lakehead University
Thunder Bay, ON, Canada

Professor Akito Y. Kawahara
Florida Museum of Natural History
University of Florida
Gainesville, FL, USA

Dr Raynald Harvey Lemelin
School of Outdoor Recreation, Parks and Touirsm
Lakehead University
Thunder Bay, ON, Canada

Professor Jeff Lockwood
Department of Philosophy
University of Wyoming
Laramie, WY, USA

Professor Forrest L. Mitchell
Texas AgriLife Research
Stephenville, TX, USA
and
Department of Entomology
Texas A&M University
College Station, TX, USA

Professor Tim R. New
Department of Zoology
La Trobe University
Bundoora, Australia

Dr David L. Pearson
School of Life Sciences
Arizona State University
Tempe, AZ, USA

Dr Robert M. Pyle
The Xerces Society for Invertebrate Conservation
Portland, OR, USA

Dr Jessica J. Rykken
Museum of Comparative Zoology
Harvard University
Cambridge, MA, USA

Professor Michael J. Samways
Department of Conservation Ecology and Entomology
Stellenbosch University
Stellenbosch, South Africa

Mr Matt Shardlow
Chief Executive
Buglife: The Invertebrate Conservation Trust
Peterborough, UK

Mr Edward M. Spevak
Curator of Invertebrates
St Louis Zoo
St Louis, MO, USA

Professor Arnold van Huis
Laboratory of Entomology
Wageningen University
Wageningen, The Netherlands

Mr Ko Veltman
Natura Artis Magistra (Artis Royal Zoo)
Amsterdam, The Netherlands

Ms Kristen M. Vinke
Department of Biology
University of Prince Edward Island
Charlottetown, PE, Canada

Dr Alan L. Yen
Department of Primary Industries
Victoria & La Trobe University
Victoria, Australia

1

Introduction

RAYNALD HARVEY LEMELIN

The recreational dimensions of insects have been described in entomological texts, edited volumes (see *Grzimek's Animal Life Encyclopedia: Insects* by Hutchins (2003)), cultural entomology (Chevancy *et al.* 2004; Hogue 1987), and various books such as Berenbaum's (1995) *Bugs in the System*, Kritsky and Cherry's (2000) *Insect Mythology* and Mitchell and Lasswell's (2005) *A Dazzle of Dragonflies*. More recently, social scientists like Preston (2006) and Sleigh (2003) have examined human–insect encounters by focusing on specific insect species, while Brown's (2006) *Insect Poetics*, Motte-Florace and Thomas' (2003) cultural study of insects, Parikka's (2010) examination of technology and animals, Raffles' (2010) anthropological study, and Rodger's (2008) examination of social theory and social insects have examined human–insect interactions from various disciplinary perspectives. This book is an attempt to build on these works by incorporating dimensions of leisure and tourism in human–insect interactions from an interdisciplinary perspective, as this topic is sorely lacking in leisure and tourism research (Fennell 2012).

While the link between insects and leisure may at first glance appear to be dubious, these links, as Klein (2007) explains, are deeply embedded within the socio-economic fabric of human history, for we have domesticated insects, exploited their products (e.g., silk, beeswax, honey, cochineal, lacquer), and deliberately introduced them in biocontrol measures for agriculture. These contributions were estimated at US $57 billion annually (Losey and Vaughan 2006). These figures do not, however, include the contributions of insects to forensic

The Management of Insects in Recreation and Tourism, ed. Raynald Harvey Lemelin. Published by Cambridge University Press. © Cambridge University Press 2013.

entomology (the use of insects in crime scene investigations), entomophagy and leisure and tourism. In other instances, we have polluted or radically altered their habitats in attempts to attract, repel or extinguish certain species of insects (Russell 2001). Some insects, in return 'use our homes for shelter, parasitize our bodies, spread pathogens and allergens, and feed on our resources and excrement and remains' (Klein 2007: 1).

From socio-cultural perspectives, insects have long inspired writers and poets (Tennyson 2004), musicians (see Nikolai Rimsky-Korsakov's *Flight of the Bumblebee*) and artists (van Gogh and Picasso) (Laufer 2009). In contemporary times, artists and craftsmen still use insects as fabric motifs, models in painting, sculpture, jewellery, furniture, household items, toys and tattoos (Samways 2005). In the sciences, dragonflies have inspired 'ornithopters' (Thomas *et al.* 2004), while termite mounds have inspired the exoskeletal skyscraper designed for the city of Cheonga in South Korea (Ball 2010).

One of the oldest documented recreational activities associated with insects in Europe is the collection of specimens. Dürer's watercolour of the stag beetle (1505) is regarded as the first formal portrayal of the insect as specimen, setting the convention for many subsequent portrayals (Neri 2011). Early evidence of the promotion of insect collection as a valuable leisure pursuit was noted in 1766 when Drury, an insect enthusiast, wrote, 'if you have curiosity enough to employ an hour in this amusement [capturing and examining insects], permit me to say you will have a scene of wonders opened to you in the insect world, you will have such a number of objects of speculation present themselves, that will amaze you' (Drury 1837: viii). Six decades later, documenting the first insect craze in Europe in 1836, John James Audubon noted that 'the world is all agog… for *Bugs* the size of *Water Melons*' (Clark 2009: 8, original emphasis).

Modern art exhibits, movies, festivals and books all dedicated to insects are testaments to this long, diversified and ongoing relationship between humans and insects (Laufer 2009). In the United States alone, US $46 billion annually is spent on hunting, fishing, and wildlife viewing (US Fish and Wildlife Service 2006). A substantial proportion of this spending goes directly into observing insects which have been facilitated in part through landscape transformation (pollinator gardens, dragonfly ponds) and by easier access to information through the web and various technological innovations.

Figure 1.1   Katydid. Photo by Elaine Wiersma.

THE ROLE OF TECHNOLOGY AND ASSOCIATIONS
IN HUMAN–INSECT ENCOUNTERS

As Dodd (this volume) explains, the invention of the microscope in Victorian England brought about a scalar revolution in which the world of insects and insects themselves were now visible and subject to human curiosity and investigation. More recent advances in technology have stimulated on-line chat groups (e.g., Odonata Central; The North American Butterfly Association), on-line verification of specimens (e.g., Digital Dragonflies). Personal Digital Assistants (PDAs) and Global Positioning System (GPS) have further facilitated experiential interactions with insects, and even in some cases, encouraged non-researchers to participate in citizen science (Johansen and Auger, this volume). Digital images suggest Mitchell, Rykken and Farrell (this volume) have helped to capture and depict the beauty and the deadly appeal of some insects. These digital images have in some instances replaced specimen harvest and could result in a visual renaissance in insect interests (Cheesman and Key 2007).

Further facilitating this emergence in insect interest, Pearson (this volume) argues, is the availability of books, field guides and various associations such as the Amateur Entomologists' Society (AES) in the UK, and the Young Entomologists' Society in the United States. The Xerces Society, the largest insect organisation in North America, is estimated at 6000 members while Buglife: The Invertebrate Conservation Trust in the UK has over 1192 members. While the number of organizations and participants affiliated with entomological associations has been difficult to estimate, authors Pearson and Shetterly (2006) suggest that according to the 1999 Directory of Entomological Societies (1999: 246–247):

> 514 organized associations worldwide have insects and spiders as their primary focus; of these associations, 194 are interested in general entomology. The others have more focused missions: pest control (99), honeybee business (79), forensics (1), or a single taxonomic group (Lepidoptera, 45; spiders, 38; Odonata, 15; Coleoptera, 11; and fewer for Diptera, Orthoptera, Hymenoptera, Isoptera, Heteroptera, Neuroptera and Ephemeroptera). According to their Websites and mission statements, 107 of the general entomology associations are primarily for professionals, 85 are for professionals and amateurs, and 2 are expressly for amateurs. The membership of most of the 130 associations with a single taxon as the focus is a combination of professionals and amateurs, but amateurs make up the bulk of the membership.

Many of these national and regional entomological societies also have researchers and staff, or committees promoting insect awareness through educational strategies, festivals, citizen science projects and conservation initiatives.

AIM OF THE BOOK

This book was written in order to illustrate human–insect interactions from historical to contemporary perspectives, and highlight the opportunities and contributions from the realms of leisure, tourism and experiential education. It is an attempt to focus on the innovators, the educators, the dedicated researchers and activists who, through various collaborative approaches, have brought insects from the recreational fringes to the forefront, in many cases, of conservation and leisure initiatives. Through the work of these individuals and their supporters, the importance of bees and the migratory patterns of butterflies and some dragonflies are now known; in some instances, policies and conservation strategies to protect these animals and their habitats have been enacted (see New, Samways and Spevak, this volume).

Figure 1.2   Youth interacting with insects at an insectarium. Photo by Elaine C. Wiersma.

The book presents the first synthesis of insect–human interactions in various and international settings, and highlights ways in which leisure, education and tourism can provide greater protection of insects. As Lewis *et al.* (2007: 432–433) suggest,

> public perceptions of insects are a key facet of conservation – much of the value of butterflies as flagships stems simply from their intrinsic appeal for many people. Conversely, many people admit to not liking moths, cockroaches and other insects, and any moves to redress this perception imbalance through education (particularly of young people) are important.

By illustrating a myriad of human–insect encounters from interdisciplinary perspectives, the book is also meant to challenge the notions that animals lacking anthropomorphic features have little to no appeal for humans. Two central questions aim to be answered in this volume: (1) what are the biggest challenges to human-insect interactions? (2) what techniques, strategies, and/or activities have insect proponents (researchers, amateur entomologists, managers) used to encourage human–insects encounters in leisure and tourism settings?

A critical overview of the social constructs of entomophobia (the fear of insects) and anthropomorphism (the human predisposition to favour human-like animals); taxonomic bias and arthropod discourse disorder (the intentional or unintentional exclusion of arthropods and insects in human–animal studies (Leather 2009); the extinction of experience or nature deficit disorder (the loss of ecological knowledge through urbanization) (Louv 2008) is provided, along with a discussion of how these concepts actually deter our understandings of insects. Also examined are approaches and situations that hold the best prospects for advancing our understanding of insects in leisure and recreational settings.

COMING TO TERMS WITH A FEW TERMS: DEFINING
THE CONTEXT

Insects are members of the animal kingdom; in 'Latin, animals are referred to as (*animans)* 'animate beings', because they are 'animated' (animare) by life and moved by spirit' (Fennell 2012: 12). An animal is a 'living organism which feeds on organic matter, typically having specialized sense organs and a nervous system and are able to respond rapidly to stimuli' (Fennell 2012: 12). Insects or *insectare* which literally means cut into pieces, are members of the arthropod phylum, and usually have an exoskeleton, a three-part segmented body (head, thorax and abdomen), three pairs of jointed legs, compound eyes and two antennae. Insects are among the most diverse groups of animals, and can be found in nearly all environments (Morris 1998).

A bug is loosely defined as a creeping, crawling invertebrate with many legs, which can in some instances disturb or annoy people (Thone 1949). Bugs, from an entomological perspective, refer to insects in the order Hemiptera. Derived from the Greek, hemiptera means half-winged, and is descriptive of typical true bugs like stink-bugs, assassin-bugs and bed-bugs, which have shorter fore-wings than hind-wings, and specialized mouthparts which have evolved into a 'piercing sucking beak, which can be sunk' (Thone 1949: 334) into food.

The 'pest' label is often used to incorporate invasive creatures (cockroaches, flies), biting creatures (mosquitoes, horse flies) or parasites (lice, ticks, bed-bugs). An animal can be considered a pest when it causes damage to agricultural, silvicultural and structural processes, parasitizes livestock or becomes a vector of human disease. The irony is that insect pests represent a minute portion of all insects, yet it is this tiny minority which is responsible for spreading numerous patho-gens and causing untold damage to food crops and property (Fichter *et al.* 1987; Gurr *et al.*, 2012). In a study of public perceptions of pests and subsequent use of pesticide use in American homes, Baldwin *et al.* (2008) suggest that the role of the pesticide industry is quite pervasive and influential. The power of this industry in fuelling various fears and phobias associated with insects in leisure and tourism cannot be underestimated (Russell 2001). While all insects are animals, not all insects in the true entomological sense are bugs, nor are all insects pests. In this volume, the discussion, apart from a few examples of spiders, is largely relegated to insects and a few bugs, but no pests.

Many researchers within the realms of social science use a process called researcher disclosure. This process is used to situate objectivity and contextualize the research approach. I am not an entomologist, I have no training in biology nor do I have any ecological background. My training is in the social sciences, more specifically sociology, anthro-pology and the human dimensions of wildlife management. Like my uncles in the prologue, I also, in my youth, chased butterflies and accumulated grasshoppers and fireflies, but somewhere something changed, and I began to fear and dislike insects, and often indiscrim-inately killed them for no particular reason. This abhorrence of insects continued unabated until I was conducting my master's research and a Mohawk Elder, in his taciturn way, pointed out how irrational my behaviour was. Henceforth, my interactions with these animals were tempered somewhat (i.e., usually through avoidance). Approximately 6 years ago, my wife and I participated in the Great Lakes Odonata Meeting (GLOM) which was held in Fort Francis, Northern Ontario, Canada. Armed with a net, identification guides and GPS, I chased one of the most elusive creatures I have ever met. When I captured my first emerald and I gazed into those multi-faceted eyes, I was mesmerized, and enthralled. In many ways, I had an epiphany that would trans-form my own research and activism, and persuade me to research and educate others on the wonders of these animals. Part of this epiph-any was facilitated by the fact that the entomological realm was very supportive. Coming from the world of polar bears, where research is politically charged, territorial and competitive, I found the passion, collaboration and dedication of the volunteers, professional amateurs and researchers at the GLOM, a refreshing change of pace.

Since then I have been learning about insects and researching human–insect encounters in various settings. In addition to visiting eight insectariums and butterfly pavilions throughout North America and Western Europe, and participating in five insect festivals (two of which I have organized), I have published three peer-reviewed publications, two conference proceedings, one encyclopaedia entry and a book chapter on human–insect interactions (Lemelin, 2007, 2009, 2012). Through a mixed-method approach consisting of participant observations and 20 semi-structured interviews, I describe the findings acquired from attending two dragonfly symposiums held in Northern Canada (2007, 2009). Consisting of 47 structured internet surveys, two structured telephone interviews, and four structured face-to-face interviews with managers, researchers and professional amateurs, my (2011) study examined current insect management and conservation strategies from an international perspective. The 2011 study revealed that few respondents mentioned or attributed any value to recreation and leisure as opportunities to increase our awareness or knowledge of insects. With the help of the Personal Meaning Map (PMM), I am currently conducting inductive studies on human perceptions of insects; so far over 50 PMMs with university undergraduate and graduate students have been administered.

I cannot call myself an entomophile, for I do not love and revere all insects. Some, like dragonflies, tiger beetles, wasps and mantises, fascinate me; some like ants and bees, I admire; and others, like blackflies, ticks and mosquitoes, I loathe. This doesn't mean I want to eradicate all of the latter, but I prefer to minimize my contact with them whenever possible. Living in Northern Canada, this is virtually impossible, so I have learnt to tolerate them.

A criticism of the book may be that it focuses too much on the charismatic micro-fauna of the insect world, that is, bees, beetles, butterflies and dragonflies. Considering that insects and other invertebrates are often overlooked in favour of vertebrates, a process that Leather (2009) labelled as taxonomic bias, or that arthropods are deliberately or inadvertently excluded from human–animal narratives since they are not, in the words of one social scientist, 'animals', a symptom that I named Arthropod Discourse Disorder (ADD) (Lemelin, 2012), I make no apologies for the focus on the charismatic micro-fauna of the insect world. The fact that these animals, as highlighted next, have been featured in books and documentaries, and incorporated in leisure and conservation strategies, suggest that large, conspicuous, colourful, mostly diurnal insects such as bees, beetles, butterflies and dragonflies are excellent subjects for conservation strategies, nature interpretation and public education (Cardoso *et al.* 2011; Snaddon and Turner 2007).

HUMAN–INSECT INTERACTIONS

A chalk drawing dated 8000 BP depicts a woman gathering honey (Cave of the Spider, Spain) (Klein 2007). Similar images of bee-handling were painted on rocks in Africa and advanced beekeeping was depicted in an ancient Egyptian tomb (2626 BP.) (Klein 2007). More recently, tourism activities featuring bees include beekeeping museums found throughout North America, South America, Europe, Australia, Israel and Japan, and incorporating the art of beekeeping in tourism strategies at larger apiaries (see the Galil Apiary in Israel) and at various hotels and resorts throughout Canada, the United States and Kenya (Lemelin 2012). Tourists have joined the Mowalis, the honey gatherers of Sunderbans (India, Bangladesh), and the Rai, the honey collectors of the Himalayas in Nepal on their honey hunting excursions for the Giant Rock bee (*Apis dorsata*) (Kakani 2011; Shahwahid *et al.* 2008; Spevak, this volume).

In China, the domestic silkworm (*Bombyx mori*) has been cultivated for silk production since at least 2700 BC (New, this volume), while, as the painting of 'La chasse aux papillons' by Manet depicts, viewing and collecting butterflies is a well-established leisure activity practised by numerous cultures across the world (Russell 2003). Butterflies in the words of Samways (2005) continue to be the biggest draw of insect-related leisure and tourism activities. In fact:

> Many tourists now wish to see butterflies in the wild and increased leisure and wealth in the 'affluent world' may engender tropical butterfly-watching safaris in addition to the more widespread bird-watching and 'wildlife' vacations which have proliferated in recent years (New 1997: 207).

Butterfly tourism opportunities offered in Costa Rica and in Taiwan attract nearly 500 000 butterfly tourists per year (Samways 2005), while the Sierra Madre Biosphere Reserve in Mexico, home of the monarch butterfly (*Danaus plexippus*) aggregations, is visited by 250 000 people per season (Barkin 2000). A more specific example of a micro-fauna being used as flagship species for tourism strategies includes the rare and threatened Karkloof blue (*Orachrysops ariadne*) found in southern Africa. A reserve, complete with a Karkloof blue logo, was created especially for this particular flagship species (Samways 2005). Elsewhere, butterfly festivals such as the 9th Annual Texas Butterfly Festival of Mission, Texas, and the El Cieolo Butterfly Festival in Cd. Mante, Tamaulipas, Mexico, attest to the popularity of these animals in recreation and tourism activities (Lemelin 2009). As stated earlier, there are over 250 insectariums and butterfly pavilions and 20 butterfly ranches worldwide. The butterfly exhibit at the Oakland Mall, Michigan, the National Butterfly Center in Mission, Texas, the butterfly garden at the Changi Airport in Singapore, the Penang Butterfly Farm in Malaysia (the world's 'first tropical live butterfly and insect sanctuary'), the Butterfly Park and Insect Kingdom Museum in Sentosa, Singapore, the Museum of World Insects and Natural Wonders in Chiang Mai, Thailand, and the insect exhibition featuring a butterfly garden and farm and insectarium in Phuket, Thailand, are all multi-purpose facilities showcasing butterflies and other insects (Lemelin 2012).

One of the most remarkable initiatives involving butterflies is the butterfly gardens of Batticaloa, Sri Lanka. The programme, which began in 1994, offers children affected by war the opportunity to interact and learn about butterflies through play and art activities (music, arts and crafts) (Chase 2000). Similar therapeutic approaches involving butterfly gardens have also been used for people living with Alzheimer's disease (Tyson 2002).

Recently, in order to meet the growth in popularity of collecting, butterfly exhibits and even the release of live butterflies at weddings, several butterfly farms and insect ranches have been established, catering to these demands (Laufer 2009). As Veltman (this volume) explains, butterfly farming is now practised throughout the world since it requires little investment, uses locally available tools and materials and the skills to raise butterflies can be learned relatively quickly (Gordon and Ayiemba 2003). A review of the International Association of Butterfly Exhibitors and Suppliers revealed that 250 insect and butterfly exhibits (more than 60 of these are located in North America alone), 20 butterfly ranches and 107 festivals and special events celebrating insects and butterflies, are located throughout the world (Hvenegaard *et al.* and Veltman, this volume). According to Parsons (1992) and Slone *et al.* (1997), the worldwide retail sales of butterflies may be as high as US $100 million per annum. The butterfly industry is also estimated to employ 20 000 people (Parsons 1992).

Each year thousands of visitors travel to rural areas in North America, Japan and Malaysia to view fireflies, while glow-worm cave aggregations in Australia and New Zealand attract 63 000 visitors (Hall, this volume). The glow-worm visitation figures, argues Hall, are comparable to whale-watching operations in Hervey Bay, Australia. In Japan, South Africa and Taiwan, visitors can explore the Nakamura Dragonfly Reserve, the dragonfly awareness trails at the National

Figure 1.3   Butterflies in a butterfly pavilion. Photo by Elaine C. Wiersma.

Botanical Gardens in Pietermaritzburg (Suh and Samways 2001) and the Taomi Ecological Village (Lemelin 2007).

Humans have, and continue to pursue, collect, identify and photograph insects in various recreational settings. In some cases, insects are brought into our homes as specimen collections, as pets (i.e., stick insects, praying mantis) and for entertainment (i.e., crickets in Japan) (Raffles 2010; Suga 2006). As of 2010, more than 20 million formicaries (ant colonies usually enclosed in a transparent box made of glass or plastic), more commonly known as Ant Farms, have been sold around the world (Lemelin 2012). Throughout Asia, certain species like stag beetles, rhinoceros beetles and goliath beetles have been pitted in combat against each other for centuries. Today these competitive bouts remain quite popular in China, Laos, Myanmar and Thailand (Rennesson *et al.* 2011). The market for the stag beetle alone has been estimated at US $100 million (Goka *et al.* 2004). However, it is not only insect enthusiasts that seek these encounters, for five of the world's largest insect and butterfly exhibits (The Magic Wings Butterfly Conservatory and Gardens, Butterfly World, the Butterfly House at the San Diego Wild Animal Park, the Insectarium of New Orleans in the United States and the Insectarium of Montreal) alone

Figure 1.4   Damselfly photographed at a dragonfly symposium. Photo by Elaine C. Wiersma.

hosted 3.3 million visitors in total and generated millions in revenues (Lemelin 2012).

CHALLENGING ENTOMOPHOBIA AND ANTHROPOMORPHISM

The word science is derived from the Latin word *scientia*, meaning to know and to learn. The goals of science are to acquire knowledge about the world, to empower through knowledge and to dissipate the thousands of prejudices and dangerous superstitions which mislead innocent people and harm innocent creatures (Haraway 1991; Harding 2008; Lakatos 1970).

Despite this quest for knowledge there is a general assumption in human–animal studies that preferred animals are those perceived to be most human-like (Arluke and Sanders 1996; Kellert 1993; Lorenz 1971). In fact, according to Morris and Morris (1965, 1966) and Stokes (2007), physical traits (large size, posture similarity to human shape) and aesthetics (surface texture and colour) influence human preference for certain animals such as pandas and lions over ants and most other invertebrates. In situations where we do incorporate insects into our popular culture (A Bug's Life, Fern Gully, Bee Movie), we anthropomorphize these creatures into human-like protagonists, thereby reinforcing the notion that we can only admire those creatures most like us (Stokes 2007). Critics note that much of this 'alien-ness', this fear of the other, is often promoted by a very powerful pesticide industry (Russell 2001; Van Hook 1997), and through inaccurate depictions of insects in media and urban myths (Bixler *et al.* 1994; Van Velsor and Nilon 2006). In other cases, much of our understanding of human–insect encounters is derived from self-reported behaviours of fear and disgust of insects (Plous 1993; Woods 2000). From this perspective, it comes as no surprise so many researchers have accepted the argument that it is this lack of anthropomorphic quality that provokes dislike, disgust and intense phobias towards insects (entomophobia) (Arrindell 2000; Hardy 1998; Hillman 1988).

Lemelin (2009, 2012), Lorimer (2007), Raffles (2010) and most of the contributors in this edited volume suggest that the fear of insects is not universal, nor does it always result in disgust and the fright response. Indeed, some of these encounters are met with indifference and apathy, others with curiosity, while still others can actually result in awe, jouissance and epiphanies. The appeal of the negative sublime (the attraction of 'creepy-crawlers'; the 'yuck' factor, see Lockwood, this volume) has been promoted successfully in various educational strategies for children in parks and protected areas (Rykken and Farrell, this volume), citizen science projects (Johansen and Auger, this volume) and festivals (Hvenegaard *et al.*, this volume), casting further doubt on the notion that insects have no appeal because they possess few if any anthropomorphic features.

BOOK OVERVIEW

A multi-authored, interdisciplinary volume, this book features some of the leading experts and upcoming researchers in the fields of entomology (Chris Buddle, Jaret Daniels, Brian Farrell, Donna Giberson, Yupa Hanboonsong, Akito Kawahara, Forrest Mitchell, Tim New, David Pearson, Jessica Rykken, Michael Samways, Arnold Van Huis, Ernst Vinke, Alan Yen); cultural history (Adam Dodd); leisure and tourism (Alaine Auger, Kathleen Brager, Tom Delamere, Michael Hall, Glen Hvenegard, Kelsey Johansen); management (Matt Shardlow); museum and zoo curators (Ed Spevak and Ko Veltman); philosophy (Jeff Lockwood); and sociology (Adrian Franklin). The book also discusses the role of experiential education and interpretation in increasing our understanding of human–insect interactions. In order to present each of these ideas, the book is divided into four sections.

Part I: Human–insect encounters provides a socio-historical overview describing the philosophical and social dimensions of human–insect interactions. Adam Dodd examines the role of scalar anthropocentrism in our understanding and subsequent interactions with insects and describes the importance of the microscope in entomological research and insect collections. Never one to shy from challenging existing assumptions or expanding philosophical boundaries of entomology, Jeff Lockwood's chapter examines the appeal of the negative sublime and suggests creatures that negatively impact people (e.g., biting insects, perceived ugliness or pests that consume crops) can generate emotions translatable into visitor interest. This appeal of the negative sublime is explored in much greater detail in Parts II, III and IV of the book. The next four chapters provide an overview in alphabetical order of four charismatic micro-fauna: Coleoptera (beetles) by Pearson; Hymenoptera (bees) by Spevak; Lepidoptera (butterflies and moths) by New; and Odonata (dragonflies and damselflies) by Samways. Pearson, in particular, also discusses the valuable contributions made by professional amateurs interested in insects.

Part II: Insects and leisure focuses on the long history of insects in recreational activities. Franklin's chapter on the contributions of fly-fishing to aquatic entomology highlights how this knowledge was, and remains, essential to this leisure activity. Today, the abilities of fly-fishers in the UK to identify and monitor 'riverflies' (caddisflies, mayflies and stoneflies) have been incorporated into modern-day citizen science projects (Cheesman and Key 2007). Nakamura and Pyle describe how collecting and general interaction with insects have forged various relationships with humans in diverse socio-cultural settings. More importantly they also provide possible solutions to overcoming the challenge of urbanization and changing recreational patterns through leisure experiential education and citizen science. Specific landscape modifications such as butterfly gardens aimed at attracting certain types of arthropods as Daniels' chapter explains, are increasing in popularity and are crucial to introducing a largely urban society to insects in an aesthetically pleasing environment. Since eating is one of the world's most popular leisure activities, Yen, Hanboonsong and Van Huis provide a discussion on entomophagy. Edible insects, they suggest, are becoming part of the culinary experience for tourists in some countries. This can involve snacking on insects in busy tourist locations or eating them as part of an ecotourism holiday.

Part III: Insects and tourism provides an overview of the attraction and appeal of insects in tourism. Authored by one of the pioneers of butterfly conservatories, the chapter by Ko Veltman provides an historical evolution of insect pavilions and the interdependence of these attractions with butterfly ranches in Papua New Guinea and Kenya. Butterfly viewing, when done properly, suggests Veltman, creates awareness and generates social and environmental justice for local inhabitants. Hvenegaard, Delamere, Lemelin, Brager and Auger discuss how hundreds of insect festivals and fairs celebrating various insects generate economic and educational capital. Hall's chapter on glow-worms in Australia illustrates the importance of insects, not only as an economic generator, but also as a vehicle of conservation and protection.

Part IV: Conservation frontiers highlights how many entomologists, professional amateurs and naturalists have been essential in creating protection (sanctuaries), conservation (through legislation) and educational strategies. Mitchells begins this section by discussing how new forms of technology are increasing our understanding of insects. Johansen and Auger discuss the contributions of citizen science to science and experiential education. Shardlow discusses the increasing awareness of the importance of certain insects and the role that insect associations such as Buglife can play in this increasing recognition. Projects like the Northern Biodiversity Program in Canada, like the Boston Harbor Islands All Taxa Biodiversity Inventory (ATBI) suggest Ernst, Vinke, Giberson, and Buddle and Rykken and Farrell, respectively, have produced collaborative educational efforts between university researchers, teachers and school children. The goal of these educational programmes is to provide participants with entomological learning opportunities, while increasing their understanding of insects from biological and cultural perspectives. The conclusion draws together key themes and lessons synthesized from the contributions in this volume and outlines various insect awareness strategies that are currently being implemented to promote insect awareness.

In many ways this book is a tribute to all those insect enthusiasts and innovators who have brought insects from the recreational fringes to the limelight of leisure and tourism; it is also a reminder and a challenge to the nay-sayers who are too blinded by their anthropocentrism to realize that insects are indeed an attraction, and have been so for millennia.

REFERENCES

Arluke, A. and Sanders, C. R. (1996) *Regarding Animals*. Philadelphia, PA: Temple University Press.

Arrindell, W. A. (2000) Phobic dimensions: IV. The structure of animal fears. *Behavior Research & Therapy*, **38**, 509–530.

Baldwin, R. W., Koehler, P. G., Pereira, R. M. and Oi, F. M. (2008) Public perceptions of pest problems. *American Entomologist*, **54**(2), 73–79.

Ball, P. (2010) Bright lights, bug city. *New Scientist*, **205**(35), 2748.

Barkin, D. (2000) The economic impacts of ecotourism: conflicts and solutions in Highland Mexico. In *Tourism and Development in Mountain Areas*. London: CABI, pp. 157–172.

Berenbaum, M. R. (1995) *Bugs In The System: Insects and Their Impacts On Human Affairs*. Cambridge, MA: Helix Books.

Bixler, D. B., Carlisle, C. L., Hammitt, W. E. and Floyd, M. F. (1994) Observed fears and discomforts among urban students on field trips to wildland areas. *The Journal of Environmental Education*, **26**(1), 24–33.

Brown, E. C. (ed). (2006) *Insect Poetics*. Minneapolis, MN: University of Minnesota Press.

Cardoso, P., Erwin, T. L., Borges, P. A. V. and New, T. R. (2011) The seven impediments in invertebrate conservation and how to overcome them. *Biological Conservation*, **144**(11), 2647–2655.

Chase, R. (2000) The butterfly garden. Batticaloa, Sri Lanka: Final report of a program development and research project (1998–2000). Ratmalana Sri Lanka: Vishva Lekha Printers.

Cheesman, O. D. and Key, R. S. (2007) The extinction of experience: a threat to insect conservation? In *Insect Conservation Biology*, ed. A. J. A. Stewart, T. R. New and O. T. Lewis. Wallingford, UK: CABI, pp. 322–350.

Chevancy, G., Goyons, J-C., Goyon, M. *et al.* (2004) *Des cultures qui racontent une histoire des insectes et des hommes : Ethnoentomologie*. Lyon: EMCC.

Clark, J. F. M. (2009) *Bugs and the Victorians*. New Haven, CO: Yale University Press.

Directory of Entomological Societies (1999) Available at: www.sciref.org/links/EntSoc/index.htm.

Drury, D. (1837) *Illustrations of Exotic Entomology*, new edition. London: Henry G. Bohn.

Evans, A. V. (2008) *What's Bugging You? A Fond Look at the Animals we Love to Hate*. Charlottesville, VA: University of Virginia Press.

Fennell, D. A. (2012) *Tourism and Animal Ethics*. New York: Routledge.

Fichter, G., Zim, S., Spencer, H. and Strekalovsky, N. (1987) *Insect Pests*. New York: Golden Press.

Goka, K., Kojima, H. and Okabe, K. (2004) Biological invasion caused by commercialization of stag beetles in Japan. *Global Environmental Research*, **8**(1), 67–74.

Gordon, I. and Ayiemba, W. (2003) Harnessing butterfly biodiversity for improving livelihoods and forest conservation: the Kipepeo Project. *The Journal of Environment Development*, **12**(1), 82–98

Gurr, G. M., Wratten, S. D. and Snyder, W. E. (eds.) (2012) *Biodiversity and Insect Pests: Key Issues for Sustainable Management*. Oxford: Wiley-Blackwell.

Haraway, D. J. (1991) *Simians, Cyborgs and Women: The Reinvention of Nature*. New York: Routledge.

Harding, S. (2008) *Sciences from Below: Feminisms, Postcolonialities, and Modernities*. Durham, NC: Duke University Press.

Hardy, T. (1998) Entomophobia: the case for Miss Muffet. *Bulletin of the Entomological Society of America*, **34**, 64–69.

Hillman, J. (1988) Going bugs. *Spring: A Journal of Archetype and Culture*, 1988, 40–72.

Hogue, C. (1987) Cultural entomology. *The Annual Review of Entomology*, **32**, 181–199

Hutchins, M. (2003) *Grzimek's Animal Life Encyclopedia: Insects*, Vol. 3, 2nd edn. Farmington Hills, MI: Gale Cengage.

Kakani, M. (2011) Preserving the biodiversity of wild bees and supporting the traditional Honey Gatherers. Available at: http://murielkakani.hubpages.com/hub/support-the-honeygatherers.

Kellert, S. R. (1993) Values and perceptions of invertebrates. *Conservation Biology*, **7**, 845–855.

Klein, B. A. (2007) Insects and humans: a relationship recorded in visual art. In *Encyclopedia of Human–Animal Relationships*, ed. M. Bekhoff. Portsmouth, NH: Greenwood Publ.

Kristky, G. and Cherry, R. (2000) *Insect Mythology*. Bloomington, IN: Writers Club Press.

Lakatos, I. (1970) History of science and its rational reconstructions. *PSA: Proceedings of the Biennial Meeting of the Philosophy of Science Association*, 1970, 91–136.

Laufer, P. (2009) *The Dangerous World of Butterflies: The Startling Subculture of Criminals, Collectors, and Conservationists*. Guilford, CT: Lyons Press.

Leather, S. R. (2009) Taxonomic chauvinism threatens entomology. *Biologist*, **56**, 10–13.

Lemelin, R. H. (2007) Finding beauty in the dragon: the role of dragonflies in recreation, tourism, and conservation. *Journal of Ecotourism*, **6**(2), 139–145.

Lemelin, R. H. (2009) Goodwill hunting? Dragon hunters, dragonflies and leisure. *Current Issues in Tourism*,**12**(3), 235–253.

Lemelin, R. H. (2011) Beetlemania: insects take leisure world by swarm. Conference Proceedings from the 13th Canadian Congress on Leisure Research. An evolving tapestry: weaving together the threads of leisure. May 18–21, 2011, Brock University, St. Catharines, ON.

Lemelin, R. H. (2012) To bee or not to bee: whether 'tis nobler to revere or to revile those six-legged creatures during one's leisure. *Leisure Studies*, doi:10 .1080/02614367.2011.626064

Lemelin, R. H. and Williams, G. (2012) Blossoms and butterflies, waterfalls and dragonflies: integrating insects in the hospitality and tourism industries through swarm supposition. In *Sustainable Hospitality and Tourism as Motors for Development: Case Studies from Developing Regions of the World*, ed. P. Sloan, C. Simons-Kaufmann and W. Legrand. New York: Routledge, pp. 198–212.

Lewis, O. T., New, T. R. and Stewart, A. J. A. (2007) Insect conservation: progress and prospects. In *Insect Conservation Biology*, ed. A. J. A. Stewart, T. R. New and O. T. Lewis. Wallingford, UK: CABI, pp. 431–436.

Lorenz, K. (1971) Part and parcel in animal and human societies: a methodological discussion. In *Studies in Animal and Human Behavior*, Vol. 2. Cambridge, MA: Harvard University Press, pp. 115–195.

Lorimer, J. (2007) Nonhuman charisma. *Environment and Planning Development : Society and Space*, **25**(5), 911–932.

Losey, J.E. and Vaughan, M. (2006) The economic value of ecological services provided by insects. *Bioscience*, **56** (4), 331–323.

Louv, R. (2008) *Last Child in the Woods: Saving Our Children From Nature-Deficit Disorder*. Chapel Hill, NC: Algonquin Paperbacks.

Mitchell, F. L. and Lasswell, J. L. (2005) *A Dazzle of Dragonflies*. College Station, TX: Texas A&M University Press.

Morris, B. (1998) *The Power of Animals*. Oxford: Berg.

Morris, D. and Morris, R. (1965) *Men and Snakes*. New York: McGraw-Hill.

Morris, D. and Morris, R. (1966) *Men and Pandas*. New York: McGraw-Hill.

Motte-Florace, E. and Thomas, J. M. C. (2003) *Les 'insectes' dans la tradition orale/ Insects in oral literature and traditions*. Paris: Peeters.

Neri, J. (2011) *The Insect and the Image: Visualizing Nature in Early Modern Europe, 1500–1700*. Minneapolis, MN: University of Minnesota Press.

New, T. R. (1997) *Butterfly Conservation*, 2nd edn. New York: Oxford University Press.

Parikka, J. (2010) *An Archaeology of Animals and Technology*. Minneapolis, MN: University of Minnesota Press.

Parsons, M. J. (1992) The butterfly farming and trading industry in the Indo-Australian region and its role in tropical forest conservation. *Tropical Lepidoptera*, **2**, 1–31.

Pearson, D. L. and Shetterly, J. A. (2006) How do published field guides influence interactions between amateurs and professionals in entomology? *American Entomologist*, **52**, 246–252.

Plous, S. (1993) *The Psychology of Judgement and Decision Making*. Philadelphia, PA: Temple University Press.

Preston, C. (2006) *Bees*. London: Reaktion Books.

Raffles, H. (2010) *Insectopedia*. New York: Pantheon Books.

Rennesson, S., Grimaud, E. and Césard, N. (2011) Jeu d'espèce. Quand deux scarabées se rencontrent sur un ring. In *Humains, non-humains: Comment repeupler les sciences socialise*, ed. S. Houdart and O. Thiery. Paris: La découverte, pp. 30–39.

Rodgers, D. M. (2008) *Debugging The Link Between Social Theory and Social Insects*. Baton Rouge, LA: Louisiana State University Press.

Russell, E. (2001) *War and Nature: Fighting Insects with Chemicals from World War I to Silent Spring*. Cambridge: Cambridge University Press.

Russell, S. A. (2003) *An Obsession with Butterflies*. Cambridge, MA: Perseus Publishing.

Samways, J. M. (2005) *Insect Diversity Conservation*. Cambridge: Cambridge University Press.

Shahwahid, H. O. M., Yong, P. L., and Sius, T. (2008) Comparative valuation of eco-tourism attractions between honey bees museum and wild honey collection. *The Malaysian Forester*, **71**(1), 21–28.

Shepardson, D. P. (2002) Bugs, butterflies and spiders: children's understanding about insects. *International Journal of Science Education*, **24**(6), 627–643.

Sleigh, C. (2003) *Ants*. London: Reaktion Books.

Slone, T. H., Orsak, L. J. and Malver, O. (1997) A comparison of price, rarity and cost of butterfly specimens: implications for the insect trade and for habitat conservation.' *Ecological Economics*, **21**(1), 77–85.

Snaddon, J. L., and Turner, E. C. (2007) A child's eye view of the insect world: perceptions of insect diversity. *Environmental Conservation*, **34**(1), 33–35.

Stokes, D. L. (2007) Things we like: human preferences among similar organisms and implications for conservation. *Human Ecology*, **35**(3), 361.

Suga, Y. (2006) Chinese cricket-fighting. *International Journal of Asian Studies*, **3**(1), 77–93.

Suh, A. N. and Samways, M. J. (2001) Development of a dragonfly awareness trail in an African botanical garden. *Biological Conservation*, **100**, 345–353.

Tennyson, A. (2004) *Tennyson: Poems*. New York: Everyman's Library.

Thomas, A. L. R., Taylor, G. K., Srygley, R. B., Nudds, R. L. and Bomphrey, R. J. (2004) Dragonfly flight: free-flight and tethered flow visualizations reveal a diverse array of unsteady lift-generating mechanisms, controlled primarily via angle of attack. *Journal of Experimental Biology*, **207**, 4299–4323.

Thone, F. (1949) Nature ramblings: what are bugs? *Society for Science and The Public*, **55**(21), 334.

Tyson, M. (2002) Treatment gardens: naturally mapped environments and independence. *Alzheimer's Care Quarterly*, **3**(1), 55–60.

US Fish and Wildlife Service (2006) 2006 National survey of fishing, hunting, and wildlife-associated recreation. Available at: http://www.census.gov/prod/2008pubs/fhw06-nat.pdf.

Van Hook, T. (1997) Insect coloration and implications for conservation. *Florida Entomologist*, **80**(2), 193–209.

Van Velsor, S. W. and Nilon, C. H. (2006) A qualitative investigation of the urban African-American and Latino adolescent experience with wildlife. *Human Dimensions of Wildlife*, **11**, 359–370.

Woods, B. (2000) Beauty and the beast: preferences for animals in Australia. *Journal of Tourism Studies*, **11**, 25–35.

Part I Human–insect encounters

2

Minding insects: scale, value, world

ADAM DODD

Most, if not all, invertebrates lack the human interest which birds and other animals high in the evolutionary scale often have. You cannot extract a meaningful expression from a beetle's eye, and a spider is always a spider, a creature as remote in all its physical attributes from the human animal as it is possible to be... No matter how beautifully formed or how colourful they may be, these lowly creatures can never be, as birds and mammals sometimes are, mirrors to ourselves. (Clark 1977: 168–169)

INTRODUCTION

Sir Kenneth Clark's rather defeatist view of insect–human relations reflects a deeply engrained species bias, especially prominent in Western cultures, that emphasizes what are routinely presented as irreconcilable differences between human beings and 'lowly creatures' such as insects. These differences are often figured in ways that maintain an asymmetric power balance: our inability to extract a meaningful expression from a beetle's eye, for example, is regarded as a shortcoming on the part of the beetle – the onus is thus placed on other species to 'rise to the occasion' and interact with human beings on our own terms. Tellingly, it is the insect, and not the human being, that is positioned as 'remote' in this relationship. Although the morphological, behavioural and cognitive differences between human beings and insects are both evident and significant, it does not logically follow that these differences must translate into a hierarchy of species. While such hierarchies undoubtedly assist in the comprehension of

The Management of Insects in Recreation and Tourism, ed. Raynald Harvey Lemelin. Published by Cambridge University Press. © Cambridge University Press 2013.

nature's otherwise overwhelming diversity, it should not be forgotten that, in doing so, they demonstrably embody the values and interests of the human societies which have produced them, rather than the properties of the natural world those societies seek to describe. As Marc Bekoff passionately argues in *Minding Animals*, 'we must seek to understand each and every individual in his or her own world and be extremely cautious of thinking of differences in terms of their being 'good' or conferring more 'value' on an individual's life' (Bekoff 2003: 13). Insects, of course, represent a particular challenge in this sense, and may even exemplify the practical limits of our ethical, mindful interaction with nonhuman species. In this chapter, I examine the ways in which perceptions of scale, attribution of value and 'world-making' intersect in the appreciation and conception of insects.

THINKING WITH INSECTS

That Western societies have historically exhibited a range of attitudes towards insect–human relations may come as a surprise to those who assume such outlooks have been uniformly consensual and fixed in place since time immemorial. To assert flatly that micro-faunae can never be seen as mirrors to ourselves, as Clark does in his analysis of animals in Western art, is to disregard an extensive history of empath-etic reflections on insect life and overlook their crucial, ongoing role in fostering meaningful insect–human relations. Ants and bees (see Spivek, this volume), for example, have often been considered as 'mirrors' of human beings and our societal organization, even if such considerations are inherently anachronistic – ants and bees exhibited forms of social organization long before *Homo sapiens* existed as a spe-cies. Moreover, in early modern Europe, bees were seemingly com-municated with by their keepers, who would whistle, clap their hands and ring bells to settle down a swarm, a practice dating back to Roman times and universally observed as recently as the eighteenth century (Thomas 1984: 96). Although the practice seems to have stemmed from a legal obligation to notify a neighbour that one's bees were to be followed onto their property, and hence to secure permission in advance, 'by early modern times the noise was widely regarded by country people as a means of addressing the bees themselves' (Thomas 1984: 96).

Whether we believe it is possible for human beings to address insects or not, the central problem with opting to emphasize or exag-gerate the difference of insects is that such a gesture obscures the fact that all animal species, no matter how dissimilar we may seem from one another, are profoundly united in our shared cellular ancestry, cohabitation of the planet, and various cooperative and competitive roles in the sustenance of the Earth's ecosystem. Far from being irreconcilable, the perceived differences between human beings and insects are largely sustained by discursive conventions which work to ensure particular rapports between human beings and other species. To the extent that such conventions enable and enact deeply ideological perspectives on the place of human beings in relation to insects, the rapports they establish may seem natural and immutable. Yet, discursive conventions can also be readily employed to encourage the appreciation of resemblances, parallels and correspondences where none were previously thought to exist, and may even allay the fundamental assumption that *difference* between species is tantamount to *distance* between species.

A major obstacle for socially, psychologically and environmentally sound insect–human relations, however, is the tension arising from insects' known ecological importance, on the one hand, and their apparent indifference to human beings and our affairs, on the other. This indifference is often misconstrued as an absence of 'animality', and hence the very term 'animal' is regularly used in a way that implicitly excludes insects. As Myers and Saunders (2002: 152) have observed, 'one of the reasons animals are so fascinating to us is that they are highly responsive and offer many dynamic opportunities for interaction. We are social creatures and animals appeal to our propensity to interact socially.' Insects, of course, are fascinating animals, too, although in general they offer very few opportunities to us for what could be called responsive, social interaction. Exchanges with various mammals, birds, fish, reptiles and amphibians are recognized as fostering positive social, psychological and environmental sensibilities in adults and children alike. Insects, however, are often and perhaps inevitably seen as creatures primarily 'to be looked at', and in a particularly modern denial of their animality, are frequently regarded as more robot than animal. While interacting with a domestic dog, for example, will in most cases be more immediately gratifying (to both parties) than interacting with an insect, the simple refusal to seek or accept engagement with insects on their own, admittedly challenging, terms can lead to an inflexibly normative framework of human–animal interactions, one which ultimately reserves a privileged role for vertebrates. This is not to deny the existence of real, visceral bonds between mammalian species, or the presence of certain evolutionary antipathies towards insects. It is merely to point out that succumbing to an allegedly insurmountable chasm between humans and insects – which usually results in a diminished status of animality for the insect – may inhibit comprehension of the fact that insects, for all their indifference, apparent lack of empathy and seemingly automated behaviours, are the most abundant and well-established animals on Earth. Indeed, considered outside the anthropocentric province of species bias, insects essentially typify what can be meant by 'animal life' in the broadest possible sense. To marginalize the insect, then, is to marginalize the animal itself and, ultimately, to alienate the human being.

If we accept, firstly, that there is considerable room for improvement in Western insect–human relations and that, secondly, these improvements must address broader problematic attitudes to the natural world in general, what might be some possible ways forward? If insects offer few opportunities for responsive, social interaction, on what basis do they then fascinate us? Although they mostly focus on vertebrates, Myers and Saunders (2002: 168) note that 'even connecting to remote, nonanthropomorphic, microscopic, ugly, inert, or other 'nonsocial' animals is possible but may require encapsulating them in narratives'. As many popular nineteenth-century natural history books about insects attest, encapsulating insects in narratives is indeed a useful method with which to establish or encourage 'connections' to insects. Bernard Lightman has shown how popularizers of science in the nineteenth century 'seemed to acknowledge that they were storytellers and that there was a narrative structure to their work' (Lightman 2000: 2). In the sections that follow, I provide an overview of where insect narratives have been historically located: in the so-called 'insect world'. The importance of scale and value in entomological 'worldmaking' and the mindful appreciation of insect life is explored, with a view to opening up the productive, though often neglected 'ways of looking' at, and 'thinking with' insects that we have at our disposal – tools which may ultimately enhance and enrich our interactions with the natural world in which we are embedded.

WAYS OF LOOKING AT INSECTS: HUMAN VISION AND THE SUBJECTIVITY OF SCALE

Armed with that sixth sense which man has achieved for himself, I can move forward, at pleasure, in any direction. It is in my power to track out, to reach, to compute the spheres, and gravitate with them in their vast orbits. But I feel much more strongly attracted towards the other abyss – that of the infinitely little. And I myself, what am I but an atom? Neither Jupiter nor Sirius, those enormous globes so great a distance from, and possessing so little sympathy with me, will teach me the secret of terrestrial life. But these, on the contrary, surround and press upon me, injure me or lend me their assistance. If they are not of my own kind, they are at all events associated with me. Ay, fatally associated. (Michelet 1883: 143–144).

In the absence of opportunities for responsive, social interaction with insects, our fascination with them is principally derived from how they appear to us when we look at them. Again, we should resist the temptation to regard this as some kind of shortcoming on the part of the insect. After all, while human beings harbour a deeply mammalian proclivity for social interaction, we are also highly visual animals. The cultivation of vision and the desire to extend it into the natural world has led to the development of many sophisticated visioning technologies designed to assist engagement with insects; these have, as Michelet's passage exemplifies, profoundly shaped the ways in which we have come to think about, not only insects, but also ourselves. Popular nineteenth-century natural history books about insects, such as Michelet's (1883) *The Insect*, are replete with extended panegyrics on the wonders of the so-called 'insect world' (an eighteenth-century concept I will come back to later). This was the historical period that gave birth to *Alice's Adventures in Wonderland*, when amateur microscopy reached its zenith, and lay folk were everywhere amazed by the elaborate complexity of the novel miniature world beneath their very feet that, even within itself, seemed to have its giants and miniatures. And although many of these texts may appear exceedingly quaint to the contemporary amateur, and scientifically defunct to the expert, there is much we can still learn from them. In particular, their often vibrant articulations of the flexible subjectivity of scale, obligatorily produced in response to the new way of looking enabled by affordable microscopes, continue to serve as humbling reminders of what might be called our own 'scalar anthropocentrism', particularly as it relates to our perception of the insects around us.

The perception of size is, to a significant degree, a phenomenon governed by the principles of geometric optics: the apparent size of an object is determined by the light it reflects, which forms a pattern on the retina of the eye relative to the object's physical distance from the eye itself. But these images, rather than being taken purely at face value, are intimately bound up with value judgments. That which appears large is 'proximate' while that which appears small is 'distant'. The eye is not, as was held by Empedocles, Plato, Euclid and, apparently, a significant portion of modern-day children, 'a kind of window through which we see out into the world' (Winer and Cottrell 2004: 97). It is, rather, a kind of screen onto which images of the world are projected. This means that many objects appear small to us, not because 'they are small', but simply by virtue of the unique physical circumstances governing the production of their image on the retina of the human eye. As N. R. Hanson (1977: 6–7) succinctly observes in *Patterns of Discovery*: 'People, not their eyes, see. Cameras, and eye-balls, are blind. Attempts to locate within the organs of sight (or within the neurological reticulum behind the eyes) some nameable called 'seeing' may be dismissed… there is more to see-ing than meets the eyeball.' This is a point worth bearing in mind when considering the perception of insects. We are not slaves to the images of scale formed on our retinas; indeed, in our increasingly media-saturated environment – and especially with the proliferation of microscopic imagery – awareness of the credulity of human vision has become a prerequisite for the coherent consumption of increas-ingly sophisticated forms of visual representation, many of which exploit the dynamism of scalar relations (Rykken and Farrell, this volume). In short, just because something *looks* small to us does not require that we *think* of it as being small in any totalizing or objective sense. Susan Stewart (1993: 55) captures this sentiment most aptly in *On Longing*, pointing out that: 'There are no miniatures in nature; the miniature is a cultural production, the product of an eye performing certain operations, manipulating and attending in certain ways to, the physical world.'

The subjectivity of scale and its role in the evaluation of insect life was first highlighted in the Western philosophic tradition by Pliny the Elder, in his *Natural History*, around AD 77. This momentous and endur-ingly influential work was translated into English in 1601, coinciding with a turn towards the thoughtful observation and representation of insects in Western Europe. Here, Pliny (1601 [1967]: 435) implores the reader to overcome a scalar anthropocentrism that appraises the life of nonhuman species based largely on their physical size in relation to that of a fully grown, adult human body:

> But we marvel at elephants' shoulders carrying castles, and bulls' necks and the fierce tossings of their heads, at the rapacity of tigers and the manes of lions, whereas really Nature is to be found in her entirety nowhere more than in her smallest creations.

Writing more than 1500 years before the appearance of the earliest microscopes, which radically reformed the evaluation of insects, Pliny was addressing the widespread assumption that the 'biggest' animals are most worthy of our admiration, while the 'smallest' are of little interest or consequence. But he also intimated the philosophical observer's innate ability to circumvent this assumption through the close and attentive examination of animals much smaller than ourselves – especially insects. When the microscope emerged in the early seventeenth century, and was immediately turned towards insects, it did not create this kind of reassessment of scale, but rather encouraged, facilitated and extended it. Even without a microscope, indeed without a magnifying lens of any kind, Pliny was sensitive to the marvels of insect life as they appear to the naked human eye. For Pliny (1601 [1967]: 433), the flexible vertebrae of the insect showed

> a craftsmanship on the part of Nature that is more remarkable than in any other case: inasmuch as in large bodies or at all events the larger ones the process of manufacture was facilitated by the yielding nature of the material, whereas in these minute nothings what method, what power, what labyrinthine perfection is displayed!

Pliny was probably not the first person to marvel at the minutiae of nature, but he was one of the earliest to advocate it with authority. His way of looking at and describing insects, oriented by a novel, non-anthropocentric revaluation of insect life, established something of a blueprint for the revaluation of micro-faunae in general that surfaced in early modern European natural history. From the final quarter of the sixteenth century, insects became seen as both valuable and worthy of our attention precisely because of the complexity they exhibit in the face of their minuscule size, a complexity which was seen to ultimately speak to the immense creative power of nature and/or God. In the absence of a theory of evolutionary biology that could incorporate micro-faunae into a methodical account of the natural environment as 'ecosystem,' the smallest animals were primarily significant as aesthetic catalysts of a deep and lasting reverence.

This notion was perhaps most intensely illustrated in modern times by the pioneering seventeenth-century Dutch microscopist, Jan Swammerdam (1637–80), who began his *The Book of Nature; or, The History of Insects* with the declaration that: 'After an attentive examination of the nature and fabric of the least and largest animals, I cannot but allow the less an equal, or perhaps superior degree in dignity' (Swammerdam 1758: 1). Specializing in the dissection of insects and the investigation of insect anatomy at the microscopic level, Swammerdam was drawn ever deeper into the world of the miniature in nature. Significantly, as the quote above attests, it was a new way of looking at insects that engendered a new way of evaluating them – the microscope allowed the observer to see insects as if they were as large as an elephant, a bull or a lion. The visual preeminence of insects as seen through the microscope tends to leave a mark upon the observer in the form of a mental image; once an insect has been seen in greatly magnified detail, it is thereafter perceived with such detail in mind, even when witnessed with the naked eye. This allows new evaluations, not only of the insect's form, but also of its behaviour, its dignity and its virtue. Of ants, Swammerdam wrote that

> nothing hinders our preferring them to the largest animals, if we consider either their unwearied diligence, their wonderful strength, or their inimitable propensity to labour; or, to say all in one word, their amazing and incomprehensible love to their young, whom they not only carry daily to such places as may afford them food, but, if by accident they are killed, and even cut into pieces, they, with the utmost tenderness, will carry them away piecemeal in their arms. Who can show such an example among the largest animals, which are dignified with the title of perfect? (Swammerdam 1758: 1–2)

This kind of favourable comparison of the smallest animals to the largest, a rhetorical move derived from Pliny and empowered by the microscope, was repeated widely throughout microscopical discourse of the seventeenth century and beyond, especially where practitioners felt obliged to convince others that paying close attention to insects and other small animals, was a worthwhile pursuit at all. And although microscopy, with entomology, has traditionally presented itself as a rational pursuit, consistent with the operations of the modern, empirical sciences, contained within both fields of inquiry is the terminal point of logic itself, as it relates to the experience of size. As Gaston Bachelard writes in *The Poetics of Space*: 'Platonic dialectics of large and small do not suffice for us to become cognizant of the dynamic virtues of miniature thinking. One must go beyond logic to experience what is big in what is small' (Bachelard 1994: 150). Here we enter precarious territory. What happens when we go beyond logic in order to experience what is 'big' in insects?

BEYOND LOGIC: IMAGINING THE INSECT WORLD

In his classic work on human geography, *Space and Place*, Yi-Fu Tuan describes 'mythical space' as a 'conceptual extension of the familiar workaday spaces given by direct experience… a fuzzy area of defective knowledge surrounding the empirically known; it frames pragmatic space' (Tuan 1977: 86). I suggest that Tuan's articulation of mythical space sheds much light on what is meant by 'the insect world', a term which has maintained currency in both popular and scientific accounts of insect life since at least the early eighteenth century, despite having no clear referent. Where is this so-called 'insect world', what are its limits, its rules, its history? Is it an actual 'place?' It seems clear that the insect world is anything but literal. One need not turn any substantial amount of critical scrutiny to the term in order to realize that it describes what can, with very little effort, be understood as a kind of mythical space. It is consistent also with Joseph Campbell's observation that 'myth is the secret opening through which the inexhaustible energies of the cosmos pour into human cultural manifestation' (Campbell 2004: 3). Like so much mythology, 'the insect world' exists in order to make sense of the infinite; this may at least partly explain its endurance in both scientific and popular conceptions of insect life.

I would suggest, however, that the insect world, as traditionally described, goes further than this – that it has come to function as a kind of spatial metaphor for the unconscious mind. It is a modern emblem of how we think, and don't think, about insects. Turning our attention to 'the insect world' tells us something about not only insects and their environment, but about our own imagination as it corresponds with the natural world. 'Imagination', as Patrick Harpur writes in *Daimonic Reality* (1995),

> always imagines itself spatially. Like… the collective unconscious and *Anima Mundi*, it is in itself non-spatial, just as it is timeless. Like the traditional definition of God, it is 'an intelligible sphere whose centre is everywhere and whose circumference is nowhere.' But in order to speak about these models at all, we are forced to fall back on spatial metaphors, to talk about realms or domains, worlds or realities, oceanic repositories or storehouses of images. They invite such prepositions as 'below' or 'behind' – they 'underlie' consciousness or 'lie behind' our world… (Harpur 1995: 173–174)

Harpur's focus is the Otherworld, a realm populated with supernatural beings which exhibits a kind of non-literal reality, and yet his description suitably accounts for what is known as 'the insect world': it is both here and there, with us, surrounding us, beneath or behind our lived, conscious reality, influencing our lives in ways of which we are normally only vaguely aware. It is this kind of framing that positions insects as quasi-supernatural, or at least occult, animals – sharing connotations with spectres, fairies and even UFOs. One need think only of the insectoid forms of the nineteenth-century fairy, or the twentieth-century extraterrestrial, to bring the insect's modern otherworldliness to mind. The inhabitants of the insect world sometimes infringe the parameters of our conscious reality – in the form of a sudden insect intruder in our home, for example. Such moments, which are usually unexpected and sometimes shocking, remind us of the deeper biological reality in which we are embedded, but which we usually disavow in the maintenance of our increasingly symbolic, human-centred social reality. Modern domestic spaces, in particular, are largely defined by their selective inclusion of nonhuman organisms and successful exclusion of insects – but it would be shortsighted to define such spaces in purely literal terms. They are, like the insect world, also symbolic constructs, predicated on what Slavoj Žižek has discussed at length as fetishistic disavowal, which follows the structure of: 'I know very well that [insects are all around us], but I choose to act as if I don't know it.' The insect, inhabiting the fringes of our sensory parameters, exposes the limits of our consciousness, and thus retains the power to invert our everyday, symbolic order, since the mythical space of the insect world, rather than constituting an immaterial fantasy, actually refers to a reality more 'real' that the space we inhabit both physically and cognitively. This transgressive quality makes the insect, potentially, monstrous.

In *The Philosophy of Horror*, Noël Carroll distinguishes between the monsters of horror narratives and the monsters of fairy tales, as a way of articulating the differing effects such creatures have on audiences. He writes that

> the monsters of horror… breach the norms of ontological propriety
> presumed by the positive human characters in the story. That
> is, in examples of horror, it would appear that the monster is an
> extraordinary character in an ordinary world, whereas in fairy tales
> and the like the monster is an ordinary creature in an extraordinary
> world… they can be accommodated by the metaphysics of the
> cosmology that has produced them. (Carroll 1990: 16)

This distinction can be usefully transferred to the insects of many – especially popular – entomological texts that seek to create an otherworldly status for their subjects. Consider, for example, that famous popularizer of amateur entomology, Jean-Henri Fabre (1823–1915) who, in the very title of his *Social Life in the Insect World* (1912), contextualizes

his insect subjects as the inhabitants of another world. The reader is
to presume that, while some things might be familiar in this other
world, important differences will be apparent, differences that will
extend to that world's inhabitants themselves. Thus, the fairytale cos-
mology described by Carroll can be seen to shape many descriptions
of insects, that is, as ordinary creatures in an extraordinary world, a
world in which the radically different parameters of time and space
are what primarily set it apart from our own and make it extraordinary. Within the insect world, humans enjoy no ontological propriety,
and insects are not extraordinary. However, such a configuration again
belies a disavowal: insects do in fact inhabit 'our' world. Indeed, many
of the insects described by Fabre inhabited his own garden, in which
case they could be validly described as extraordinary creatures in an
ordinary world; if an insect were to enter the familiar space of the
home of its own accord, this effect would be magnified further still,
not by any transformation of the insect, but by the ordinariness of its
surroundings – its context.

The spatially capricious nature of insects themselves, then,
makes their cosmological status uncertain – their status is in fact
intimately tied to their surrounding environment. Observed candidly
from 'above' or 'without' they appear as ordinary creatures in an
extraordinary world, yet when appearing in domestic spaces, they are
quite the opposite. Fabre seemed to recognize this aspect of insects (he
captured mantids and kept them in his home) and so his description
of the mantis contains intimations of both horror (the vampire) and
fairytales (the ogre):

> Good people, how very far astray your childlike simplicity has led
> you! These attitudes of prayer conceal the most atrocious habits; the
> supplicating arms are lethal weapons; these fingers tell no rosaries,
> but help to exterminate the unfortunate passer-by… It is the tiger of
> the peaceful insect peoples; the ogre in ambush which demands a
> tribute of living flesh. If it only had sufficient strength its bloodthirsty
> appetites, and its horrible perfection of concealment would make it the
> terror of the countryside. The *Prègo-Diéu* [creature which prays to God]
> would become a Satanic vampire. (Fabre 1912: 69)

In a move typical of nineteenth-century worldmaking, Fabre attempts
to resolve the intercosmic quality of the insect by claiming that

> the habits of the Mantis cannot be continuously studied in the freedom
> of the fields; the insect must be domesticated. There is no difficulty
> here; the Mantis is quite indifferent to imprisonment under glass,

> provided it is well fed. Offer it a tasty diet, feed it daily, and it will feel
> but little regret for its native thickets. (Fabre 1912: 71)

For Fabre, then, the world of the artificial terrarium is not significantly different (at least for the mantid) than the world of the garden. In its very otherworldliness, the mantis embodies the intersection of multiple worlds, simultaneously connoting vampires, ogres and tigers.

Perhaps insects lend themselves so readily to myth and imagination (think of Psyche and the Imago) because the complete reality of the insect world, and indeed of the microscopic dimensions of nature in their entirety, is ultimately inaccessible to us via empirical methods. While we may feel secure in our ability to produce facts about what happens at the microscopic level – and we should not for a moment dismiss the utility of microscopic and entomological knowledge – it is a level of reality which seems inherently possessed by the fantastic. 'The insect world' trope effectively contains this tension between microscopic fact and microscopic fancy, while speaking to the human imagination and its propensity – we might even say compulsion – to think about, to imagine, other worlds. Consider what is perhaps the earliest appearance of the term in printed English, Henry Brooke's (1706–1783) 'The Reptile and Insect World,' the title of a passage in Book V of his six-book philosophical poem, *Universal Beauty* (1735 [1875]). In this exemplary account, all the essential qualities of the insect world as subsequently configured are present: it is a fairy-like lifeworld that strangely mirrors our own, brimming with action and purpose, yet beyond our normal, conscious awareness, and articulated by a wry distinction between fantasy (or falsehood) and the true wonder of miniature as revealed by observation and reflection:

> No debt to fable, or to fancy due,
> And only wondrous facts reveal'd to view…
> Though numberless these insect tribes of air,
> Though numberless each tribe and species fair,
> Who wing the noon, and brighten in the blaze,
> Innumerous as the sands which bend the seas;
> These have their organs, arts, and arms, and tools,
> And functions exercised by various rules,
> The saw, the ax, auger, trowel, piercer, drill;
> The neat alembic, and nectareous still:
> Their peaceful hours the loom and distaff know;
> But war, the force and fury of the foe,
> The spear, the falchion, and the martial mail,
> And artful stratagem, where strength may fail.

Each tribe peculiar occupations claim,
Peculiar beauties deck each varying frame;
Attire and food peculiar are assign'd,
And means to propagate their varying kind. (in Campbell 1875:607)

CONCLUSION

What I have attempted in this chapter is a description of the imaginative dimensions of insect–human relations in modern Western societies that avoids both the condescending diminution of the insect to sub-animal status, and the dismissal of imaginative conceptions as lacking in utility or purpose. Perhaps the reader may feel that the perspectives I have outlined are simply 'too esoteric' to have any real-world value. Yet simply by acknowledging that we are psychologically, as well as ecologically bound up with insects can, I hope, constitute a step towards the breakdown of psychological barriers to ecological realities. The ways in which we think about insects say as much about us as they do about the insects themselves, yet it would appear that our ability to imagine other, self-contained worlds has indeed been facilitated by the mere presence of insects. This is not to advance a teleological argument about insect–human relations, but merely to grant nature, if not an agency, then at least an active role in how we have come to think about it. The human mind, and its imagination, does not exist in some vacuum, divorced from nature. As Edith Cobb observes in *The Ecology of Imagination in Childhood*: 'The energies of nature impinging upon the nervous system meet in the open-system, questioning attitude of child or man. The search for perceptual form and meaning in the perceptually unknown evokes a conjugation of mind and nature' (Cobb 2004: 45). It is in this sense, I suggest, that we should critically reflect on our perception of scale in, and our attribution of value to, the insect world.

REFERENCES

Bachelard, G. (1994) *The Poetics of Space*, trans. M. Jolas. Boston, MA: Beacon Press.
Bekoff, M. (2003) *Minding Animals: Awareness, Emotions, and Heart*. Oxford: Oxford University Press.
Brooke, H. (1735 [1875]) Universal Beauty. In *Cyclopedia of English Poetry. Specimens of the British Poets: Biographical and Critical Notices, and an Essay on English Poetry*, ed. T. Campbell. New Edition. Philadelphia, PA: J.B. Lippincott and Co.
Campbell, J. (2004) *The Hero with a Thousand Faces*, Princeton and Oxford: Princeton University Press.

Campbell, T. (1875) *Cyclopædia of English Poetry. Specimens of the British Poets: Biographical and Critical Notices and an Essay on British Poetry*. Philadelphia, PA: J.B. Lippincott and Co.

Carrol, N. (1990) *The Philosophy of Horror, or Paradoxes of the Heart*, New York/London: Routledge.

Carroll, L. (2009) *Alice's Adventures in Wonderland; and, Through the Looking Glass: and what Alice Found There*, ed. P. Hunt. Oxford: Oxford University Press.

Clark, K. (1977) *Animals and Men: Their Relationship as Reflected in Western Art from Prehistory to the Present Day*. New York: William Morrow.

Cobb, E. (2004) *The Ecology of Imagination in Childhood*, Putnam, CT: Spring.

Fabre, J. H. (1912) *Social Life in the Insect World*, trans. B. Miall. London: T. Fisher Unwin.

Hanson, N. R. (1977) *Patterns of Discovery: An Inquiry into the Conceptual Foundations of Science*. London: Cambridge University Press.

Harpur, P. (1995) *Daimonic Reality: Understanding Otherworldly Encounters*, London: Penguin.

Lightman, B. (2000) The story of nature: Victorian popularizers and scientific narrative. *Victorian Review*, **25**(2), 1–29.

Michelet, J. (1883) *The Insect*, trans. W. H. Davenport Adams. London: T. Nelson and Sons.

Myers, O. E. and Saunders, C. D. (2002) Animals as links toward developing caring relationships with the natural world. In *Children and Nature: Psychological, Sociocultural and Evolutionary Investigations*, ed. P. Kahn and S. R. Kellert. Cambridge/London: MIT Press.

Pliny (1601 [1967]) *Natural History: with the English Translation in Ten Volumes*, Vol. 3, trans. H. Rackham. Cambridge, MA: Harvard University Press.

Stewart, S. (1993) *On Longing: Narratives of the Miniature, the Gigantic, the Souvenir, the Collection*. Durham, NC: Duke University Press.

Swammerdam, J. (1758) *The Book of Nature; Or, The History of Insects: with the Life of the Author*, trans. T. Flloyd. London: C. G. Seyffert.

Thomas, K. (1984) *Man and the Natural World: Changing Attitudes in England 1500–1800*. London: Penguin.

Tuan, Y-F. (1977) *Space and Place: The Perspective of Experience*. Minneapolis, MN: University of Minnesota Press.

Winer, G. A. and Cottrell, J. E. (2004) The odd belief that rays exit the eyes during vision. In *Thinking and Seeing: Metacognition in Animals and Children*, ed. D. T. Levin. Cambridge/London: MIT Press.

3

The philosophical and psychological dimensions of insects: tourism, horror and the negative sublime

JEFF LOCKWOOD

There is only one kind of ugliness that cannot be presented
in conformity with nature without obliterating all aesthetic pleasure,
hence artistic beauty: namely the ugliness arousing disgust.

Moses Mendelssohn

SETTING THE UNPLEASANT STAGE

In this chapter, I make a case for the potential of insects (and their relatives, particularly spiders) to evoke the sublime – a response to nature that tourists often seek and value (e.g., the majesty of the Canadian Rockies or the awe of the Grand Canyon) (Knudson *et al.*2003; Kozak and Decrop 2009; Woodside and Martin 2008). The twist to this conventional formulation of ecotourism is that while beauty underpins conventional features of natural settings, insects often trigger profound repulsion. But I'll argue that horror and disgust can align with (some) touristic motives. In an important way, this discussion aligns with ongoing reconceptualization of historical and cultural 'dark tourism' (and its various manifestations such as war and battlefield tourism).

THE ENTOMOLOGICALLY HORRIFIED TOURIST

Not all insects elicit negative responses. The glow-worm caves of New Zealand and Australia (see Hall, this volume), the overwintering grounds of monarch butterflies and butterfly pavilions around the world are

The Management of Insects in Recreation and Tourism, ed. Raynald Harvey Lemelin. Published by Cambridge University Press. © Cambridge University Press 2013.

popular because the creatures are enchanting (see Veltman, this volume). And there are a few arachnophiles among us (Schnoeker-Shorb and Shorb 1996). However, for the vast majority of people insects and spiders evoke negative responses (Barlow 2002; Olatunji and McKay 2009). Nonetheless, tourists are sometimes enthralled by arthropods which trigger profoundly negative emotions and cognitions (Lorimer 2007).

An exemplary case of this phenomenon unfolded in 2007, when spiders constructed an immense, communal web in a Texas state park (Figure 3.1; Quinn 2007). This unusual structure covered trees along 200 m of trail and was inhabited by millions of spiders. Given the frequency of aversive responses, tourists clearly found the spider web to be horrifying (e.g., '[the web] sort of draws you in, if you are not creeped out because some people just want to run'; Quinn 2007). However, the web became a tourist destination over the Labor Day weekend with more than 3000 people coming to see the 'creepy' spectacle.

The desire for disturbing experiences reflects an important feature of modern leisure. At zoos and aquaria, people flock to feeding times – and these often entail unpleasant scenes of carnivores consuming meat or live prey (e.g., small fish and invertebrates). As Orkin (2011) explains, exhibits of invertebrates themselves are also popular, as exemplified by the Insect Zoo at the Smithsonian which receives more than a million people a year (for a greater discussion on the appeal of insect pavilions, see Veltman, this volume).

Museums tap into our paradoxical attraction to repulsive experiences. On my recent visit to the Arizona Science Center, the featured exhibit was Gunther von Hagens' *Body Worlds* (Body Worlds 2011). This display of human anatomy uses plasticized cadavers variously dissected and posed. Overheard conversations often expressed macabre fascination. The two items at the entrance of the museum gift shop consisted of kits featuring 'Slime' and 'Gross Science'. Within the store, the majority of shelf space was devoted to animal-based products. Of these, 39% involved insects and spiders, 29% pertained to reptiles and amphibians, 12% were based on dinosaurs (including fossilized 'dinosaur poop') and only 18% involved cuddly simulacra of mammals and birds.

Videographic media make extensive use of repulsive images and themes. Many 'reality programs' exploit elements of fear and disgust. 'Fear Factor' challenges contestants to engage in often abhorrent activities (e.g., eating live insects) and 'Dirty Jobs' features difficult and often disgusting labours (e.g., maggot farming, forensic entomology

Figure 3.1   Communal spider web at Lake Tawakoni State Park, Texas, 15 August 2007. Over the Labor Day weekend, more than 3000 visitors came specifically to see this natural phenomenon which most found horrifying – and enticing. Photo by Donna Garde.

and dung beetle research). Likewise, 'Billy the Exterminator' follows a pest control operator who encounters dangerous and repulsive creatures. Finally, horror films are premised on evoking negative emotions, with the 'Saw' series exemplifying the continuing popularity of this genre.

Thus, many people seek negative experiences in nature, museums and media. In each of these contexts, and particularly in natural settings, insects and spiders are often the offensive stimuli (Miller 1997; Miller 2004). To make sense of this, I consider the elements of psychology (emotions), philosophy (aesthetics), tourism and insects – and how these components interact to create a touristic experience.

In the following section, I focus on emotions, with an eye toward linking the qualities of insects and the opportunities for nature interpretation. The next section considers aesthetics and particularly the negative sublime. This is followed by a discussion of interpreting entomological phenomena for tourists. In the next section I propose some insects and related creatures that might engage tourists seeking a disturbing experience of nature. I conclude with a synthesis of my lines of argument using an example from my own teaching.

THE PSYCHOLOGY OF NEGATIVE EMOTIONS

To comprehend why people are drawn to disturbing natural phenomena, we must delve into the nature of emotions. And understanding emotions is relevant to tourism researchers and managers (Gnoth and Zins 2009). To begin, most people respond to large aggregations of insects and their kin with disgust and fear. Miller (1997: 24) explains that, 'We have a name for fear-imbued disgust: horror' (see also Smith and Korsmeyer 2004). To analyse entomogenic horror, I will begin with the foundations of disgust and then consider the modifying emotion of fear.

The nature of disgust

Disgust is a topic of intense theoretical and empirical inquiry (Miller 1997; Miller 2004; Olatunji and McKay 2009; Smith and Korsmeyer 2004), but its relevance to nature-based tourism is largely unexplored. Miller (1997: 11) contends that, 'To feel disgust is human and humanizing.' Only humans are capable of disgust – a potentially powerful context for a naturalist's interpretation.

Disgust plays a protective function for the body or the more inclusive 'self' (Barlow 2002; Miller 2004; Smith and Korsmeyer 2004). At the corporeal level, disgust allows us to avoid ingesting or absorbing infectious or toxic substances. Some psychologists offer a further interpretation, proposing that this emotion protects the self. Not just our bodies but our identity, integrity and mental wellbeing can be contaminated. As Miller (1997: 47) puts it, 'Nature routinely challenges us with regard to matters of personal identity and self-boundary.'

Given the role of disgust in protecting the individual's boundaries – whether physical or psychic – it is not surprising that the senses associated with close contact are most important in eliciting this response (Miller 1997; Miller 2004; Smith and Korsmeyer 2004). Understanding these sensory connections is important to meaningful interpretation. Etymologically, 'disgust' is derived from distaste; having something offensive in one's mouth triggers rejection. The senses of smell and touch are similarly associated with close contact and play important roles in disgust. Vision and hearing are much less important but become relevant when a sight or sound triggers an imagined or remembered repulsive encounter.

Disgust is elicited in seven domains: food, body products, sexual acts, hygiene, violations of the body envelope, death and animals

(Davey and Marzillier 2009). This emotion is triggered by creatures that we perceive to be dominated by base drives, that exhibit teeming profusion and that have the capacity to enter or attach to our bodies. Miller (1997: 40) maintains that, 'What disgusts, startlingly, is the capacity for life,' echoing the observation of Kolnai (in Smith and Korsmeyer 2004). This ironic feature bridges the domains of disgust. First, sexual fecundity is offensive: 'Every swarming thing that swarms upon the earth is an abomination' (Leviticus 11: 41). Second, bodily secretions and excretions (often associated with base indulgences) are filthy – and insects represent a surfeit of viscid, squishy, sticky and wiggly alien life forms (Miller 1997). Third, rampant life evokes a sense of rank excess or generative rot which we associate with the inconsequence of our own lives, the mindless death of the profuse beings and decomposition through which the dead are converted into the living – including ourselves (Smith and Korsmeyer 2004).

If disgust were simply repulsive, there'd be no touristic appeal. But this paradoxical emotion involves both a sense of being drawn toward and pushed away. In short, 'one almost savours its object while being revolted by it in a kind of perverse magnetism' (Smith and Korsmeyer 2004: 9). This macabre allure may represent the reality that to be nourished, physically or psychically, we must approach 'the other' closely enough to put ourselves at risk (Miller 2004). As Miller (1997: 17) notes: 'Disgust shocks, entertains by shocking, and sears itself into memory' – and these qualities provide a potent basis for naturalists' interpretations.

Knowing the origins of disgust are important to effectively engaging people via this emotion. In terms of development, disgust is absent from very young children, although they are capable of distaste. At 4 to 8 years of age, disgust begins to emerge. This emotion becomes particularly keen during adolescence with intensified awareness of one's body (Miller 1997).

With regard to social differences – an important touristic consideration – disgust is largely invariant across cultures, except for dietary prohibitions (Smith and Korsmeyer 2004). Cultures expand what constitutes disgust (e.g., eating dogs or insects) but have little capacity to exclude biologically fundamental experiences (e.g., touching a carcass or dung pile seething with maggots) (Miller 1997). Because disgust easily transitions into moral judgment, interpreters must clarify that objects (and people) evoking disgust should not be targets of destruction (Smith and Korsmeyer 2004).

What we find disgusting is not simply explained by either social constructionism or sociobiology. Rather, we are evolutionarily prepared to readily learn that particular stimuli are disgusting. Furthermore, there is a strong relationship between disgust and fright (Barlow 2002), especially with regard to insects (Mulkens *et al.* 1996). This connection leads to a consideration of fear.

The nature of fear

Both disgust and fear entail aversion, but while disgust focuses on the removal of an offending stimulus, fear leads to flight from a dangerous stimulus (Miller 1997). And both emotions may be in play at once. For example, while maggots evoke disgust and pit bulls generate fear, spiders stimulate both fear and disgust (van Overveld *et al.* 2006) – and perhaps stinging insects such as hornets also evoke both emotions.

Fear reflects the primal fight–flight–freeze response to immediate danger (Barlow 2002). A cross-cultural study found that animals fell into three groups: harmless (e.g., rabbits and cows), fearsome (e.g., tigers and sharks) and feared but not predatory (usually invertebrates) (Davey *et al.* 1998). We are evolutionarily primed to acquire particular fears (Barlow 2002), and we readily learn to fear insects and their relatives. Although most phobic individuals cannot recall a traumatic encounter, parental and social modelling are etiologically important (Barlow 2002). Biting and stinging arthropods may have been life threatening to our ancestors, so we readily acquire a fear of spiders and bees (Davey and Marzillier 2009).

Aside from agoraphobia (fear of open or public places), the most common phobias involve animals, illness (which provides a link to disgust), thunderstorms and psychosocial trauma (Beck *et al.* 1985). Animals are the source of nearly half of all clinical cases of specific phobias (Chapman 1997; Marks 1987). Feared animals typically include rodents, reptiles and invertebrates such as maggots, slugs, worms, cockroaches, beetles, moths and spiders (Davey 1994). Indeed, arachnophobia is the experimental 'white rat' of psychologists investigating specific phobias (Olatunji and McKay 2009), and it is arguably the most common fear in Western societies with 55% of women and 18% of men reporting a fear of spiders (Davey 1994). Perhaps the most important feature of fear in terms of tourism is that intense fears are most prevalent in women, African Americans and religious practitioners (Davey *et al.* 1998).

While fear typically evokes flight, it shares with disgust a push–pull quality such that we can't quite leave alone that which we are terrified to confront (Asma 2009). So while Gnoth and Zins (2009: 196) argue that, 'tourism is all about the search for pleasure', it is evident that pleasure is a complex state that can arise in unexpected ways (including historically and culturally disturbing sites).

AESTHETICS: POSITIVE AND NEGATIVE

Our aesthetic sense interacts with our emotional state to generate powerful encounters with natural and built objects. Ultimately, I wish to contend that entomological phenomena can evoke the (negative) sublime. But it is necessary to first consider beauty as the foundation for complex aesthetic experiences.

The nature of beauty

The standard philosophical debate regarding beauty is whether this quality is subjective or objective (Graham 2005). If it is purely subjective, why do we express our judgments in objective terms and why do we argue about beauty at all? But if purely objective, why is there so much disagreement and why are we moved by beauty?

In *The Critique of Judgement*, Immanuel Kant provided an elegant solution, arguing that beauty is both sentimental and rational (Frierson and Guyer 2011). Aesthetic pleasure is a feeling that transcends individual preferences through the 'common sensitive nature' of humanity. I'd suggest that this shared sensibility is most plausible for that which is ugly, as there is considerable cross-cultural agreement on what is repulsive (Miller 1997; Miller 2004; Olatunji and McKay 2009).

There is an equally challenging debate regarding what constitutes a proper object of aesthetic appreciation (Graham 2005). Artworks can be judged, but what about natural objects? Mountains, rivers, trees and locust swarms are formed without the intentional meaning or expressive properties of art. However, various theories have been proposed as to how aesthetic evaluation of nature might be conducted (Carlson 2002).

The closest theories to standard artistic judgment treat nature as a sculpture (the object model) or a painting (the landscape model). The old Kodak overlooks reflected the picturesque aesthetic which lives on through postcards but garners little serious attention in contemporary aesthetics. Psychological and spiritual theories of aesthetics include the engagement model (appreciation via immersion in nature), the arousal model (appreciation via emotional response), the mystery model (appreciation via nature's incomprehensibility) and the metaphysical imagination model (appreciation via insights about the meaning of our existence). Cultural theories include the postmodern model (appreciation via cultural consensus) and the pluralist model (appreciation via each culture's particular natural context). And finally, the analytical theories include the non-aesthetic model (appreciation is appropriate only for art) and the natural environment model (appreciation via scientific knowledge).

Perhaps the most viable model in terms of nature-based tourism is Carlson's (2002) version of the natural environment model. He maintains that by understanding ecology we develop a 'thick sense' of beauty (Hospers 1946), rather than a 'thin sense' of prettiness – a contention consistent with Aldo Leopold's (1949) aesthetics. Carlson contends that when we comprehend the unity, harmony, interdependence and stability of nature via science, our aesthetic is deepened such that we see nature as it is, rather than in naïve or anthropocentric terms.

Carlson (2002) advocates a positive aesthetic such that with knowledge nothing in nature is seen as ugly. I would propose that informed and authentic comprehension includes the observer's relationship with a natural object. As such, a negative aesthetic can be appropriate – some things are truly disgusting, even genuinely horrifying, in light of our biological nature.

The nature of the sublime

A synthetic definition of the positive sublime is: the greatness of beauty, scale, goodness or brilliance which draws us closer for its virtue but terrifies us with its power and supremacy. According to Morley (2010), the sublime is a transformative experience in which reason falters. Kant was the first philosopher to systematically explore the sublime and differentiate it from beauty (Frierson and Guyer 2011). His analysis involved a series of conceptual pairings of beauty/sublimity: day/night, short/long, small/large, comedy/tragedy, youth/age, cunning/bold, romance/friendship, wit/understanding, decorated/simple and charming/touching. Kant maintained that experiencing the sublime required disinterest so that we could detach from and contemplate our terror. But modern scholars (Graham 2005) have questioned whether such objectivity is psychologically possible or normatively desirable, suggesting that detachment precludes an aesthetic experience. Although scholars have revised Kant's ideas (Graham 2005; Morley 2010), the paradoxical quality of attraction and repulsion remains central.

The Janus-faced quality of the sublime reflects the push–pull feature of both disgust and fear – including the psychological dialectic of awesome and awful creatures (Lorimer 2007). In describing the communal spider web in Texas, the park manager said, 'You just can't believe what you are seeing, and then it sort of draws you in' (Quinn 2007). Edmund Burke noted the perverse pleasure of mixing fear and delight in the sublime (Morley 2010). Kant's concept of the sublime involved the experience of that which is both attractive and repulsive in its excess. And a modern framing of the sublime posits that, 'something rushes in and we are profoundly altered' (Morley 2010: 12) – suggesting a breaching of our barriers.

Conventionally, the sublime entails terror which is converted into a positive aesthetic via amazement or wonder (Carlson 2002). However, one might substitute 'horror' (fear-imbued disgust) for 'terror' and thereby develop a concept of the negative sublime: The greatness of ugliness, badness or destruction which tempts us to draw closer but repulses us with its depravity. In opposing a negative aesthetic, Carlson (2002: 95) maintains that, 'the positive aesthetic appreciation of previously abhorred life forms, such as insects and reptiles, seems to have followed developments in biology.' This is plausible for our cognitive capacities but is less convincing for our emotional responses. Although his argument endorses natural beauty, the 'thick sense' that comes with scientific understanding also validates the negative sublime. That is, if science is the key to aesthetic appreciation, then as we learn about our psychological and evolutionary links to that which is transformatively horrifying, we have a basis for deep engagement. And insightful guides could provide a cogent interpretation of this authentic interaction with nature.

THE INTERPRETATION OF NATURE FOR TOURISTS

The motivations of tourists

While people might accidentally encounter natural phenomena that evoke disgust, horror (fear-imbued disgust), or the negative sublime (overwhelming awfulness), perhaps the central question is whether tourists would seek such an experience as either a destination or a part of a larger venture. Various motivation models of tourists' behaviour suggest that there may be a place for such opportunities (Hsu and Huang 2008) for both nature-based and cultural-historical travel.

According to Maslow's 'Hierarchy of Needs', self-actualization is the highest level which can be achieved through personal growth and fulfilment – qualities that would seem to entail experiences with the full range of emotions including horror. Indeed, Maslow (1970) also maintained that aesthetics and curiosity are basic human needs, and an encounter with the negative sublime might qualify in both regards.

'The Career Ladder' and 'Travel Career Patterns' models propose that travel motivations change with experience (TCL; Pearce 1988) and that tourists increasingly seek satisfaction of higher-level needs (TCP; Pearce 2005). The 14 motivational factors in TCP include novelty, nature and stimulation. And these needs might be satisfied with communal spider webs or masses of cicadas, for example. Similar developments from macro-fauna to birds to insects in animal observations, have been suggested by professional amateurs (see Pearson this volume for a discussion on professional amateurs).

Models based on 'Push–Pull Factors' cast motivations in terms of internal (push) and external (pull) factors, with the former important to initiating travel and the latter key to choosing destinations (Gnoth 1997). Different researchers have identified a range of push factors that could motivate engagement with a horrifying natural phenomenon, including escape from a mundane environment, transcending the everyday, self exploration and evaluation, regression, novelty, education, curiosity, adventure and excitement (e.g., Crompton 1979; Dann 1977; Jamrozy and Uysal 1994; Yoon and Uysal 2005). Pull factors are perhaps less readily associated with horror, although television and movies and the *Annual Insect Fear Film Festival* at the University of Illinois would suggest there is potential. It should be noted, however, that a disturbing encounter in nature would be avoided in light of many push and pull factors.

The 'Escaping and Seeking Dimensions' of Mannell and Iso-Ahola (1987) involve a two-dimensional model in which tourists seek optimal arousal by escaping from either routine or stressful environments to experience mastery, challenge, learning and exploration. As such, some people (i.e., those escaping from drudgery) might choose to encounter the negative sublime. Of course, allocentric individuals who are venturesome and self-assured would be far more likely to do so than psychocentric individuals who are anxious and insecure (Plog 2001).

Interpreting disgust and the negative sublime

Whatever model of tourist motivation one adopts, the rubber hits the road when interpreting occasions of disgust, horror and the negative sublime arising from experiences of superabundant insects and their relatives. Knudson *et al.* (2003: 3) maintain that interpretation, 'translates or brings meaning to people about natural and cultural environments'. Others have proposed that interpretation, 'forges emotional and intellectual connections between the interests of the audience and the inherent meanings in the resource' (Brochu and Merriman 2000). In this light, it seems possible to forge meaningful connections with disturbing entomological phenomena.

Pushing a bit further, the generally accepted purposes of interpretation conceivably accommodate a horrifying mass of spiders, locusts, bees or maggots. The classic view that interpretation relies on firsthand experience (Tilden 1967) accords well with such opportunities. In the field of interpretation, there is attention to both the object and the viewer such that the guide, 'communicates the traits common to the human species' (Knudson *et al.* 2003: 8). Surely these shared traits include the basic emotions derived from our evolutionary experience with things that can harm us. Indeed, understanding why we respond with horror and how disgust functions in human biology constitute 'larger truths' lurking beyond 'mere facts' (Tilden 1967).

Much of the discussion of interpretation concerns the desire of tourists to see natural beauty, and the sublime is taken to be rooted in beauty (Knudson *et al.* 2003). However, scholars of interpretation also assert that guides should lead visitors to places of unusual aesthetic interest including unappreciated landscapes such as prairies and tidal flats (Knudson *et al.* 2003). So why not hordes of insects? However, tourism advocates pull up short: 'death follows growth, decay follows death, and life follows decay [the beauties of which are] unattached to vulgar, restricting concepts of what constitutes beauty in nature' (Porter and Gleick 1990: 1). Why must interpretation convince people that nature is beautiful? To deny the overwhelming majority of visitors their primal feeling of disgust seems counterproductive to the goal of genuine engagement. While we want them to stop short of condemning or destroying the offending objects or beings, perhaps we should honour authentic emotions if we want visitors to take away deep and lasting memories.

The disconnection between our urbanized lives and our sense of the natural world is cause for concern (Louv 2005). Knudson *et al.* (2003: 47) contend that, 'Our environmental handicaps arise from our incompleteness, our lack of understanding, our loss of contact with what we are and what supports us.' But then they warn interpreters against using a snake to draw attention to broader themes because the information is lost when people are absorbed with their fear of the creature. Perhaps the savvy interpreter understands that the big picture is not about the supposed balance of nature provided by predators or some other ecological principle. Rather it is about our evolutionarily deep, psychologically profound and culturally compelling relationship to serpents. And that in this broader context, fear-imbued disgust and the chill of the negative sublime in response to a mass of wriggling maggots is part of our becoming complete, gaining understanding and making contact with the natural world.

Knudson *et al.* (2003) suggest that the goal of interpretation is not instruction but provocation, and one might contend that a meadow overrun with grasshoppers is more provocative than one overrun with wildflowers. They admonish interpreters to: 'Formulate it well. Enunciate it clearly. Hide it not' (Knudson *et al.* 2003: 60). This is good advice – and we should not hide those parts of nature that elicit disgust, horror and the negative sublime.

HORRIFYING TOURISTIC OPPORTUNITIES WITH INSECTS

Having considered both how an experience of the negative sublime accords with tourists' motives and how interpretation might provide deeper understanding of such experiences, I'd like to briefly consider some exemplar cases. This sampling reflects my own experiences with insects, and I have no doubt that there are other opportunities noted in this book with similar or greater potential.

None of the instances that I describe have been exploited as touristic opportunities to my knowledge, but it should be apparent how an innovative interpreter might frame these occasions. The examples that I offer include both destination-based phenomena in which tourists travel to a location to encounter the arthropods (e.g., driving to a state park to see a communal spider web) and opportunistic phenomena in which tourists decide to encounter insects in the course of other travels (e.g., going on a nature walk in a National Park to see a black widow spider web).

Grasslands and pastures

I have previously considered the potential of *Orthoptera* (e.g., grasshoppers, crickets and locusts) to elicit the negative sublime (Lockwood 2002a, 2004a, b). On the North American steppe, populations of rangeland grasshoppers can exceed 100 individuals per square metre over hundreds hectares. I've described the sense of walking into such aggregations: 'Rather than waves of movement parting in my path, there was sheer pandemonium. Grasshoppers ricocheted off my face and chest, clung to my legs and boiled in every direction' (Lockwood 2004b). Such infestations are not likely to be destination phenomena, but they are readily visited in the course of travelling across the western states and provinces.

Bands of Mormon crickets (*Anabrus simplex*) can also be encountered in areas of the western United States (e.g., Dinosaur National Monument; Gildart and Gildart 2005). Few people could fail to be awed by millions of these highly cannibalistic, thumb-sized insects that give the impression that, 'a sadistic fisherman had dumped a net-full of melanic prawns on the prairie' (Lockwood 2002a:45).

The people of the southern United States may have become inured to fire ants of the genus *Solenopsis*, but for those who have not witnessed the fury of thousands of these insects pouring out of a metre-high mound of soil, the experience is stunning. A hectare of infested pasture can support 100 million fire ants in a massive network of interconnected mounds (Tvedten 2002).

Forests

Outbreaks of forest insects can elicit the negative sublime, particularly via the scale of their damage. Stands of eastern forests denuded by gypsy moths exemplify the destructive capacity of nature (Gerardi and Grimm 1979), and vast swaths of western pine forests are being decimated by the bark beetles (US Forest Service 2011). Given the role that climate change is playing in the latter awe-inspiring entomological phenomenon (Bentz *et al.* 2010), there are valuable opportunities for interpretation.

Deciduous forests support breathtaking emergences of the periodical cicada (Lockwood 2004c):

> [T]he cicada carnival – a bacchanalian celebration of the flesh – reminds
> us that despite our tidy gardens and parks, vestiges of nature remain
> untamed … the excrement that these insects are going to rain down in
> backyards and parks [would] fill six hundred Olympic-sized swimming
> pools a day … a single square mile of the forest [will] have as many eggs
> as there are stars in the Milky Way.

I have proposed that the emergence of cicadas could constitute a sublime phenomenon worthy of national recognition:

> In a sense, the periodical cicadas are like our National Parks. Maybe
> the US Department of Interior should declare a new category: National
> Events. These would be natural happenings that define the character
> of our nation, occasions that warrant our attention, or processes that
> merit celebration.

Finally, swarms of bees and paper wasp nests (which reach impressive sizes) can be opportunistically located by attentive interpreters, and these insect aggregations have the potential to evoke both aversion and fascination.

Deserts

As the annual *Invertebrates in Education and Conservation Conference* and other activities hosted by the *Sonoran Arthropod Studies Institute* (SASI) in Arizona demonstrate, deserts host spatiotemporally erratic outpourings of arthropods that would serve as ideal opportunities for the negative sublime. For example, the migration of tarantulas in the desert southwest is a remarkable phenomenon (Janowski-Bell and Horner 1999). Likewise, mass emergences of winged (reproductive) ants and termites following desert rains are spectacular events. Even individual scorpions, whip scorpions and sun spiders have the capacity to evoke horror – and black widow spiders are readily located.

Aquatic ecosystems

Mass emergences of mayflies are remarkable outbursts of insect life. Billions of insects can swarm from an aquatic habitat in a single night, blanketing an area (Fremling 1968). The individual insects are unlikely to evoke disgust, but there are few natural phenomena that better capture the appalling fecundity of the animal world. A predictable surfeit of insects occurs around the Great Salt Lake, where clouds of brine flies reach 220 million insects/km of shoreline (Great Salt Lake Ecosystem Program 2011).

As Rykken and Farrell (this volume) demonstrate, aquatic habitats also produce enormous numbers of fierce micro-predators and blood-feeding flies that can evoke a dark sense of wonder. Interestingly, the classic view of interpretation has framed these insects as negating the experience of nature: 'An onslaught of mosquitoes can distract from even the best group discussion of wetland values' (Knudson *et al.* 2003: 112).

It might also be noted that in some regions, people have become familiar with regular emergences of insects. And with environmental changes, the disappearance of these harbingers of summer is perceived with angst and a sense of loss (Césard, 2010). Hence, even the absence of insects can elicit negative emotions such as melancholy.

Decomposing materials

In addition to the disgust that is elicited by carcasses and faeces, the presence of insects fosters the negative sublime. For example, parks could include a site where visitors view a dead deer replete with maggots and other necrophagic insects. The opportunity to witness the process of decomposition would need sensitive interpretation, but the experience could be enormously compelling. Perhaps the 'hook' to forensic entomology (e.g., the popular show 'CSI: Crime Scene Investigation') could make the encounter cognitively, as well as emotionally, engaging. Likewise, a guided tour of the insect fauna of faeces (cattle and bison dung pats being particularly rich habitats) could evoke a sense of profound disgust – and conceptual wonder. The marvel of dung beetles making and rolling perfect balls of faeces is intriguing, and when one realizes the biological context – that the dung balls will become nurseries for their offspring – a 'thick' aesthetic appreciation makes the phenomenon utterly captivating.

CONCLUSION

In my effort as an entomologist to understand the sometimes disturbing encounters that I've had with insects – experiences of horror, disgust and the negative sublime leading to enchantment – I have written several essays (Lockwood 2002a, b, 2004b, c, 2006). What follows is adapted from Lockwood (2012).

A disgusting education

Within minutes, our hands are covered in vomit and faeces. Our quarry, the plains lubber grasshopper, is a disgusting creature to subdue. The largest of all insects on the Wyoming grasslands, *Brachystola magna* is reminiscent of the chewed, cigar butts that I encountered jammed into ashtrays when I was a kid. They're about the same size and equally appealing.

Every summer, I take my students to collect a few dozen of these grasshoppers from patches of roadside sunflowers, where the insect lives in obscene, orgiastic abundance. They make outstanding specimens for dissections in my 'Insect Anatomy and Physiology' laboratory. But this is not the primary reason for bringing students into intimate contact with these repulsive beasts.

Upon capture, this grasshopper regurgitates copiously, smearing itself and its handler with dark brown fluid. Then it begins to defecate, producing mushy, mouse-like turds in quick succession. And when a humongous insect pukes and craps on you, no amount of cultivated sensitivity saves this from being disgusting. Evolution decrees that we twist our faces into masks of revulsion.

As a kid, I turned over my fair share of decomposing critters and poked at the edges of many fetid puddles. But in school, the science of biology was ironically sterile. High school biology relied on models and pickled specimens, and college teaching laboratories were neat and sanitary back then – just as they are today.

I doubt that my students are any more ethical for their experiences in the field. Rather than the soft lap of a mythic mother, they understand that nature usually doesn't give a crap about us, and when it does, it is as likely to crap on us as to embrace us. Students with a Disneyfied view of the world may be compassionate and gentle, but theirs is an insipid, secondhand morality. For better or worse, my students arrive at values of their own making through authentic experiences with the natural world.

Shock artists appreciate the power of real materials to evoke disgust: dancing in the midst of dismembered animals, painting with human blood and sculpturing with dung. The purpose of these artworks, according to Robert Wilson (avant-garde playwright and director), is to force the observer to face the 'sad recognition of how much has been missed'. Perhaps I am a shock biologist. I want my students to see what they miss in nature documentaries, college textbooks and computer simulations. With apologies to Rene Descartes, in today's videographic virtual world: *Fastidium movet, ergo est*: It is disgusting, therefore it is real.

REFERENCES

Asma, S. T. (2009) *On Monsters: An Unnatural History of Our Worst Fears*. New York: Oxford University Press.

Barlow, D. H. (2002) *Anxiety and its Disorders: The Nature and Treatment of Anxiety and Panic*. New York: Guilford.

Beck, A. T., Emery, G. and Greenberg, R. L. (1985) *Anxiety Disorders and Phobias: A Cognitive Perspective*. New York: Basic.

Bentz, B. J., Regniere, J., Fettig, C. J. *et al.* (2010) Climate change and bark beetles of the Western United States and Canada: direct and indirect effects. *BioScience*, **60**, 602–613.

Body Worlds (2011) Gunther von Hagens' Body Worlds [Online]. Available at: http://www.bodyworlds.com/en.html.

Brochu, L. and Merriman, T. (2000) *Interpretive Guide Training Workbook*. Ft. Collins, CO: National Association for Interpretation.

Carlson, A. (2002) *Aesthetics and the Environment: The Appreciation of Nature, Art and Architecture*. New York: Routledge.

Césard, N. (2010) Vie et mort de la Manne blanche des riverains de la Saône. *Études rurales*, **185**, 83–98.

Chapman, T. F. (1997) The epidemiology of fears and phobias. In *Phobias: A Handbook of Theory, Research and Treatment*, ed. G. C. L. Davey. Chichester, UK: Wiley, pp. 415–434.

Crompton, J. (1979) Motivations for pleasure vacation. *Annals of Tourism Research*, **6**, 408–424.

Dann, G. (1977) Anomie, ego-enhancement and tourism. *Annals of Tourism Research*, **4**, 184–194.

Davey, G. C. L. (1994) Self-reported fears to common indigenous animals in an adult UK population: the role of disgust sensitivity. *British Journal of Psychology*, **85**, 541–554.

Davey, G. C. L. and Marzillier, S. (2009) 'Disgust and animal phobias' in *Disgust and its Disorders*, eds. B. O. Olatunji and D. McKay, Washington DC: American Psychological Association, pp. 169–190.

Davey, G. C. L., McDonald, A. S., Hirisave, U. *et al.* (1998) A cross-cultural study of animal fears. *Behaviour Research and Therapy*, **36**, 737–750.

Fremling, C. R. (1968) Documentation of a mass emergence of *Hexagenia* mayflies from the Upper Mississippi River. *Transactions of the American Fisheries Society*, **97**, 278–281.

Frierson, P. and Guyer, P. (2011) *Immanuel Kant: Observations of the Feelings of the Beautiful and Sublime and Other Writings*. Cambridge: Cambridge University Press.

Gerardi, M. H. and Grimm, J. K. (1979) *The History, Biology, Damage, and Control of the Gypsy Moth*, Porthetria dispar *(L.)*. Cranberry, NJ: Associated University Presses.

Gildart, B. and Gildart J. (2005) *A Falcon Guide to Dinosaur National Monument*. Guilford, CT: Falcon.

Gnoth, J. (1997) Tourism motivation and expectation formation. *Annals of Tourism Research*, **23**, 283–304.

Gnoth, J. and Zins, A. H. (2009) Emotions and affective states in tourism behavior. In *Handbook of Tourist Behavior*, ed. M. Kozak and A. Decrop. New York: Routledge.

Graham, G. (2005) *Philosophy of the Arts*. New York: Routledge.

Great Salt Lake Ecosystem Program (2011) Brine flies [Online]. Available at: http://wildlife.utah.gov/gsl/brineflies/index.php.

Hospers, J. (1946) *Meaning and Truth in the Arts*. Chapel Hill, NC: University of North Carolina Press.

Hsu, C. H. C. and Huang, S. (2008) Travel motivation: a critical review of the concept's development. In *Tourism Management*, eds. A. Woodside and D. Martin. Wallingford, UK: CABI.

Jamrozy, U. and Uysal, M. (1994) Travel motivation variations of overseas German visitors. *Journal of International Consumer Marketing*, **6**, 135–160.

Janowski-Bell, M. E. and Horner, N. V. (1999) Movement of the male brown tarantula, *Aphonopelma hentzi* (Araneae, Theraphosidae), using radio telemetry. *Journal of Arachnology*, **27**, 503–512.

Knudson, D. M., Cable, T. T. and Beck, L. (eds.) (2003) *Interpretation of Cultural and Natural Resources*, 2nd edn. State College, PA: Venture.

Kozak, M. and Decrop, A. (eds.) (2009) *Handbook of Tourist Behavior: Theory and Practice*. New York: Routledge.

Leopold, A. (1949) *A Sand County Almanac and Sketches Here and There*. New York: Oxford University Press.

Lockwood, J. A. (2002a) *Grasshopper Dreaming: Reflections on Killing and Loving*. Boston, MA: Skinner House.

Lockwood, J. A. (2002b) Lessons of the Rocky Mountain locust. *Orion*. Summer, 88–93.

Lockwood, J. A. (2004a) *Locust: The Devastating Rise and Mysterious Disappearance of the Insect that Shaped the American Frontier*. New York: Basic.

Lockwood, J. A. (2004b) The joyful terror of oneness. *Wild Earth*, Spring/Summer, 54–56.

Lockwood, J. A. (2004c) The orgy in your backyard. *New York Times*, Op-Ed page, May 20, 2004.

Lockwood, J. A. (2006) The nature of violence. *Orion*. January/February, 14–19.

Lockwood, J. A. (2012) 'The joy and wonder of fear and loathing' in *Trash Animals*, eds. K. Nagy and P. D. Johnson II. Minneapolis: University of Minnesota Press, in press.

Lorimer, J. (2007) Nonhuman charisma. *Environment and Planning Development: Society and Space* **25**, 911–932.

Louv, R. (2005) *Last Child in the Woods: Saving Our Children from Nature-Deficit Disorder*. Chapel Hill, NC: Algonquin.

Mannell, R. C. and Iso-Ahola, S. E. (1987) Psychological nature of leisure and tourism experience. *Annals of Tourism Research*, **14**, 314–331.

Marks, I. M. (1987) *Fears, Phobias and Rituals*. New York: Academic.

Maslow, A. (1970) *Motivation and Personality*. New York: Harper and Row.

Miller, S. B. (2004) *Disgust: The Gatekeeper Emotion*. Hillsdale, NJ: Analytic.

Miller, W. I. (1997) *The Anatomy of Disgust*. Cambridge, MA: Harvard University Press.

Morley, S. (ed.) (2010) *The Sublime*. Cambridge, MA: MIT Press.

Mulkens, S. A. N., de Jong, P. J. and Merckelbach, H. (1996) Disgust and spider phobia. *Journal of Abnormal Psychology*, **105**, 464–468.

Olatunji, B. O. and McKay, D. (eds.) (2009) *Disgust and its Disorders: Theory, Assessment, and Treatment Implications*. Washington DC: American Psychological Association.

Orkin (2011) O. Orkin Insect Zoo [Online]. Available at: http://www.orkin.com/l earningcenter/o-orkin-insect-zoo/.

Pearce, P. L. (1988) *The Ulysses Factor: Evaluation Visitors in Tourist Settings*. New York: Springer-Verlag.

Pearce, P. L. (2005) *Tourism Behaviour: Themes and Conceptual Schemes.* Clevedon, UK: Channel View.

Plog, S. C. (2001) Why destination areas rise and fall in popularity: an update of a Cornell quarterly classic. *Cornel Hotel and Restaurant Administration Quarterly,* **42**, 13–24.

Porter, E. and Gleick, J. (1990) *Nature's Chaos.* New York: Viking Penguin.

Quinn, M. (2007) Giant Spider Web in an East Texas State Park – 2007 [Online]. Available at: http://www.texasento.net/Social_Spider.htm.

Schnoeker-Shorb, Y. A. and Shorb, T. L. (1996) *The Spiders and Spirits of Petunia Manor.* Prescott, AZ: Native West Press.

Smith, B. and Korsmeyer, C. (eds.) (2004) *On Disgust: Aurel Kolnai with Introduction.* Chicago, IL: Open Court.

Tilden, F. (1967) *Interpreting Our Heritage.* Chapel Hill, NC: University of North Carolina Press.

Tvedten, S. (2002) About the fire ant [Online]. Available at: http://www.getipm.com/thebestcontrol/fireants/factoids.htm.

US Forest Service (2011) Regional bark beetle impact and information [Online]. Available at: http://www.fs.usda.gov/wps/portal/fsinternet/!ut/p/c4/04_SB8K8xLLM9MSSzPy8xBz9CP0os3gjAwhwtDDw9_AI8zPwhQoY6BdkOyoCAPkATlA!/?ss=110299&navtype=BROWSEBYSUBJECT&cid=FSE_003853&navid=091000000000000&pnavid=null&position=BROWSEBYSUBJECT&ttype=main&pname=Rocky%20Mtn.%20Bark%20Beetle-%20Home.

van Overveld, M., de Jong, P. J. and Peters, M. L. (2006) Differential UCS expectancy bias in spider fearful individuals: Evidence toward an association between spiders and disgust relevant outcomes. *Journal of Behavior Therapy and Experimental Psychiatry,* **37**, 60–72.

Woodside, A. G. and Martin, D. (eds.) (2008) *Tourism Management: Analysis, Behaviour and Strategy.* Wallingford, UK: CABI.

Yoon, Y. and Uysal, M. (2005) An examination of the effects of motivation and satisfaction on destination loyalty: a structural model. *Tourism Management,* **26**, 45–56.

4

Tiger beetles: lessons in natural history, conservation and the rise of amateur involvement

DAVID L. PEARSON

INTRODUCTION

Ecologists, medical researchers, behaviourists, natural resource managers, conservation biologists, policy makers and the judiciary represent some of the vast array of disciplines that have come to rely heavily on basic documentation of insects (Bale et. al 2008; Pimental *et al.* 1997), what otherwise can be called natural history studies. Areas that are critical to natural history advances include consistent names (taxonomy), reliable identification, descriptions of behaviour, geographical distribution and population trends (Wilson 2000). Support for professional biologists to pursue these types of data has decreased considerably in the past few decades. Salaries and administrative support have shifted to areas that are considered more sophisticated, hypothesis-driven and with greater opportunities to obtain funding, such as molecular genetics, mathematical modelling and population regulation (Cotterill and Foissner 2010; Felsenstein 2004; Wheeler *et al.* 2004).

As a result, not only are many areas of research at risk for lack of supporting descriptive data but also the availability of information on which we rely for fields such as medicine, agriculture, ecosystem and wildlife management planning and public policy (Wilcove and Eisner 2000). If we can learn from history, we can apply established patterns from the past together with present and predicted future economic indicators to help broaden the pool of those who can contribute to

The Management of Insects in Recreation and Tourism, ed. Raynald Harvey Lemelin. Published by Cambridge University Press. © Cambridge University Press 2013.

the base of data that professionals no longer feel free to pursue. This plan draws on literature from economics and from recent work on the rise of professional amateur biologists (Pro-Ams) to argue for specific changes in the way contributors to the biological enterprise are recruited, trained and supported in their efforts. New technology and broadened goals present an opportunity for natural history to expand its boundaries by attracting and actively supporting a range of participants far beyond the classical model that is centred on professional biologists. My goal here is to look into what factors may affect the dynamics between amateur and professional insect workers.

AN HISTORICAL OVERVIEW OF NATURAL HISTORY

One of the most reliable techniques for answering historical questions and testing for patterns is by using insights from one field to tell us something about another – a philosophical technique called consilience by some historians (Wilson 1999). In so doing, we can make sense of the past and perhaps anticipate the future (Gaddis 2002). Historical development of such varied fields as military science, astronomy, archaeology, geology, physics, computer science, horticulture, ichthyology, malacology, ornithology, mammalogy, entomology and herpetology share a common pattern with respect to the relationship between experts and amateurs (Leadbeater and Miller 2004). The earliest stages of these fields were dominated by amateurs – there often being no established professional experts at the time – who described distribution patterns and named component parts such as species, rocks, stars or bytes of information. As these fields of study became solidified, power was transferred from expert amateurs to trained professional scientists and graduate training for employment in the field became available. In these cases, the maturation of these fields led to research that was increasingly conceptual and theoretical, including systems analysis and the use of formal models. As a result, technical terminology and methodology in these fields have become so refined that most of what is known about these topics is now accessible by a narrow audience of trained professionals (Pearson and Cassola 2007).

This trend suggests that theoretical and institutional development in the sciences will lead reliably or even inevitably to the exclusion of expert amateurs. Indeed, the twentieth century saw a great deal of this, first with the rise of population genetics in the early decades of the century, then with the advent of molecular biology in the 1950s, followed by the push toward mathematical approaches to ecology in the 1970s. Even basic natural history studies have become quite technical and specialized over the last few centuries. They require access to specimens, to libraries of historical literature, to multiple technical vocabularies, to the expertise of colleagues, to computers for analysing data and to wet labs for acquiring genetic information. The trend is toward continued professionalization and the exclusion of amateurs. Interested amateurs, however, may again become a crucially important part of the study of natural history (Pearson *et al.* 2011).

There is little doubt that expert amateurs play a smaller role in natural history than they once did. In Europe and North America, for instance, a significant part of the infrastructure for natural history was built by amateurs. It was common for those who were assembling research collections in the late nineteenth and early twentieth centuries to rely heavily on large networks of amateur botanists, entomologists, birders and mammalogists. Joseph Grinnell, founding Director of the Museum of Vertebrate Zoology, C. Hart Merriam, first Chief of the Division of Economic Ornithology and Mammalogy of the USDA (later renamed the Bureau of Biological Survey) and Spencer Baird, Secretary of the Smithsonian Institution, are all examples (Kohler 2006). These administrators made extensive use of commercial collectors who were not trained biologists and who today we would call parataxonomists. They contributed tens of thousands of specimens to science and had long careers as freelance naturalists and some eventually obtained museum or university appointments.

Professionals eventually became more significant in some areas and for some taxa largely supplanted the amateurs. Yet, in other fields the amateurs remain a significant part and may even dominate the progress and agenda of the discipline. These amateurs can range in level of interest from devotees and dabblers to skilled amateurs. The most committed can become quasi-professional and operate at professional levels without receiving pay. According to the Directory of Entomological Societies, worldwide there are 514 organized associations that have insects and spiders as their primary focus. Of these associations, 194 are interested in general entomology. The rest have a more focused mission including: pest control (99), honeybee business (79), forensics (1) or a single taxonomic group (Lepidoptera, 45; spiders, 38; Odonata, 15; Coleoptera, 11; and smaller numbers for Diptera, Orthoptera, Hymenoptera, Isoptera, Heteroptera, Neuroptera and Ephemeroptera). From their websites and mission statements, 107 of the general entomology associations are primarily for professionals, 85 are for both professionals and amateurs, and two are expressly

Figure 4.1 The big sand tiger beetle (*Cicindela formosa*), a species common in sandy areas throughout much of North America. Copyright R. Planck.

for amateurs. The membership of the majority of the 130 associations with a single taxon as the focus is made up of a combination of professionals and amateurs, but with amateurs comprising the bulk of the membership.

TIGER BEETLES AS A MODEL FOR UNDERSTANDING
HISTORICAL PATTERNS

Tiger beetles (Figure 4.1) are a small but distinct group of nearly 2800 species whose historical background of studies is relatively well known (Pearson 2006; Pearson and Vogler 2001). These beetles are attractive, fast-flying and fast-running insect predators that occur in many diverse habitats around the world. Many of their adaptations, such as for thermoregulation, competition and avoiding their own enemies, are well studied (Pearson and Vogler 2001).

For tiger beetles, the near monopoly of a single expert, Walther Horn, had great influence on the direction of natural history studies (Horn 1926). Beyond his tight control of tiger beetle publications, however, a few other professional biologists began to publish scientific articles using tiger beetles as test organisms for geological history (Wickham 1904), behaviour (Shelford 1902), physiology (Shelford 1913) and ecology (Shelford 1907).

At the same time, the few graduate studies on tiger beetles focused on their colouration (Schultz and Rankin 1983a, b), ecology (Hori 1982; Mury Meyer 1987), physiology (Zerm *et al.* 2004), neural anatomy (Strausfeld *et al.* 2009) and use in conservation efforts, such as a search for bioindicators (Andriamampianina *et al.* 2000; Arndt *et al.* 2005, Rodríguez *et al.* 1998), local extinction (Knisley *et al.* 1987; Satoh 2008; Spomer and Higley 1993) and reintroduction (Fenster *et al.* 2006; Omland 2002).

Tiger beetle researchers showed little socialization well into the twentieth century. There were no organized peer groups, meetings or associations of those interested in tiger beetles, and only in the 1990s did field guides and general books on the biology of tiger beetles appear (Acorn 2001; Choate 2003; Pearson *et al.* 2006). Before this time, only those with time and interest to search through often obscure journals and arcane terms could acquire the basic knowledge to do research using tiger beetles.

In 1969, an informal correspondence among tiger beetle enthusiasts developed into a journal called *Cicindela*. Its publication goals were to provide a forum to share observations, collecting sites, natural history, distributional data, identification help and taxonomic insight of tiger beetles. The subscriber list to this journal quickly rose to about 200 but stayed at that level for the next 40 years, and they included primarily enthusiastic amateurs from North America and Europe. Small groups of subscribers would go on collecting trips together, but there were few attempts to organize meetings or symposia where these people could interact face to face. A continued aura of exclusivity among tiger beetle workers was apparent in the paucity of support programmes for active recruitment of new and especially young enthusiasts. Because of this, there are likely fewer than 3000 tiger beetle professional and amateur enthusiasts in the world.

The rapid growth of such highly sophisticated fields as molecular biology, statistical modelling and satellite imagery introduced many technical words and concepts that quickly limited comprehension to a narrow array of associated professionals. Among tiger beetle workers there is a growing separation between professionals and amateurs, especially in complex fields, such as molecular studies (Vogler and DeSalle 1994; Vogler *et al.* 1997), physiology (Irmler 1973; Toh and Mizutani 1994; Yager *et al.* 2000) and mathematical modelling (Carroll and Pearson 1998; Pearson and Carroll 2001; Pearson and Juliano 1993).

Paradoxically, the growing sophistication among many tiger beetle professionals has manifested itself as a rejection of taxonomy and natural history as valid pursuits (Acorn 2009), and a general decline in funding for these areas of research has created a paucity of these critical data (Bossart and Carlton 2002; Pearson *et al.* 2011). In addition, beginning in the last third of the twentieth century, governments in many countries listed several tiger beetle species as endangered or threatened (Berglind *et al.* 1997; Dangalle *et al.* 2011; Diogo *et al.* 1999). Legislators, economists, sociologists, foresters, politicians, land owners and many members of the public, who had little or no previous interest in these taxa, suddenly needed to know about them (Schlesinger and Novak 2011). Often by default, the pursuit of these basic taxonomic, distributional and natural history data has fallen to individuals searching for an avocation but can only do so during their free-time. Their investment of free time and level of interest in an avocation or hobby depends on social class, educational level, household income and gender (66% for men and 50% for women) (Leadbeater and Miller 2004).

The sources of these amateurs lie in an understanding of economics. As the economy of a country or region rises, its middle class grows. Families will have fewer children and invest more time and money into each child including increased support of higher education (Barro 2001), donations to private organizations (NGOs), time and money for avocations (Leadbeater and Miller 2004) and concern for the environment (Bhattarai and Hammig 2004). Increased access to the internet (Godfray 2007) and published field guides (Pearson and Shetterly 2006) are especially significant factors in attracting and training Pro-Ams into biology and conservation.

The primary underlying social and demographic factors that best explain the emergence and growth of all levels of amateur interest in society include: expanding life span; growing levels of education; increasing search for individual fulfilment; more social mobility; changing occupational patterns that encourage second careers early in life; and more affluence to allow trading income for a better quality of life (Leadbeater and Miller 2004).

Although there appears to be a general decline in the numbers of professional natural historians and taxonomists (Hopkins and Freckleton 2002), amateurs or citizen scientists are notably active and numerous among those studying tiger beetles, especially in relation to their conservation and protection. The influence of these active amateurs appears to be impacting not only scientific studies but economics as well. Based on notices of regional and national meetings and ecotourist tours published on line, amateurs studying such insect groups as dragonflies, butterflies and tiger beetles are investing more time and money into domestic and international tours that focus on searches for insect species and observations of their behaviours. Some local Audubon Society-sponsored field trips in Virginia, Florida and Arizona focused specifically on tiger beetles. The appearance of published field guides for tiger beetles in North America (Pearson *et al.* 2006), Thailand (Naviaux and Pinratana 2004), Colombia (Vítolo 2004) and other parts of the world was quickly followed by a notable increase in the number of local amateurs and professionals interested in tiger beetles as a hobby or research organism (Pearson and Shetterly 2006). In addition, local tiger beetles have become regional icons for areas such as Colorado's Great Sand Dunes National Park and Reserve and the Lincoln, Nebraska area (Tweit 2010)

AREAS OF POTENTIAL CONFLICT BETWEEN AMATEURS
AND PROFESSIONALS

The history of some areas of study can sometimes help foresee problems and provide solutions for other fields. Because studies of birds, and to some degree Lepidoptera and Odonata, have such a long history, perhaps they can serve as paths along which studies of other groups such as tiger beetles can choose to follow. These past experiences could clarify and anticipate many potential interactions, including those between amateurs and professionals in entomology.

To collect specimens or not to collect

The world of bird studies endured a bitter battle of philosophies in the late 1800s regarding the collection of specimens (Barrow 1998). Professionals and many enthusiastic amateurs argued that a specimen in the hand was the only incontrovertible way to establish identification and range extensions. The burgeoning number of amateur bird enthusiasts, however, most of whom were attracted to birds through field guides, argued that with training, observational skills could adequately substitute for most sightings of new and unusual species. Among ornithologists today, laws and legal restrictions as well as a change in philosophy about collecting specimens have made a general non-collecting philosophy the norm, in developed as well as developing nations (Vuilleumier 1998). As Lemelin (2009) revealed, this debate is repeating itself now for insects, especially among the Lepidoptera and Odonata. Much of this controversy can be traced to recently popularized field guides, such as the *Butterflies Through Binoculars* series, written and edited by Glassberg (1999) and other expert amateurs. At the same time, it is becoming more apparent that the controversy is not simply between collectors and non-collectors. The various levels of amateurs and general public interested in Nature introduce a range of philosophies that include tolerating collecting for science but not for personal use.

Because tiger beetle studies have only recently opened up to a broad amateur market, collecting specimens is still the philosophy underlying most studies of these beetles. Already, however, concerns among the old guard anticipate a growing change in this philosophy. Discussions of collecting in the most recent literature typically take an apologetic tone and advise that binoculars and cameras are legitimate alternatives to nets. Observations and photographs are often touted as equally appropriate methods of registering species and documenting range extensions, behaviour and natural history. A chapter or section on conservation and threatened species has become de rigueur and implicitly downplays the importance of collecting specimens, especially of those species with threatened populations. As their popularity grows with the availability of widely marketed field guides, history tells us that the collecting versus non-collecting philosophies may quickly evolve into a pro- and anti-collecting schism. Not only do the new initiates need to be shown that there is a time and place for collecting, but the old guard needs to change their preconceptions that a specimen is needed to document every event and site. To promote these compromises, careful and non-inflammatory communication will be essential.

Some of the growing number of new amateur devotees likely will be tempted to observe and perhaps even collect specimens from localities that previously were under little pressure from enthusiasts. The compulsion to collect or even disturb populations for observation and photography is hard to control, especially for rare or endemic species. As knowledge and interest grow, perhaps dealers will enter the field and start offering for sale taxa in which there was limited interest until the appearance of a field guide. Occasionally this interest mutates into a grotesque monster. Fads for live, pinned and fossil insects have come to dominate markets in places such as Japan and influenced mass collecting overseas to fulfil the demand. During a fad, amateur collectors will blanket accessible areas and deplete the population of their target insects (Brock 2006). The potential for growing misuse of collecting is likely to influence legislation and controls that will affect amateurs and professionals as well as the rate of legitimate data accumulation for tiger beetles.

Common English versus scientific names

Again in the late1800s, as field guides for birds became more popular, a personal and volatile disagreement arose between many professionals and most amateurs over the use of scientific and common names. On one side, arguments were made that scientific names were stable and made communication easier. In addition, proponents of scientific names argued there is no inherent barrier preventing the public from embracing scientific names. Nevertheless, publishers, the press and much of the public perceived scientific names as elitist (Barrow 1998). A subsequent mushrooming of often-duplicated common names made communication difficult. Finally a committee, made up of professionals and expert amateurs, chose a single common name for each bird species in North America. This committee has since become the final arbiter of these names. Among insect students, odonates and butterflies have repeated this history. There are now at least two competing 'official' lists of common names of species for butterflies in North America.

The community of tiger beetle enthusiasts has already laid the groundwork for an uncomplicated transition to common names. Largely under pressure from publishers of field guides, newspaper reporters and government agencies dealing with threatened species, the use of common names was increasingly discussed in the 1990s. There was a good deal of opposition to the introduction of common names from many entomologists, professional and amateur alike, who argued that some of the common names proposed were obscure, more complicated than the Latin names, and unlikely ever to be used by many (Pearson and Shetterly 2006).

Even more problematic was the bourgeoning problem of duplicated common names. In response to this issue, ten professionals and amateurs formed an ad hoc committee in 2003 to guide a selection process that would establish standardized English names for tiger beetle species in Canada and the United States. Using the history of how bird, Lepidoptera and Odonata groups got their common names, the committee tried to avoid historical pitfalls in the process by including as many as possible from the North American community of tiger beetle enthusiasts. The entire readership of the journal *Cicindela* was drafted, and after a year of solicitations, winnowing and voting, a cautious consensus of common names was chosen (Pearson 2004). No claim was made that these were the official names, but they were used in a field guide to the tiger beetles of the United States and Canada that appeared soon after (Pearson *et al.* 2006). Whether or not this procedure adequately anticipates and avoids the problems of applying common names remains to be seen, but for the committee, the past was an important ingredient in developing the steps used for naming tiger beetles.

Writing style, terminology and methodology

More subtly, distinctive writing styles emerge that indicate levels of expertise and establish levels of authority that can separate professionals from amateurs. Some examples of writing devices that are preferred by professionals include reduced use of personal pronouns, reliance on passive voice, a decrease in the number of simple sentences, the presence of technical terminology, an emphasis on reliability of evidence and the use of citations (Carter 1990; Chafe 1986; Lakoff and Johnson 1980). Amateurs writing articles for popular consumption frequently mix science and sentiment, use the personal and active tense, and describe species in technical detail and telegraphic phrases or Romantic prose in full sentences. The various writing styles often make articles and field guides more readable to amateurs, but less scientifically respectable for some professional biologists.

The most appropriate writing style for articles and books on tiger beetles is usually an individual choice, not a committee decision. The journal *Cicindela* originally evolved from a newsletter in 1969, and it has been an important form of communication among amateurs and professionals for more than 40 years. Originally informal, primarily un-refereed, and largely written by amateurs, the writing style has become more formal and professional in terms of data presentation, citations, acknowledgements sections and writing style. This journal publishes information on range extensions, descriptions of new species and unusual natural history and behaviour – subjects that are unlikely to be published by professionals. For more experimental and broader philosophical subjects, expert amateurs and professionals publish in a broad range of refereed international journals, a pattern similar to that seen in studies of other taxa.

Tiger beetle field guides have experimented with a range of writing styles to try and attract a broader readership and initiates into tiger beetle studies. Some, such as Acorn's (2001) *Field Guide to the Tiger Beetles of Alberta*, combine a readable style of species accounts and distributions interspersed with calculated informality and poetry. Most of the others have a range of style from a stilted discourse usually associated with monographs to a compromise of limited jargon and uncomplicated sentences that will reassure professionals but not intimidate readers new to the tiger beetles. This latter style is the one that the majority of successful field guides have adopted, but it is a difficult tight-rope to walk (Pearson and Shetterly 2006). A style that is too simplistic and colloquial will be perceived patronizing and unchallenging, but a style that is too laden with jargon and theoretical concepts will likely be considered stilted or elitist. In addition, choosing the right level of discourse when amateurs are just entering a field will also be different for that same field as it matures and becomes more sophisticated (Pearson and Shetterly 2006).

Segregation of areas of study and expertise

Vuilleumier (2003) and others have argued that the phenomenal rise in availability of bird field guides and interactive websites has revolutionized and vitalized field studies in region after region, and both amateurs and professionals have gained from them. In the last 30 years, these modern sources of information have become useful and attractive to a growing public. Their market success is largely due to authors that have more than just museum experience with organisms. Extensive experience in the field helps these authors understand what is important both for recognizing the characters to use for separating similar species as well as what information the inexperienced readers most need to become skilled. The majority of these modern field guides and websites are authored by expert amateur ornithologists rather than by academic professionals. A common criticism by professionals is that the veracity of amateurs is inherently questionable. However, many professional scientists have been caught in plagiarism and other intentionally deceitful uses of data. Peer review and peer pressure are the most valuable and powerful tools for controlling misleading data and interpretation of patterns for both professionals and amateurs.

Once these popular sources of information are made available, they attract more professionals and amateurs to go to the field and study organisms there in greater depth. Their studies then provide more detailed information and data on the species that need to be incorporated into the next and more sophisticated field guide or website. Thus, these identification guides often reflect the stage of development of the study of organisms and influence its activity directly. They also accelerate skills that in turn help basic and applied knowledge grow (Pearson and Shetterly 2006).

As mentioned previously, natural history observations, geographical distributions and seasonal records of occurrence and dispersion of many taxa, have by default been turned over largely to amateurs. Largely because of funding pressures and changing perceptions of acceptable levels of sophistication, professionals rarely publish these types of basic data. However, because of language use and perceived low standards of scientific rigour, acceptance of the resultant data by professionals is not universal.

Expert amateurs have influenced studies of tiger beetles throughout history (Pearson and Cassola 2005). But now as this taxon becomes more available and attractive to graduate students, academicians and other professionals seeking a study organism, a division of labour is becoming evident. As in many other taxa, field studies of distribution, declining populations and descriptions of new species of tiger beetles have been taken over almost completely by amateurs (Knisley 2011). Professionals dominate more technical fields of spatial modelling, molecular studies, physiological adaptations and conservation policy. Presently the cooperation and mutual respect between professionals and amateurs is remarkable, but it may be largely due to the small cadre on both sides that find they need each other to advance. Popular publications and websites are likely to throw off this delicate balance. As they introduce larger numbers of participants to the thrills and joys of tiger beetles, ground-swell changes are almost guaranteed. As we have seen historically for studies of other taxa, a culture of cooperation is likely to decline when a rapidly growing number of enthusiasts joins in. Many of these recent initiates no longer know each other personally, and more extreme views regarding collecting and other potentially disruptive turmoil are likely to arise. How the presently small world of tiger beetle workers can prepare for major changes is not yet clear. Perhaps knowing changes are coming is the first important step.

DISCUSSION

Does the history of natural history studies follow a predictable pattern? Despite substantial differences in their biology and taxonomic level, studies of taxa as diverse as birds, butterflies and tiger beetles show similar patterns of change over their histories at large-scale time intervals. The most obvious divergence among them is the speed with which some steps were completed and the comparable maturity of research at any given time, differences similar to those suggested by theories of paradigm shifts (Kuhn 1996).

For instance, amateurs initiated natural history and taxonomic studies of several taxa, such as birds and tiger beetles, at the same time in the eighteenth century. In the nineteenth century amateurs, especially bird enthusiasts, were instrumental in starting conservation societies and influencing legislation for protection of the environment. However, professionalization of the field became apparent much earlier in ornithology and tiger beetle studies lagged in these changes by at least 75 years. Collectors and authors working alone have been very important over the entire history of tiger beetles. Passionate adherence to ideas is the norm for creative scientists and critical for developing concepts and popularizing or socializing adherents to these concepts (Mitroff 1976). Perhaps some fields and taxa are more likely or lucky to attract these passionate pioneers.

The rise of such issues as trinomial use, biological studies and graduate education are a few additional examples of differential rates of change by bird and tiger beetle researchers. The use of common English names versus scientific names was debated among ornithologists in the nineteenth century, probably because birds had attracted so much attention from the public early on. Amateurs complained of too much dependence on scientific names in ornithology (Barrow 1998). For tiger beetles, in contrast, scientific names were retained as virtually the only nomenclature until the twenty-first century (Pearson 2004; Erwin and Pearson 2008; Wu and Shook 2010). More recently, the publication of tiger beetle field guides with English names helped recruit a huge increase in amateur involvement, most of whom eschewed scientific names of tiger beetles.

The minimization by professionals of basic but critical studies of natural history and range distributions, and in many cases descriptions of new species (Acorn 2009) impacted ornithology earlier than tiger beetle studies. Today, largely because few undescribed bird species remain in the world, descriptions of new species are so few that bird taxonomists spend little time in this effort. Instead they concentrate on refining studies such as phylogenetic and evolutionary relationships, areas that are also more likely to be funded and recognized as intellectually appropriate. Also, most professional ornithologists no longer pursue studies of long-term presence–absence data, range expansions and descriptions of natural history, even though these types of data are often critical for sophisticated modelling and hypothesis testing. Instead professional ornithologists are helping empower citizen scientists to gather long term data and basic descriptive natural history observations (Droege *et al.* 1998; Pearson *et al.* 2010). This trend is also apparent among tiger beetle professionals (Pearson *et al.* 1988), but less formally than the programmes organized by ornithologists. If tiger beetles and other insect groups are to be widely incorporated into studies and management plans for conservation biology, the history of other more studied groups, such as that of butterflies and dragonflies, demonstrates that facilitating the interaction of amateurs and professionals, not establishing further barriers, should be a high priority. These comparisons of taxa provide insight into factors that may permit some groups to advance more quickly and with greater sophistication in an accumulation of knowledge of their natural history and usefulness for conservation.

Number of species

An ideal taxon for research and conservation studies should have sufficiently small species numbers so that researchers can expect to understand regional or even global patterns of species relationships as well as details of biology and distribution. However, it should also have high enough species numbers to provide a wide range of examples and potential concepts to test.

Range of habitats

Although, for instance, butterflies, tiger beetles and birds occur over a wide range of latitudinal and altitudinal habitats, some butterfly and bird species extend into higher latitudes, altitudes and extreme habitat types not occupied by tiger beetles. Bird and butterfly studies thus provide a wider range of questions and potential biological problems to solve.

Obviousness and economic importance

Groups with attractive or colourful species are more likely to attract the attention of potential amateur natural historians. Those taxa that can be readily observed in the field will offer fewer frustrations and less discouragement to continue, especially in the initial stages of interest.

In addition, any economic significance such as hunting and domestication will produce an economic impact that is likely to attract funding and interest from professionals and amateurs.

The number of researchers

A combination of the previous three factors likely contributed to the number of researchers using each taxon. In early steps, they influenced how many aficionados could compete for recognition. Those groups that attracted few workers were likely to have less chance for socialization and formation of interactive peer groups, few specialized journals, little recruitment of additional enthusiasts, and slower and narrower development into more experimental studies.

CONCLUSION

We now have a better idea of priorities for selecting which taxa or fields will yield the most useful and broadest results for advances in natural history knowledge. But we also understand that many problems remain. With limited funds, time and personnel, how do we best balance the costs and benefits of speciose habitats and taxa, economic importance, detectability and human or natural threats (Sorenson 1995)? How do we redefine the training and support of professional biologists so that they are rewarded for developing and applying communication skills among scientists, Pro-Ams, legislators, decision makers and the public? How can we most effectively educate, recruit and mentor Pro-Ams who will probably provide the bulk of future taxonomic and natural history data for most taxa in the future? And how do we recruit enthusiasts to study more obscure groups that may not be charismatic, economically important or obvious (Russell 2010)?

The British social critics, Leadbeater and Miller (2004), proposed that the government, funding agencies and the public must become involved and together with professionals facilitate the contributions of Pro-Ams. This facilitation needs to include not only moral and financial support and direction of Pro-Ams but also a willingness of professionals to share the stage with them so that there can be mutually beneficial advances. Technology is likely to make the borders separating professionals and amateurs less defined in the future as these groups work more and more closely, but only if it is accompanied by a social change in attitude and acceptance between professionals and amateurs.

ACKNOWLEDGEMENTS

I thank my wife, Nancy, for a lifetime of support and encouragement in pursuit of tiger beetles around the world. I also thank the many Pro-Ams who have faithfully provided data, ideas and been role models over the last 50 years, especially Ronald Huber, Dave Brzoska, Fabio Cassola, Juergen Wiesner and Roger Naviaux. My professional colleagues, including Joachim Adis, Terry Erwin, Barry Knisley, Jon Paul Riodriguez and Alfried Vogler, have generously helped me develop ideas about interactions between Pro-Ams and amateurs.

REFERENCES

Acorn, J. H. (2001) *Tiger Beetles of Alberta: Killers on the Clay, Stalkers on the Sand.* Edmonton, Alberta: University of Alberta Press.

Acorn, J. H. (2009) Amateurs and abandoned science. *American Entomologist*, **55**, 127–128.

Andriamampianina, L., Kremen, C., Vane-Wright, D., Lees, D. and Razafimahatratra, V. (2000) Taxic richness patterns and conservation evaluation of Madagascan tiger beetles (Coleoptera: Cicindelidae). *Journal of Insect Conservation*, **4**, 109–128.

Arndt, E., Aydin, N. and Aydin, G. (2005) Tourism impairs tiger beetle (Cicindelidae) populations: a case study in a Mediterranean beach habitat. *Journal of Insect Conservation*, **9**, 201–206.

Bale, J. S., van Lenteren, J. C. and Bigler, F. (2008) Biological control and sustainable food production. *Philosophical Transactions of the Royal Society B*, **363**, 761–776.

Barro, R. J. (2001) Human capital: growth, history, and policy: a session to honor Stanley Engerman. *The American Economic Review*, **91**, 12–17.

Barrow, M. V., Jr. (1998) *A Passion for Birds: American Ornithology after Audubon.* Princeton, NJ: Princeton University Press.

Berglind, S.-A., Ehnström, B. and Ljungberg, H. (1997) Strandskalbaggar, biologisk mångfald och reglering av små vattendrag: exemplen Svartån och Mjällån. *Entomolgisu Tidskrift*, **118**, 137–154.

Bhattarai, M. and Hammig, M. (2004) Governance, economic policy, and the environmental Kuznets curve for natural tropical forests. *Environment and Development Economics*, **9**, 367–382.

Bossart, J. L. and Carlton, C. E. (2002) Insect conservation in America: status and perspectives. *American Entomologist*, **48**, 82–92.

Brock, R. L. (2006) Insect fads in Japan and collecting pressure on New Zealand insects. *The Weta*, **32**, 7–15.

Carroll, S. S. and Pearson, D. L. (1998) Spatial modeling of butterfly species richness using tiger beetles (Cicindelidae) as a bioindicator taxon. *Ecological Applications*, **8**, 531–543.

Carter, M. (1990) The idea of expertise: an exploration of cognitive and social dimensions of writing. *College Composition and Communication*, **41**, 265–286.

Chafe, W. (1986) Evidentiality in English conversation and academic writing. In *Evidentiality: The Linguistic Coding of Epistemology*, ed. W. Chafe and J. Nichols. Norwood, NJ: Albex Press, pp. 261–272.

Choate, P. M. (2003) *A Field Guide and Identification Manual for Florida and Eastern US: Tiger Beetles*, Gainesville, FL: University Press Florida..

Cotterill, F. P. D. and Foissner, W. (2010) A pervasive denigration of natural history misconstrues how biodiversity inventories and taxonomy underpin scientific knowledge. *Biodiversity and Conservation*, **19**, 291–303.

Dangalle, C., Pallewatta, N. and Vogler, A. (2011) The current occurrence, habitat and historical change in the distribution range of an endemic tiger beetle species *Cicindela (Ifasina) willeyi* Horn (Coleoptera: Cicindelidae) of Sri Lanka. *Journal of Threatened Taxa*, **3**, 1493–1505.

Diogo, A. C., Vogler, A. P., Gimenez, A., Gallego, D. and Galian, J. (1999) Conservation genetics of *Cicindela deserticoloides*, an endangered tiger beetle endemic to southeastern Spain. *Journal of Insect Conservation*, **3**, 117–123.

Droege, S., Cyr, A. and Larivée, J. (1998) Checklists: an under-used tool for the inventory and monitoring of plants and animals. *Conservation Biology*, **12**, 1134–1138.

Erwin, T.L. and Pearson, D.L. (2008) *A Treatise on the Western Hemisphere Caraboidea (Coleoptera): Their Classification, Distributions, and Ways of Life. Vol II (Carabidae – Nebriiformes 2 – Cicindelitae).* Sofia, Bulgaria: Pensoft.

Felsenstein, J. (2004) A digression on history and philosophy. In *Inferring Phylogenetics*, ed. J. Felsenstein. Sunderland, MA: Sinauer.

Fenster, M. S., Knisley, C. B. and Reed, C. T. (2006) Habitat preference and the effects of beach nourishment on the federally threatened Northeastern Beach Tiger Beetle, *Cicindela dorsalis dorsalis*: Western Shore, Chesapeake Bay, Virginia. *Journal of Coastal Research*, **22**, 1133–1144.

Gaddis, J. L. (2002) *The Landscape of History: How Historians Map the Past*, New York: Oxford University Press.

Glassberg, J. (1999) *Butterflies Through Binoculars: The East (Butterflies Through Binoculars Series)*. New York: Oxford University Press.

Godfray, H. C. J.. (2007) Linnaeus in the information age. *Nature*, **446**, 259–260.

Hopkins, G. W. and Freckleton, R. P. (2002) Declines in the numbers of amateur and professional taxonomists: implications for conservation. *Animal Conservation*, **5**, 245–249.

Hori, M. (1982) The biology and population dynamics of the tiger beetle, *Cicindela japonica* (Thunberg). *Physiological Ecology, Japan*, **19**, 77–212.

Horn, W. (1926) Carabidae: Cicindelinae. In *Coleopterorum Catalogus, Pars 86*, ed. W. Junk. Berlin: W. Junk.

Irmler, U. (1973) Population-dynamic and physiological adaptation of *Pentacomia egregia* Chaud (Col. Cicindelidae) to Amazonian inundation forest. *Amazoniana*, **3**, 219–227.

Knisley, C. B. (2011) Anthropogenic disturbances and rare tiger beetle habitats: benefits, risks, and implications for conservation. *Terrestrial Arthropod Reviews*, **4**, 41–61.

Knisley, C. B., Luebke, J. L. and Beatty, D. R. (1987) Natural history and population decline of the coastal tiger beetle, *Cicindela dorsalis dorsalis* Say (Coleoptera: Cicindelidae). *Virginia Journal of Science*, **38**, 293–303.

Kohler, R. (2006) *All Creatures: Naturalists, Collectors, and Biodiversity 1850–1950*. Princeton, NJ: Princeton University Press.

Kuhn, T. S. (1996) *The Structure of Scientific Revolutions*, 3rd edn. Chicago, IL: University of Chicago Press.

Lakoff, G. and Johnson, M. (1980) *Metaphors We Live By*. Chicago, IL: University of Chicago Press.

Leadbeater, C. and Miller, P. (2004) *The Pro-Am Revolution: How Enthusiasts are Changing our Society and Economy*. London: Demos.

Lemelin, R. H. (2009) Goodwill hunting? Dragon hunters, dragonflies and leisure. *Current Issues in Tourism,* **12**, 235–253.

Mitroff, I. I. (1976) Passionate scientists. *Society*, **13**, 51–57.

Mury Meyer, E. J. (1987) Asymmetric resource use in two syntopic species of larval tiger beetles (Cicindelidae). *Oikos*, **50**, 167–175.

Naviaux, R. and Pinratana, A. (2004) *The Tiger Beetles of Thailand*. Bangkok: Brothers of St. Gabriel in Thailand.

Omland, K. S. (2002) Larval habitat and reintroduction site selection for *Cicindela puritana* in Connecticut. *Northeastern Naturalist*, **9**, 433–450.

Pearson, D. L. (2004) A list of suggested common English names for species of tiger beetles (Coleoptera: Cicindelidae) occurring in Canada and the US. *Cicindela*, **36**, 31–39.

Pearson, D. L. (2006) A historical review of the studies of Neotropical tiger beetles (Coleoptera: Cicindelidae) with special reference to their use in biodiversity and conservation. *Studies on Neotropical Fauna and Environment*, **41**, 217–226.

Pearson, D. L. and Carroll, S. S. (2001) Predicting patterns of tiger beetle (Coleoptera: Cicindelidae) species richness in northwestern South America. *Studies on Neotropical Fauna and Environment*, **36**, 123–134.

Pearson, D. L. and Cassola, F. (2005) A quantitative analysis of species descriptions of tiger beetles (Coleoptera: Cicindelidae), from 1758 to 2004, and notes about related developments in biodiversity studies. *Coleopterists Bulletin*, **59**, 184–193.

Pearson, D. L. and Cassola, F. (2007) Are we doomed to repeat history? A model of the past using tiger beetles (Coleoptera: Cicindelidae) and conservation biology to anticipate the future. *Journal of Insect Conservation*, **11**, 47–59.

Pearson, D. L. and Juliano, S. A. (1993) Evidence for the influence of historical processes in co-occurrence and diversity of tiger beetle species. In *Species Diversity in Ecological Communities: Historical and Geographical Perspectives*, ed. R. Ricklefs and D. Schluter. Chicago, IL: University of Chicago Press, 194–202.

Pearson, D. L. and Shetterly, J. A. (2006) How do published field guides influence interactions between amateurs and professionals in entomology? *American Entomologist*, **52**, 246–252.

Pearson, D. L. and Vogler, A. P. (2001) *Tiger Beetles: The Evolution, Ecology and Diversity of the Cicindelids*. Ithaca, NY: Cornell University Press.

Pearson, D. L., Blum, M. S., Jones, T. H. *et al.* (1988) Historical perspective and the interpretation of ecological patterns: defensive compounds of tiger beetles (Coleoptera: Cicindelidae). *American Naturalist*, **132**, 404–416.

Pearson, D. L., Knisley, C. B. and Kazilek, C. J. (2006) *A Field Guide to the Tiger Beetles of the United States and Canada: Identification, Natural History and Distribution of the Cicindelids*, New York, NY: Oxford University Press.

Pearson, D. L., Anderson, C. D., Mitchell, B. R. *et al.* (2010) Testing hypotheses of bird extinctions at Rio Palenque, Ecuador, with informal species lists. *Conservation Biology*, **24**, 500–510.

Pearson, D. L., Hamilton, A. L. and Erwin, T. L. (2011) Recovery plan for the endangered taxonomy profession. *BioScience*, **61**, 58–63.

Pimentel, D., Wilson, C., McCullum, C. *et al.* (1997) Economic and environmental benefits of biodiversity. *BioScience*, **47**, 747–757.

Rodríguez, J. P., Pearson, D. L. and Barrera, R. (1998) A test for the adequacy of bioindicator taxa: are tiger beetles (Coleoptera: Cicindelidae) appropriate indicators for monitoring the degradation of tropical forests in Venezuela? *Biological Conservation*, **83**, 69–76.

Russell, S. A. (2010) True confessions of a citizen scientist: one woman's quest to become the world's leading expert on a bug. *OnEarth*, (Spring), 33–37.

Satoh, A. (2008) The current status and conservation of coastal tiger beetles (Coleoptera: Cicindelidae). *Japan Journal of Conservation Biology*, **13**, 103–110.

Schlesinger, M. D. and Novak, P. G. (2011) Status and conservation of an imperiled tiger beetle fauna in New York State, USA. *Journal of Insect Conservation*, **15**(6), 839–852.

Schultz, T. D. and Rankin, M. A. (1983a) The ultrastructure of the epicuticular interference reflectors of tiger beetles (*Cicindela*). *Journal of Experimental Biology*, **117**, 87–110.

Schultz, T. D. and Rankin, M. A. (1983b) Development changes in the interference reflectors and colorations of tiger beetles (*Cicindela*). *Journal of Experimental Biology*, **117**, 111–117.

Shelford, R. (1902) Observations on some mimetic insects and spiders from Borneo and Singapore. *Proceedings of the Zoological Society of London*, 1902, 230–284.

Shelford, V. E. (1907) Preliminary note on the distribution of the tiger beetles (*Cicindela*) and its relation to plant succession. *Biological Bulletin of the Marine Biological Laboratory, Woods Hole*, **14**, 9–14.

Shelford, V. E. (1913) The reactions of certain animals to gradients of evaporating power of air. A study in experimental biology. *Biological Bulletin of the Marine Biological Laboratory, Woods Hole*, **25**, 79–120.

Sorensen, W. C. (1995) *Brethren of the Net: American Entomology 1840–1880*, Tuscaloosa, AL: University of Alabama Press.

Spomer, S. M. and Higley, L. G. (1993) Population status and distribution of the Salt Creek Tiger Beetle, *Cicindela nevadica lincolniana* Casey (Coleoptera: Cicindelidae). *Journal of the Kansas Entomological Society*, **66**, 392–398.

Strausfeld, N. J., Sinakevitch, I., Brown, S. M. and Farris, S. M. (2009) Ground plan of the insect mushroom body: functional and evolutionary implications. *Journal of Comparative Neurology*, **513**, 265–291.

Toh, Y. and Mizutani, A. (1994) Neural organization of the lamina neuropil of the larva of the tiger beetle, *Cicindela chinensis*. *Cell Tissue Research*, **278**, 135–144.

Tweit, S. J. (2010) Beetle mania. *National Parks Magazine*, Fall, 1–3.

Vítolo, L. A. (2004) *Guia para la Identificación de los Escarabajos Tigre (Coleoptera: Cicindelidae) de Colombia*. Bogotá, Colombia: Instituto de Investigación de Recursos Biológicos Alexander von Humboldt..

Vogler, A. P. and DeSalle, R. (1994) Diagnosing units of conservation management. *Conservation Biology*, **8**, 354–363.

Vogler, A. P., Welsh, A. and Hancock, K. M. (1997) Phylogenetic analysis of slippage-like sequence variation in the V4 rRNA expansion segment in tiger beetles (Cicindelidae). *Molecular Biology and Evolution*, **14**, 6–19.

Vuilleumier, F. (1998) A need to collect birds in the Neotropics. *Ornitologia Neotropica*, **9**, 201–203.

Vuilleumier, F. (2003) Neotropical ornithology: then and now. *Auk*, **120**, 577–590.

Wheeler, Q. D., Raven, P. H. and Wilson, E. O. (2004) Taxonomy: impediment or expedient? *Science*, **303**, 285.

Wickham, H. F. (1904) The influence of the mutations of the Pleistocene lakes upon the present distribution of *Cicindela*. *American Naturalist*, **38**, 643–654.

Wilcove, D. S. and Eisner, T. (2000) The impending extinction of natural history. *Chronicle Review, Chronicle of Higher Education*, **47**, B24.

Wilson, E. O. (1999) *Consilience: the Unity of Knowledge*. New York: Alfred A Knopf, Inc.
Wilson, E. O. (2000) On the future of conservation biology. *Conservation Biology*, **14**, 1–3.
Wu, X.-Q. and Shook, G. (2010) Common English names and Chinese names for tiger beetles of China. *Journal of the Entomological Research Society*, **12**, 71–92.
Yager, D. D., Cook, A. P., Pearson, D. L. and Spangler, H. G. (2000) A comparative study of ultrasound-triggered behaviour in tiger beetles (Cicindelidae). *Journal of Zoology*, **251**, 355–368.
Zerm, M., Zinkler, D. and Adis, J. (2004) Oxygen uptake and local PO_2 profiles in submerged larvae of *Phaeoxantha klugii* (Coleoptera: Cicindelidae), as well as their metabolic rate in air. *Physiological and Biochemical Zoology*, **77**, 378–389.

5

A is for apiculture, B is for bee, C is for colony-collapse disorder, P is for pollinator parks: an A to Z overview of what insect conservationists can learn from the bees

EDWARD M. SPEVAK

BEES AND CONSERVATION

Species conservation is a multi-faceted endeavour that includes education (Cardoso *et al.* 2011), engagement (Novacek, 2008), advocacy (Brussard and Tull 2006), developing constituencies (Pritchard 2011) and often behavioural change (Schultz 2011) to implement. Invertebrates as a group are an excellent example for wildlife conservation to illustrate the multiple roles and interactions that animals have with the environment (Samways 1994, 2005), the numerous ways that conservation has been and can be implemented (Samways, 1994, 2005) and reasons for failures or impediments to further conservation efforts (Cardoso *et al.* 2011; New 2011). However, it is difficult for many people to embrace invertebrate conservation due to a variety of factors, including lack of knowledge about their roles in the environment, little understanding of their importance to humans through ecosystem services, and a lack of awareness of conservation problems (Cardoso *et al.* 2011; New 2011). What may be needed is a model or surrogate species or group of species (Wiens *et al.* 2008) that can touch on a variety of areas in the animal/environmental and human/environmental realms to promote invertebrate conservation and potentially other conservation agendas. The following is a discussion of how bees might be used

Figure 5.1   Potter wasp (*Euodynerus* sp.) on *Rudbeckia* sp. Photo by
Edward Spevak 2008.

as a model by conservationists to develop and promote a variety of
education and conservation initiatives.

BEES: WHAT ARE THEY AND WHY SHOULD WE CARE?

Bees, from a scientific point of view, are members of the superfamily
Apoidea, within the insect order Hymenoptera, which also includes
wasps, ants and their relatives (Michener 2007; O'Toole and Raw 1991)
(see Figure 5.1). Bees comprise a group of around 20,000 species (more
than all mammal and bird species combined) (Michener 2007) that
feed predominantly on pollen and nectar. It is this last fact that has
helped tie the bee to human culture and existence and makes them an
ideal candidate for promoting conservation of biodiversity and educating the public about the environment.

Most people know what a bee is compared to many invertebrates that we study or try to conserve. Unfortunately, for most people
when you mention the word bee they think of the European honeybee (*Apis mellifera*). They do not realize that the European honeybee is
just one species of this incredibly diverse, beautiful and fascinating
group of animals. Many do not know that bee species range in size

Figure 5.2   Black and gold bumble bee (*Bombus auricomus*) on blue false udigo (*Baptisia*). Photo by Edward Spevak 2012.

from the diminutive *Peridita minima* of the American Southwest at 2 mm in length to the 'enormous' Wallace's giant bee (*Megachile pluto*) of Indonesia reaching 39 mm in length with a wingspan of 63 mm and that they range in colours from black and yellow to orange and red to metallic greens and blues with eyes of turquoise and emerald. Bees also exhibit social structures from the truly social honeybees (*Apis* spp.), bumblebees (*Bombus* spp.; Figure 5.2) and stingless bees (Apidae, subfamily Meliponini) to completely solitary for the majority of bees (Michener 2007; O'Toole and Raw 1991). Few realize that most bees are solitary, having heard of or maybe even having experienced a honeybee hive, they believe all bees are social. Historically, this notion has been reinforced through numerous comparisons of human societies for various political agendas with 'the hive' (Preston 2006).

Biologically, bees are endlessly fascinating and would appeal to many inquisitive students and members of the public, but it is as pollinators and producers of honey and wax that bees first became inextricably linked with humans (Buchmann and Repplier 2005; Ellis 2004; Wilson 2004). Honey, i.e., concentrated and modified nectar, was first collected from wild colonies of honeybees, as it is still collected in various parts of the world as a sweet addition to the human diet (Buchmann and Repplier 2005). Over time people discovered how to domesticate honeybees in order to provide themselves with a constant supply of honey as well as beeswax for use in art, candles, cosmetics, medicine, sealants, etc. (Buchmann and Repplier 2005; Ellis 2004; Wilson 2004). The seemingly miraculous products of sweet honey and the unique lifestyle of honeybees, compared to other insects with which humans were familiar, caused various cultures to incorporate the honeybee into the mythologies and beliefs of humans (Buchmann and Repplier 2005; Kritsky and Cherry 2000; Wilson 2004). Bees came to symbolize diligence, organization, sociability, purity, chastity, cleanliness, spirituality, wisdom, courage, abstinence, sobriety, creativity and vigilance (Kritsky and Cherry 2000). Bees and honey also became important in the beliefs of the three Abrahamic religions; Judaism, Christianity and Islam; for example as symbolizing a promised land of bounty (Exodus 3: 7–9), symbols and representations of the virgin birth and Christ (Preston 2006), and a drink of healing for mankind (Qur'an 16: 69).

From a bee/human cultural or even a faith-based creation-care narrative one can develop a number of messages with complementary actions for conservation (O'Brien 2011; Pritchard 2011). As always it is a matter of understanding your message and knowing your audience (Novacek 2008). However, for many people a motivational element such as a concrete ecological or economic cost or benefit for conservation is necessary for a behavioural change (O'Brien 2011; Schultz 2011).

As bees collect pollen, as a protein source, to feed their developing young, they provide the keystone ecosystem service of pollination as primary pollinators for wild and domesticated plants (Abrol 2012; Michener 2007; O'Toole and Raw 1991). Wind-pollinated crops and tuber species represent the major source of energy in the human diet (Ghazoul 2005; Prescott-Allen and Prescott-Allen 1990; Tilman *et al.* 2002) but bee and insect pollination is essential for the supply of vegetable proteins (soybean, oil palm, rape seed, beans, peas), and dietary fibres (vegetables) (Abrol 2012; Delaplaine and Mayer 2000; Kremen and Chaplin-Kramer 2007) and the majority of the lipids and several micronutrients required for human health (Eilers *et al.*, 2011). Some fruits and vegetables require bee and insect pollination services for the production of the fruit or vegetable itself, e.g., almonds, apples, apricots, blueberries, cantaloupes, citrus, cucumbers, peach, plum, squash, sunflower and watermelon (Abrol 2012; Delaplaine and Mayer 2000). For other fruits or vegetables, insect pollination is not a strict requirement for fruit production, but it substantially increases yields (e.g., coffee) (e.g. Klein *et al.* 2007). In addition, bee-pollinated crops are widely used for cattle feeding (e.g. alfalfa, clover). Economically pollination of agricultural crops by bees in the United States has an estimated value of US $29 billion (Calderone 2012) and worldwide animal pollination is valued at $217 billion (Gallai *et al.* 2008). It has been estimated that one of every three bites of food or drink is dependent upon the services of pollinators, with the majority of these being bees (Buchmann and Nabhan 1996). Bee pollination is essential for global agriculture and human food security and as the planting of additional pollinator dependent crops increases, bees and other insect pollinators will become even more important (Klein *et al.* 2007; Kremen 2008; Kremen and Chaplin-Kramer 2007). However, these ecosystem services provided by bees may be endangered as they like many other species are showing population declines and are disappearing.

BEEKEEPING: APICULTURE AND COLONY COLLAPSE DISORDER

Buchmann and Nabhan (1996) are probably rightly credited with sounding the clarion call for pollinators. Through their book *The Forgotten Pollinators*, Buchmann and Nabhan illustrated beautifully and clearly the diversity and importance of pollinators to the environment, agriculture, and for the survival of plants, animals and humans. They also brought into focus the loss or decline of many pollinators and the lack of conservation efforts. Since, then additional work has explored this issue (Allen-Wardell *et al.* 1998), documented pollinator declines and loss of pollinator services for wild plants and crops (Biesmeijer *et al.*, 2006; Klein *et al.* 2007; National Resources Council, 2007) and identified further areas of research and conservation action (National Resources Council, 2007).

In 2006, a new and potentially more serious issue arose with the collapse and disappearance of honeybee colonies in the United States (Oldroyd 2007). The phenomenon has been termed Colony Collapse Disorder or CCD (Oldroyd 2007; vanEngelsdorp *et al.* 2008, 2009). No definitive cause has been identified, as of yet, but there are numerous hypotheses as to its cause or contributing factors (vanEngelsdorp *et al.* 2008, 2009), including, viruses (e.g., Bromenshenk *et al.* 2010; Singh *et al.* 2010), parasites (e.g., Bromenshenk *et al.* 2010; Core *et al.* 2012; vanEngelsdorp *et al.* 2009) and pesticides (e.g., Chauzat *et al* 2006; Krupke *et al.* 2012). CCD or CCD-like syndromes have now been seen across Europe and in other parts of the world (UNEP 2010). CCD and the loss of honeybees has become a cause célèbre with the production of feature length movies (e.g., Bee Movie), numerous documentaries and books (Benjamin and McCallum 2008; Jacobsen 2009; Schacker 2009). It is believed that the loss of the major managed pollinator for agricultural crops around the world could have devastating implications (Gallai *et al.* 2008; Klein *et al.* 2007; UNEP 2010)

The United States Department of Agriculture (USDA) has estimated that there are between 139 600 and 212 000 beekeepers in the United States. Most beekeepers (approximately 94%; National Resources Council 2007) are hobbyists with fewer than 25 hives. In the United States, CCD has mostly affected the commercial beekeepers which supply the majority of pollination services to commercial crops (vanEngelsdorp *et al.* 2008, 2009). With the advent of CCD and the concerns about pollinator loss, apiculture or beekeeping among the general public is on the rise. There are hundreds of local beekeeping associations across the United States (National Resource Council 2007) and many of them have seen dramatic increases in the number of individuals signing up for beekeeping workshops and those interested in keeping bees, in part to assist with the 'decline' of honeybees.

It must be remembered that other species of bees such as bumblebees (*Bombus* sp.), as well as the honeybee, have exhibited population declines (Biesmeijer *et al.* 2006, Cameron *et al.* 2011; Otterstater and Thomson 2008) and are equally if not more important for pollination services for some crops than honeybees (Kremen 2008). However, with the current interest in honeybees and the importance of pollination for agriculture, conservationists and educators can develop messages and programmes that expand upon public knowledge and encourage behavioural change. Though there is always a concern that a focus on honeybees to the neglect of other bee species can be detrimental for overall bee conservation (Aebii *et al.* 2012; Ollerton *et al.* 2011) many of the actions that will help honeybees will also help other species (Winfree 2010).

GARDENING FOR BEES AT HOME AND IN THE COMMUNITY

Bee conservation is a matter of stewardship of our surroundings and increasing our connections to the natural environment. In the United States, there is a growing number of people that are interested in some relationship with plants and the natural environment as demonstrated through the popularity of gardening (National Gardening Association 2011). An increased understanding of pollination and other ecosystem services and the role that invertebrates play in the garden and the environment by the general public will help to overcome some of the issues preventing effective invertebrate conservation (Cardoso *et al.* 2011; see Rykken, this volume). In addition, the loss of experience with bees and other invertebrates by many individuals, whether adult or child, has been identified as a possible threat to invertebrate conservation action (Cheesman and Key 2007) and could be improved through specific educational initiatives. The interconnection between bees and gardening is also a natural fit for improving science learning and developing conservation behaviours among people (Krasny and Tidball 2009; Mader *et al.* 2011). Developing focused, achievable conservation actions through more limited messages of 'protect the bees for your garden' as opposed to larger messages such as 'save the environment' may have a greater chance of success in changing individual behaviours (Schultz 2011). There is also evidence to suggest that promoting positive behaviour alternatives, such as gardening for wildlife, is more likely to induce change than attempts to curtail or prevent a certain behaviour (Schultz 2011; see Daniels, this volume).

Among the reasons cited for the loss of bees are habitat loss and fragmentation (Williams and Osborne 2009; Winfree, 2010) and pesticide misuse (Chauzat *et al* 2006; Jacobsen 2009: Krupke *et al.* 2012; Winfree 2010). Human actions, through development, urbanization and agriculture, reduce habitat sizes and connections. Gardening, by its very nature, is usually an attempt to create a habitat for wildlife or one that is conducive to growing plants which often attract wildlife. Gardens can act as additional habitats and corridors for bees as illustrated by high abundance and diversity of garden bees in cities like San Francisco (McFrederick and LeBuhn 2006) and New York City (Matteson and Langellotto 2009; Matteson *et al.* 2008). In the UK, higher nest densities of bumblebees were found in home gardens than in rural habitats (Osborne *et al* 2008). Gardens can be an important habitat for bumblebees by providing a greater density and diversity of floral resources (Goulson *et al.* 2002).

Actions to protect bees from pesticides can be focused through the concern of people about the effects of pesticides on human health (Damalas and Eleftherohorinos 2011). Many people are looking for safer alternatives for themselves and their foods. The promotion of alternatives to pesticides or limiting their use in gardens will benefit bees and other wildlife as well as people (Mader *et al.* 2011)

The popularity of gardening can be seen in the number of people actively involved in this activity. In the United States, according to the

National Gardening Association (2011), 83 million households participate in lawn and garden activities with up to US $30 billion spent on these activities. Interestingly, food gardening has been the only area of lawn and garden activity that has seen a significant increase in household participation over the years. For example, between 2008 and 2009 participation in food gardening in the United States increased by 5 million to 41 million households (National Gardening Association 2010) and accounts for around US $3 billion in spending (National Gardening Association 2011). Food gardening includes growing vegetables, fruit, berries and herbs. It is estimated that a well-maintained food garden can yield half a pound of fresh produce per square foot of garden area (National Gardening Association 2010). Reasons given for food gardening are better tasting food, reducing household food bills, producing better quality food and growing food known to be safe (National Gardening Association 2009) as well as to enjoy nature, and health benefits (Armstrong 2000). The importance of food gardening for education and conservation opportunities should not be underestimated as food gardening cuts across all population demographics (National Gardening Association. 2010) which can encourage an exchange of information and experiences across generations.

The majority of food gardening occurs at home but around 1 million households garden in community garden plots with an additional estimated 5 million households expressing interest in having a garden plot in a community garden (National Gardening Association. 2009). Surveys document a wide variety of populations served by, and participating in, community garden programmes. A number of community gardens have a particular socio-demographic or programme focus, for example, serving a particular age-group, socio-economic group or special population group. However, many community gardens serve general neighbourhoods, communities or villages (Armstrong 2000)

The top five garden fruits and vegetables grown in New York community gardens are tomatoes, sweet peppers, beans, eggplant and cucumbers (Gittleman *et al.* 2010) and the top four most popular vegetables grown nationwide in the United States are tomatoes, cucumbers, sweet peppers and beans (National Gardening Association 2009). All of these fruits and vegetables are either bee pollinator dependent or benefit from bee pollination (Abrol 2012; Delaplaine and Mayer 2000). The importance of bees to food gardening can be used to develop on-site programmes, educational materials and continuing education classes for gardeners, such as the Master Gardener programmes in the United States, on pollination and pollinators. An increase in an understanding of pollinators and pollination for fruit and vegetable gardens can also help to improve science and conservation education. When one considers that many community gardens partner with at least one school and many others would like to partner with local schools (Gittleman *et al.* 2010; National Gardening Association 2009), there is the potential for incorporating and expanding pollination studies and conservation into school science curricula.

One caveat regarding the promotion of bees in home and community gardens is the fear by some of getting stung (Anthony and McCabe 2005). These types of fears are not necessarily unique to bees and other insects (Anthony and McCabe 2005). There are practical ways of reducing and helping to alleviate the concerns. It has been shown that exposure therapy (Anthony and McCabe 2005) can reduce fears and concerns. As people learn to work around bees fears can be reduced or eliminated. Additionally, exposure therapy plus rewards (Jones and Friman 1999), which in the context of food gardening would be bountiful, well-formed, appealing and tasty fruits and vegetables, can have the desired effect of reducing fears. Another method to allay fears of bees in the garden is to focus on developing nest sites in the form of bee blocks and 'bee hotels' for solitary twig- and tunnel-nesting bees, e.g., leafcutter and mason bees (*Megachile* sp. and *Osmia* sp.) (Mader *et al.* 2011) instead of considering installing honeybee hives. In some community gardens, honeybee hives have been proposed, but have caused concerns among gardeners about working so close to an active hive (pers. obs.). Bee blocks would allow for potential increased pollination services while allowing gardeners to observe close-up the activities of these inoffensive bees.

MELITTOLOGY: BEE GUIDES AND CITIZEN SCIENCE

Melittology, the study of all bees, is distinct from apiology, which focuses only on the study of honeybees of the genus *Apis*. Along with the honeybee, it is the appreciation and study of all other bees or at a minimum species or groups of non-*Apis* bees (e.g., bumblebees, *Bombus* spp.) that we need to expand among the general public, students, conservation organizations and government agencies and thereby effect changes for conservation. However, to effect change in conservation actions people need to know what they are conserving. Field guides and citizen science projects have the potential to expand the education of individuals regarding bees. It has been argued that education alone is insufficient to instil change (Schultz 2011) yet it is undeniable that the advent of field guides for various taxa has inspired and encouraged people to explore the natural world (Cheesman and Key 2007), swelled the ranks of concerned individuals (Pearson and Shetterly 2006) and increased conservation advocacy (Brussard and Tull 2007).

Field guides abound for a variety of taxa from birds and mammals to reptiles and fish. For invertebrates there are numerous general field guides and some specific guides focused on butterflies and moths (see New, this volume), dragonflies and damselflies (see Samways, this volume) and beetles (see Pearson, this volume) but there are very few guides for bees. The majority of general field guides for bees focus on the larger more spectacular bumblebees (e.g., Colla *et al.* 2011; Kearns and Thomson 2001; Pinchen 2006) with more advanced guides also focused primarily on bumblebees (e.g., Prys-Jones and Corbet 2011). The development of additional guides for gardeners, students, farmers, amateur melittologists and many professionals to identify bees is sorely needed. There are plans to develop additional bee guides that illustrate non-*Apis* bees along with bumblebees, but they cannot be produced too soon.

Along with field guides, citizen science programmes offer the opportunity to expand education messages about the environment as well as collecting potentially valuable data for future conservation measures (Bonney *et al.* 2009; Cohn 2008; National Resources Council 2007). Several citizen science projects have been developed within North America and Europe that can serve as models for other initiatives for bee and other species conservation (see Johansen and Auger, this volume).

In 1999, the Bumble Boosters project began as a joint collaboration between the University of Nebraska's Department of Entomology, Lincoln Public School's Science Focus High School and the Lincoln's Children's Zoo (Golick and Ellis 2003). The main goal of the programme was to develop a community of high school students across the state of Nebraska monitoring the distribution and abundance of Nebraska bumblebee species and in the process 'raise student and public awareness of the environmental importance of pollinating insects, enhance students' understanding of the process of conducting scientific investigation, increase students' knowledge of insect biology and pollination ecology, and engage students' abilities in networking with other students and leaders to solve a shared problem' (Golick and Ellis 2003: 76). Along with identifying and mapping the distributions of the 20 species of bumblebees (*Bombus* sp.) found in Nebraska the programme participants gained a greater knowledge and appreciation for pollination by insects in the environment as well as scientific inquiry (Golick and Ellis 2003).

In 2008, researchers at San Francisco State University launched the Great Sunflower Project. The aim of the project is to develop a nationwide map of bee distributions by gathering information about urban, suburban and rural bee populations and empower people to learn about what was happening with the bees in their backyard. The project originally had participants plant Lemon Queen sunflowers in their gardens and during 30 minute observation periods during the bloom season identify the types and number of bees visiting the flowers. Participants upload their observations to the Great Sunflower Project website for collation and analyses by researchers. The project has since expanded its plant list to include bee balm, cosmos, rosemary, tickseed, goldenrod and purple coneflower. When the project started in 2008, 25 000 volunteers registered to participate in the project; by 2011 around 100 000 people had pledged their participation.

The Great Pollinator Project (GPP) was started in 2007 in New York between the Center for Biodiversity and Conservation at the American Museum of Natural History and the Greenbelt Native Plant Center. It was originally called the Bee Watchers project and was developed in collaboration with the Great Sunflower Project. The current goals of this project are more regional in scope than the Great Sunflower Project: to increase understanding of bee diversity in New York City and the region; to raise public awareness of native bees; and to improve park management and home gardening practices to benefit native bees. The initial goal of the GPP was to identify which areas of New York City have good pollinator service by determining how quickly bees show up to pollinate flowers at various locations. During the summers of 2007 through 2010, volunteer Bee Watchers observed the bees visiting selected flowering plant species such as coneflower, mountain mint and rough-leaved goldenrod. Observed bees were 'assigned' to one of five categories: honeybee, bumblebee, large carpenter bee, shiny green bee and other type of bee. The data from these observations is currently being analyzed.

Also in 2007, the BeeSpotter programme was started by the University of Illinois Urbana-Champaign. The goal of this project is to engage citizen scientists with the professional scientific community by observing and photographing honeybees and bumblebees. Citizen scientists photograph bees and submit them to the BeeSpotter website for identification and distribution mapping. The programme currently boasts 1000 beespotters but is limited to the state of Illinois; there are plans however to eventually expand the programme across the United States (M. Berenbaum, pers. comm.). Through the work of these bees-potter photographers, University of Illinois bee researchers have been able to identify the presence of the rusty-patched bumblebee (*B. affinis*) in Illinois, a species which has disappeared across its range in the eastern United States, and the red-belted bumblebee (*B. rufocinctus*) which was last seen in Illinois in 1949.

In 2006 in the UK, the BumbleBee Conservation Trust (BBCT) was established and now boasts over 7000 members. The BBCT has initiated several citizen science projects. Beewatch is designed to map the distributions of bumblebees, particularly the rarer species, from photographs submitted by the public. The blaeberry/mountain bumblebee (*B. monticola*) and tree bumblebee (*B. hypnorum*) surveys are focused on the presence and distribution of two specific species of bumblebees. BBCT has also partnered with the British Trust for Ornithology (BTO) to develop the Garden Beewatch programme, where participants record the bees in their garden each week, and submit the data to BTO. The Online software allows citizen scientists to view annual changes, and compare their garden to others.

Each of these programmes has the potential to educate a large and diverse group of individuals about the plight of bees and other pollinators and their importance in the environment. In the process individuals are able to be part of the scientific community by collecting valuable data, interact with scientists and ask questions. The programmes also help to empower them to learn about how the actions they and others take affect the world around them and to develop as advocates for environment and conservation (Brussard and Tull 2006). Though there are concerns about how much these citizen scientists retain about the species being studied and if it results in a behavioural change toward science and conservation (Jordan *et al.* 2011) there is little argument that getting hands-on experience is essential to shape conservation behaviours (Cheesman and Key, 2007; Thompson and Sorenson, 2005)

API-TOURISM, 'BEE-SERVES', BEE SANCTUARIES, POLLINATOR PARKS AND ZOOS

Conservation and education through ecotourism, travel and experiential venues has historically been the realm of people interested in the charismatic mega-vertebrates, e.g., whale watching or the wildebeest migration in the Serengeti. However, as Lemelin (2007, 2012) points out insects and more specifically bees have their own following and provide unique opportunities for conservation education and programmes.

Apitourism (Horn 2005) is a relatively new phenomenon and is primarily focused on the beekeeping community or those interested in apitherapy, i.e., the medicinal use of products from honeybees. Each year, 24 000 tourists visit the Bees Museum in Ayer Keroh, Malacca, and accompany Mowalis (traditional honey-collectors) who provide guided excursions throughout the forest while they seek out the nests of the giant honeybees (*Apis dorsata*) (Shahwadid *et al.* 2008). Since 2000, Trinidad and Tobago beekeepers have been involved in hosting beekeepers from Europe on 'Beekeepers' Safaris'. These are personalized beekeeping holiday/study tours. They are the only known exclusive beekeepers' tour to the Caribbean and have added a new dimension to regional tourism. Additional, apitourism excursions can be found in Malta, Turkey, Israel, Uganda, Sardinia, Bangladesh and the Ukraine (Horn 2005). Apitourism allows beekeepers, students or regular tourists, to see and experience how other people and cultures keep their bees and produce honey. Since most visitors found honeybees to be 'a worthwhile ecotourism attraction' (Shahwahid *et al.* 2008: 27), there is the potential to expand on this idea to include traditional Meliponiculture, stingless bee culture (Cortopassi-Laurino *et al.* 2006). Meliponiculture, is practised to varying degrees by people throughout the tropical regions of the world. Allowing tourists to see and experience traditional stingless bee culture, as for example practised historically by the Mayan culture of Mexico, could have economic benefits for local people and conservation benefits by encouraging stingless beekeepers to maintain traditional environmentally friendly practices (Cortopassi-Laurino *et al.* 2006).

The development of protected areas for bees and other pollinators is a very recent phenomenon. Bee reserves or 'bee-serves' can become local destinations and educational venues to promote conservation. In 2007, the BumbleBee Conservation Trust established what they call the first bumblebee sanctuary, a 20 acre meadow in Perth and Kinross, beside Loch Leven in central Scotland. The sanctuary was a habitat restoration designed specifically for bumblebees. In Canada, Pollination Guelph along with NSERC-CANPOLIN (the Canadian Pollination Initiative) is developing a 45 hectare (112 acre) Pollinator Park on a decommissioned landfill site. This project is currently the largest pollinator park initiative worldwide.

Finally, insect pavilions and zoos have the potential to greatly increase public knowledge and appreciation for bees, and develop and support research and conservation for bees and other pollinators (see Veltman, this volume). Gusset and Dick (2011) estimate that over 700 million people visit the world's zoos and that zoos spend over $350 million on wildlife conservation each year. Much of this is focused on vertebrates but there are many invertebrate field conservation programmes as well as invertebrate captive breeding programmes with a field component (Pearce-Kelly *et al.* 2007). A number of zoos have bee exhibits or pollinator gardens focused on bees but zoos are also developing, in collaboration with universities and other conservation organizations, programmes specifically devoted to native bee conservation (e.g. Saint Louis Zoo's WildCare Institute Center for Native Pollinator Conservation (CNPC)). The Saint Louis Zoo's CNPC is developing a variety of projects devoted to native bee conservation and is looking to encourage additional zoos to develop bee conservation programmes for which there is an interest. In 2011, the CNPC surveyed accredited zoos of the Association of Zoos and Aquariums (AZA) with the aim of determining current conservation and research programmes focused on bees and their interest in developing conservation, research and education and outreach programmes focused on bees. Of the 63 respondents only 8 (12.7%) institutions currently had programmes related to native bees, however, 55 (87.3%) were interested in developing native bee programmes (Spevak, unpublished data). Bee conservation is fairly new within the zoo community but it has great potential to be developed and supported.

CONCLUSION

Bees are unique species that have ties to the environment through the ecosystem services of pollination of wild flowers and trees that provide shelter and food for a myriad of wildlife and to humans through the pollination of their crops, production of honey and wax for human consumption and by their long associations with human cultures and beliefs. It has been argued by some that species should be conserved for their intrinsic value, that they should be conserved for their own sake, 'but this neglects other types of instrumental value that might contribute to the value of species and ecosystems: aesthetic, spiritual, educational, scientific, and even 'existence' value – satisfaction humans derive from knowing that species and ecosystems remain, even if they are not experienced' (Maguire and Justus 2008: 911). Bees should be conserved for both their intrinsic and extrinsic value.

As has been shown conservation messages and actions involving bees can engage people on a variety of levels. They are one of the few groups of species that could get people directly involved with conservation.

If a person wants to help conserve tigers or elephants what can they do? They can give money to a conservation organization working with these species in the wild or visit and support a zoo that supports a conservation breeding and field programme for tigers or elephants. If a person wants to conserve bees they can support the Xerces Society in the United States or Buglife in the UK, but they can also plant a garden, create suitable habitats in their backyards for pollinators, buy local and organic honey, fruits and vegetables or participate in a citizen science programme. In essence any individual can become a bee conservationist.

REFERENCES

Abrol, D. P. (2012) *Pollination Biology: Biodiversity Conservation and Agricultural Production*. New York: Springer.

Aebii, A., Vaissière, B. E., vanEngelsdorp, D. *et al.* (2012) Back to the future: *Apis* versus non-*Apis* pollination. *Trends in Ecology and Evolution*, **27**(3), 142–143.

Allen-Wardell, G., Bernhardt, P., Bitner, R. *et al.* (1998) The potential consequences of pollinator declines on the conservation of biodiversity and stability of food crop yields. *Conservation Biology*, **12**(1), 8–17.

Anthony, M. and McCabe, R. E. (2005) *Overcoming Animal and Insect Phobias: How to Conquer Fear of Dogs, Snakes, Rodents, Bees, Spiders, and More*. Oakland, CA: New Harbinger Publications.

Armstrong, D. (2000) A survey of community gardens in upstate New York: implications for health promotion and community development. *Health & Place*, **6**, 319–327.

Benjamin, A. and McCallum, B. (2008) *A World Without Bees*. London: Guardian Books.

Biesmeijer, J. C., Roberts, S. P. M., Reemer M. *et al.* (2006) Parallel declines in pollinators and insect-pollinated plants in Britain and the Netherlands. *Science*, **313**, 351–354.

Bonney, R., Cooper, C. B., Dickinson, J. *et al.* (2009) Citizen science: a developing tool for expanding science knowledge and scientific literacy. *BioScience* **59**(11), 977–984.

Bromenshenk, J. J., Henderson, C. B., Charles H. Wick, C. H. *et al.* (2010) Iridovirus and microsporidian linked to honey bee colony decline. *Plos ONE*, **5**(10), e13181.

Brussard, P. F. and Tull, J. C. (2006) Conservation biology and four types of advocacy. *Conservation Biology*, **21**(1), 21–24.

Buchmann, S. L. and Nabhan, G. P. (1996) *The Forgotten Pollinators*. Washington DC/Covelo: Shearwater Books/ California Island Press.

Buchmann, S. and Repplier, B. (2005) *Letters from the Hive: An Intimate History of Bees, Honey, and Humankind*. New York: Bantam Books.

Cameron, S., Hines, H., Sinn-Hanlon, J. and Spevak, E. (2008) *Bumblebees of Illinois and Missouri*. Champaign-Urbana, IL: Department of Entomology University of Illinois at Champaign-Urbana and the Saint Louis Zoo.

Cameron, S., Lozier, J. D., Strange, J. P. *et al.* (2011) Patterns of widespread decline in North American bumblebees. *Proceedings of the National Academy of the United States of America*, **108**, 662–667.

Calderone, N. W. (2012) Insect pollinated crops, insect pollinators, and US agriculture: trend analysis of aggregate data for the period 1992–2009. *PLoS ONE*, **7**(5), doi: 10.1371/journal.pone 0037235.

Cardoso, P., Erwin, T. L., Borges, P. A. V. and New, T. (2011) The seven impediments in invertebrate conservation and how to overcome them. *Biological Conservation*, **144**, 2647–2655.

Chauzat, M.-P., Faucon, J.-P., Martel, A.-C. *et al.* (2006) A survey of pesticide residues in pollen loads collected by honey bees in France. *Journal of Economic Entomology*, **99** (2), 253–262.

Cheesman, O. D. and Key, R. S. (2007) The extinction of experience: a threat to insect conservation? In *Insect Conservation Biology: Proceedings of the Royal Entomological Society's 23rd Symposium*, ed. A. J. A. Stewart, T. R. New, and O. T. Lewis. Wallingford, UK: CABI, pp. 322–348.

Cohn, J. (2008) Citizen Science: can volunteers do real research? *Bioscience* **58**(3), 192–197.

Colla, S. R. and Packer, L. (2008) Evidence for decline in eastern North American bumblebees (Hymenoptera: Apidae), with special focus on *Bombus affinis* Cresson. *Biodiversity Conservation*, **17**, 1379–1391.

Colla, S. R., Otterstatter, M. C., Gegear, R. J. and Thomson, J. D. (2006) Plight of the bumblebee: pathogen spillover from commercial to wild populations. *Biological Conservation*, **129**, 461–467.

Colla, S., Richardson, L. and Williams, P. (2011) *Bumblebees of the Eastern United States*. Washington DC: US Department of Agriculture/US Forest Service

Core, A., Runckel, C., Ivers, J. *et al.* (2012) A new threat to honey bees, the parasitic phorid fly *Apocephalus borealis*. *Plos ONE*, **7**(1), 1–9, e29639.

Cortopassi-Laurino, M., Imperatriz-Fonseca, V. L. Roubik, D. W. *et al.* (2006) Global meliponiculture: challenges and opportunities. *Apidologie*, **37**, 275–292.

Damalas, C. A. and Eleftherohorinos, I. G. (2011) Pesticide exposure, safety issues, and risk assessment indicators. *International Journal of Environmental Research and Public Health*, **8**, 1402–1419.

Delaplaine, K. S. and Mayer, D. F. (2000) *Crop Pollination by Bees*. Wallingford, UK: CABI.

Eilers, E. J., Kremen, C., Greenleaf, S. S. Garder, A. K. and Klein, A. (2011) Contribution of pollinator-mediated crops to nutrients in the human food supply. *PLoS ONE*, **6**(6), e21363.

Ellis, H. (2004) *Sweetness and Light: The Mysterious History of the Honeybee*. New York: Harmony Books.

Gallai, N., Salles, J.-M., Settele, J. and Vaissiere, B. E. (2008) Economic valuation of the vulnerability of world agriculture confronted with pollinator decline. *Ecological Economic*, **68**, 810–821.

Ghazoul, J. (2005) Buzziness as usual? Questioning the global pollination crisis. *Trends in Ecology and Evolution*, **20**, 367–373.

Gittleman, M., Librizzi, L. and Stone, E. (2010) *Community Garden Survey: New York City: Results 2009/2010*. New York: Grow NYC/GreenThumb, NYC Department of Parks & Recreation

Golick, D. A. and Ellis, M. D. (2003) Bumble boosters: doing science as a community of learners. *American Entomologist*, **49**(2), 76–80.

Goulson, D., Hughes, W. O. H., Derwent, L. C. and Stout, J. C. (2002) Colony growth of the bumblebee, *Bombus terrestris*, in improved and conventional agricultural and suburban habitats. *Oecologia*, **130**, 267–273.

Gusset, M. and Dick, G. (2011) The global reach of zoos and aquariums in visitor numbers and conservation expenditures. *Zoo Biology*, **30**, 566–569.

Horn, T. (2005) *Bees in America: How the Honey Bee Shaped a Nation*. Lexington, KY: The University Press of Kentucky.

Jacobsen, R. (2009) *Fruitless Fall: The Collapse of the Honey Bee and the Coming Agricultural Crisis*. New York: Bloomsbury.

Jones, K. M. and Friman, P. C. (1999) A case study of behavioral assessment and treatment of insect phobia. *Journal of Applied Behavior Analysis*, **32**, 95–98.

Jordan, R. C., Gray, S. A., Howe, D. V., Brooks, W. R. and Ehrenfeld, J. G. (2011) Knowledge gain and behavioral change in citizen-science programs. *Conservation Biology*, **25**(6), 1148–1154.

Kearns, C. A. and Thomson, J. (2001) *The Natural History of Bumblebees: A Sourcebook for Investigations*. Boulder, CO: University Press of Colorado.

Klein, A. M., Vaissière, B. E., Cane, J. H. *et al.* (2007) Importance of pollinators in changing landscapes for world crops. *Proceedings Royal Society B Biological Science*, **274**, 303–313.

Krasny, M. and Tidball, K. (2009) Community gardens as context for science, stewardship and advocacy learning. *Cities and the Environment*, **2**(1), article 8.

Kremen, C. (2008) Crop Pollination Services from Wild Bees. In *Bee Pollination in Agricultural Ecosystems*, ed. R. R. James and T. L. Pitts-Singer. Oxford: Oxford University Press.

Kremen, C. and Chaplin-Kramer, R. (2007) Insects as providers of ecosystem services: crop pollination and pest control. In *Insect Conservation Biology: Proceedings of the Royal Entomological Society's 23rd Symposium*, ed. A. J. A. Stewart, T. R. New and O. T. Lewis. Wallingford, UK: CABI, pp. 349–382

Kritsky, G. and Cherry, R. (2000) *Insect Mythology*. Bloomington, IN: Writers Club Press

Krupke, C. H., Hunt, G. J., Eitzer, B. D., Andino, G. and Given, K. (2012) Multiple routes of pesticide exposure for honey bees living near agricultural fields. *Plos ONE* **7**(1), e29268.

Lauck, J. E. (1998) *The Voice of the Infinite in the Small: Revisioning the Insect–Human Connection*. Mill Spring, NC: Swan Raven & Co.

Lemelin, R. H. (2007). Finding Beauty in the Dragon: The Role of Dragonflies in Recreation, Tourism, and Conservation. *Journal of Ecotourism*, **6**(2), 139–145.

Lemelin, R. H. (2012). To bee or not to bee: whether 'tis nobler to revere or to revile those six-legged creatures during one's leisure. *Leisure Studies*, doi: 10.1080/02614367.2011.626064.

Mader, E., Shepherd, M., Vaughan, M., Black, S. H. and LeBuhnXerces, G. (2011) *Attracting Native Pollinators: The Xerces Society Guide to Conserving North American Bees and Butterflies and Their Habitat*. North Adams. MA: Storey Publishing.

Maguire, L. A. and Justus, J. (2008) Why intrinsic value is a poor basis for conservation decisions. *Bioscience*, **58**(10), 910–911.

Matteson, K. C. and Langellotto, G. A. (2009) Bumblebee abundance in New York City community gardens: implications for urban agriculture. *Cities and the Environment*. **2**(1), article 5, 12 pp.

Matteson, K. C., Ascher, J. S. and Langellotto, G. A. (2008) Bee Richness and Abundance in New York City Urban Gardens. *Annals of the Entomological Society of America.* **101**(1), 140–150.

McFrederick, Q. and LeBuhn, G. (2006) Are urban parks refuges for bumblebees *Bombus* spp. (Hymenoptera: Apidae)? *Biological Conservation,* **129**, 372–382.

Michener, C. D. (2007) *Bees of the World,* 2nd edn. Baltimore, MD: Johns Hopkins University Press.

National Gardening Association (2009) *The Impact of Home and Community Gardening In America.* South Burlington, VT: National Gardening Association, Inc.

National Gardening Association (2010) *Hard Times Lawn and Garden Survey.* South Burlington, VT: National Gardening Association, Inc.

National Gardening Association (2011) *National Gardening Survey.* South Burlington, VT: National Gardening Association, Inc.

National Resource Council (2007) *Status of Pollinators in North America.* Washington DC: The National Academies Press.

New, T. R. (2011) Strategic planning for invertebrate species conservation: how effective is it? *Journal of Threatened Taxa* **3**(9), 2033–2044.

Novacek, M. J. (2008) Engaging the public in biodiversity issues. *Proceedings of the National Academy of Science,* **105**(suppl. 1), 11571–11578.

O'Brien, P. (2011) Reaching the US Public through their patriotism, pastors, and pockets. *Conservation Biology,* **25**(6), 1108–1111.

Oldroyd, B. P. (2007) What's killing American honey bees? *PLoS Biology,* **5**, 1195–1199.

Ollerton, J. O., Price, V., Armbruster, W. S. *et al.* (2011) Overplaying the role of honey bees as pollinators: a comment on Aebi and Neumann (2011). *Trends in Ecology and Evolution,* **27**(3), 141.

Osborne, J. L., Martin, A. P., Shortall, C. R. *et al.* (2008) Quantifying and comparing bumblebee nest densities in gardens and countryside habitats. *Journal of Applied Ecology,* **45**, 784–792.

O'Toole, C. and Raw, A. (1991) *Bees of the World.* Oxford: Blackwell Publishing Ltd.

Otterstater, M. C. and Thomson, J. D. (2008) Does pathogen spillover from commercially reared bumblebees threaten wild pollinators? *PLoS ONE* **3**(7), e2771. doi:10.1371/journal.pone.0002771.

Pearce-Kelly, P., Morgan, R., Honan, P. *et al.* (2007) The conservation value of insect breeding programmes: rational, evaluation tools and example case studies. In *Insect Conservation Biology: Proceedings of the Royal Entomological Society's 23rd Symposium,* ed. A. J. A. Stewart, T. R. New and O. T. Lewis. Wallingford, UK: CABI, pp. 57–75.

Pearson, D. L. and Shetterly, J. A. (2006) How do published field guides influence interactions between amateurs and professional in entomology. *American Entomologist,* **52**(4), 246–252.

Pinchen, B. L. (2006) *A Pocket Guide to the Bumblebees of Britain and Ireland.* Lymington, UK: Forficula Books

Prescott-Allen, R. and Prescott-Allen, C. (1990) How many plants feed the world? *Conservation Biology,* **4**, 365–374.

Preston, C. (2006) *Bee.* London: Reaktion Press.

Pritchard, L. (2011) Cultivating a constituency for conservation. *Conservation Biology* **25**(6), 1103–1107.

Prys-Jones, O. E. and Corbet, S. (2011) *Bumblebees.* Exeter, UK: Pelagic Publishing.

Ransome, H. M. (2004) *The Sacred Bee in Ancient Times and Folklore*. Mineola, NY: Dover Publications, Inc.

Samways, M. J. (1994) *Insect Conservation Biology*. London: Chapman & Hall.

Samways, M. J. (2005) *Insect Diversity Conservation*. Cambridge: Cambridge University Press.

Schacker, M. (2009) *A Spring without Bees: How Colony Collapse Disorder Has Endangered Our Food Supply*. Guilford, CT: The Lyons.

Schultz, P. W. (2011) Conservation means behavior. *Conservation Biology*, **25**(6), 1080–1083.

Shahwadid, H. O. M., Yong, P. L. and Sius, T. (2008) Comparative valuation of eco-tourism attractions between honey bees museum and wild honey collection. *The Malaysian Forester*, **71**(1), 21–28.

Singh, R., Levitt, A. L., Rajotte, E. G. *et al.* (2010) RNA viruses in Hymenopteran pollinators: evidence of inter-taxa virus transmission via pollen and potential impact on non-*Apis* Hymenopteran species. *PLoS ONE* **5**(12): e14357. doi:10.1371/journal.pone.0014357.

Thompson, S. R. and Sorenson, C. E. (2005) Hands-on bugs: bringing insects up close and personal with non-science majors. *American Entomologist* **51**(2), 74–77.

Tilman, D., Cassman, K. G., Matson, P. A., Naylor, R. and Polasky, S. (2002) Agricultural sustainability and intensive production practices. *Nature*, **418**, 671–677.

UNEP (2010) UNEP emerging issues: global honey bee colony disorder and other threats to insect pollinators. United Nations Environment Programme

vanEngelsdorp, D., Hayes, Jr., J., Underwood, R. M. and Pettis, J. (2008) A survey of honey bee colony losses in the US, fall 2007 to spring 2008. *Plos ONE*, **3**(12), 1–6 e4071.

vanEngelsdorp, D., Evans, J. D., Saegerman, C. *et al.* (2009) Colony collapse disorder: a descriptive study. *Plos ONE*, **4**(8), 1–17 e6481

Wiens, J. A., Hayward, G. D., Holthausen, R. S. and Wisdom, M. J. (2008) Using surrogate species and groups for conservation planning and management. *Bioscience*, **58**(3), 241–252

Williams, P. H. and Osborne, J. L. (2009) Bumblebee vulnerability and conservation worldwide. *Apidologie*, **40**, 367–387.

Wilson, B. (2004) *The Hive: The Story of the Honeybee and Us*. New York: Thomas Dunne Books, St Martin's Press.

Winfree, R. (2010) The conservation and restoration of wild bees. *Annals of the New York Academy of Sciences*, **1195**, 169–197.

6

The entomological and recreational aspects of interacting with Lepidoptera

T. R. NEW

INTRODUCTION

Human cultural diversity has led to insects of many kinds being incorporated into symbolism, visual and performance arts, music, religion and folklore. Butterflies and moths figure highly in this array (Hogue 1987), with their incidences on postage stamps and decorative fabrics and as jewellery all widespread. However, many people have mixed reactions to Lepidoptera, with the widespread major anomaly that 'people like butterflies' and 'people dislike moths'. Even the same individuals may react very differently to the two categories, with the common epithets evoking contrasting images: butterflies are viewed as brightly coloured, day-active and harmless insects reflecting tranquillity and harmony, and moths regarded as drab, nocturnal and pestiferous, objects for suppression. Historical positive appreciations of moths – some of them rather atypical species, such as the domestic silkworm (*Bombyx mori*) cultivated in China since at least 2700 BC for silk production – are in direct contrast to the more widespread perception of moths as pests. The divergent images are mirrored in some cultural traditions, with butterflies viewed as 'fairies' and moths as 'witches', so contrasting images of good and evil. Release of butterflies at weddings or other significant events for celebration or 'good luck' is one outcome of this perception. It contrasts with the image of some moths (notably the death's head hawk moths, *Acherontia* spp. (Sphingidae), and some neotropical Noctuidae) as harbingers of death.

The Management of Insects in Recreation and Tourism, ed. Raynald Harvey Lemelin. Published by Cambridge University Press. © Cambridge University Press 2013.

All these contrasting features break down with further examination, but the categorized stereotypes remain influential – to the extent that the well-known tourist feature 'The Valley of the Butterflies' in Rhodes, Greece, actually applies to a brightly coloured diurnal arctiid moth, *Panaxia quadripunctaria*. Other cases of such 'honorary butterflies' are known. The factors that influence this popularity include traditional imagery, size, conspicuousness through diurnal activity and bright colours. However, the images of many insects can be changed to some extent by education and the transitions across insect groups are in some cases diminishing with wider awareness and knowledge. Thus, the spectacular larger silkmoths (Saturniidae) are popular exhibits in many 'butterfly displays', and it has been suggested that the large furry body of these is an attractive feature, together with their large size – much as, for example, bumblebees have long been viewed favourably in relation to many other aculeate Hymenoptera. Size, alone, may foster positive attention through gaining notice so that, as another example, true dragonflies may be viewed more favourably than the generally smaller damselflies within the Odonata. Indeed, Odonata are, next to Lepidoptera, perhaps the most important insects in human culture, and also attractive in recreation and tourism (see discussion by Lemelin 2007, 2009).

The spectrum of Lepidoptera is, however, much wider than butterflies and silkmoths, but butterflies remain the best-known, most liked and most charismatic of all insects (see Figures 6.1 and 6.2). The status (1) flows from mythological intrigue, with the emergence from the pupa likened to a symbol of the resurrection and butterflies sometimes viewed as the souls of recently departed people and (2) is supported by a long history of collector interests, whereby butterflies have become 'desirable'. Perhaps the most familiar entomological stereotype, one beloved of cartoonists, is that of the eccentric enthusiast wielding a long-handled butterfly net in pursuit of quarry as a recreational activity. These interests have fostered production of well-illustrated identification guides facilitating access to the fauna by non-specialists, accumulation of life history and distributional information and the general philatelic appeal later having wide benefits for conservation and ecological assessments (e.g. Asher *et al.* 2001, for Britain) founded in sound scientific understanding of the relatively few species. In general, knowledge of, and interest in, Lepidoptera falls into three categories that parallel the major groups – of (1) butterflies (fewer species, relatively well-studied and popular), (2) 'macromoths' (much more diverse, also well known in many places

Figure 6.1   Painted lady butterfly (*Vanessa cardni*) at rest in butterfly garden, Thunder Bay, Canada. Photo by Elaine C. Wiersma, 2011.

and ranging from popular groups such as hawkmoths [Sphingidae] and silkmoths [Saturniidae] to more diverse and lesser documented families), and (3) 'Microlepidoptera', the daunting array of tiny and poorly known moths traditionally far less attractive to collectors but the most diverse components of the order, far less tractable for identification and study. Importantly, and in contrast to many other groups of insects, Lepidoptera study is a field to which both amateur and professional can contribute meaningfully and (as Young, 1997: xiii, put it for moths) 'a field where amateur and professional do still respect one another'. The cultural transitions of major and continuing relevance are those from collector to student, and from individual to team efforts in which networks or communities become involved, for example in conservation and inventory appraisals leading eventually to organizations founded on amateur interest and/or professional involvements.

Knowledge of how Lepidoptera 'work' flows largely from hobbyist interests, with detailed study of some pest species laying the foundations of understanding the factors affecting insect population dynamics. Much of the fundamental knowledge of insect ecology in the areas of insect–plant interactions has flowed from studies on Lepidoptera, and that knowledge has become a critical component of

Figure 6.2   White admiral butterfly (*Limenitis arthemis*) at rest in butterfly garden, Thunder Bay, Canada. Photo by Elaine C. Wiersma, 2011.

understanding needs for insect conservation in interpreting population and distributional changes and how these occur and are influenced by human activities.

Only in parts of Europe is it currently possible to evaluate conservation status of many Microlepidoptera other than in the most generalized terms, reflecting attention from substantial numbers of resident enthusiasts. Kristensen *et al.* (2007) noted that in that region the 'average amateur lepidopterist' ('having exhausted the challenges [at the collector's level!] presented by the local butterflies': Kristensen *et al.* 2007:708) tends to make the transition from butterflies to macromoths and subsequently often to micromoths, of the national fauna, rather than to butterflies of neighbouring countries. Over most of the rest of the world, however, in any practical sense Microlepidoptera are both unknown and unknowable as candidates for species-focused conservation or other individual attention. In noting that the number of described Microlepidoptera in south east Asia was around 6000 species, for example, Robinson *et al.* (1994) commented also that the number still to be discovered was likely to be 'at least the same number again'. The paucity of resident lepidopterists in many parts of the tropics contrasts markedly with the concentrations of interest and expertise in parts of the temperate regions. The outcome is clear: butterflies are the major focus of public interest, and many moths may gain sympathy by flow-on and education fostered from the flagship base of butterflies, not least to demonstrate the biology and conservation needs they manifest. This taxonomic bias has strong influences on community interest, but leaves abundant educational opportunities for it to be redressed. One important recent case, in Scotland, has involved one of the tiniest European moths, the Sorrel pygmy moth (*Enteucha acetosae*, a member of the Nepticulidae and only about 3 mm in wingspan), as a flagship for promoting conservation of a key valley bog site, Aucheninnes Moss. Acknowledgement that even such tiny insects need and merit conservation, and can become endangered through threats to a specialized habitat, is both welcome and noteworthy.

Other such novelties of structure or biology, such as the fly-catching ambush predatory geometrid caterpillars in Hawai'i (*Eupithecia*, Montgomery 1983), may appeal instantly. Any such unusual biological feature or aesthetic appeal may foster wider interest, perhaps combined with local pride if the species is a local endemic or of local conservation concern – and a local patronymic common name may be important in enhancing this interest. In southeastern Australia, awareness of the complex biology of the golden sun-moth (*Synemon plana*, Castniidae) has become important in assessing the distribution and status of this nationally listed Critically Endangered species as one of a small portfolio of animals (including two reptiles) that are key symbols for threatened native grassland habitats. Any such wider engagement, however, flows from interest and concerns arising from advocacy that can be seen to be both realistic and justified, and can occur only in human societies where such concerns are practicable, and in which people have leisure and resources to devote to matters beyond basic subsistence needs. These are societal environments in which conservation needs can be both acknowledged and attended. For much of the world, local concerns with Lepidoptera, and much other conservation, is a luxury other than by inclusion in ecotourism or other exercises that can provide livelihoods and generate revenue, and that can be driven locally rather than imposed from elsewhere.

The northern hemisphere tradition of natural history collecting, from the late eighteenth century flowing to other parts of the world through colonial explorations, proliferated as a pastime and source of income throughout the nineteenth century – exemplified by the efforts of famed explorer-collectors such as A. R. Wallace and H. W. Bates. The

'Aurelian legacy' so well summarized by Salmon (2000) was in many ways the core foundation of traditional public interest in Lepidoptera, albeit based initially in the philatelic aspects of amassing desirable specimens for the Victorian 'cabinet' as manifestations of the richness of creation, with their contributions to understanding evolution yet to be recognized. The transition from acceptance of created variety to observation-based interpretation based on curiosity about the natural world characterised the later Victorian age (Clark 2009). Novelty, however, was persistently important, whether as unusual varieties of familiar species in Europe, or spectacular and rare tropical taxa such as the larger birdwings (Papilionidae), which the proud owner could display to his friends and peers. Strong associations between rarity and value led to large prices being paid for unusual specimens, and has spawned strong links between conservation status and needs – with attendant fears of commercial over-collecting being a threat to some taxa, as scarcity increases desirability and financial value. There is considerable ambivalence between listing species for protection, as a step almost always including prohibition of specimen capture, and the inadvertent promotion of black market trade as a resulting threat. Much of the continuing debate involves perceptions of real need for protection in this way versus the widespread contention that threats from over-collecting are almost always far less than those from habitat loss and change, so that such blanket prohibitions on specimen capture are only rarely warranted – notwithstanding the conservation awareness that such steps may engender.

Novelty, again, remains important in eliciting interest in Lepidoptera in a largely media-influenced world. Two major themes in butterfly conservation are addressed elsewhere in this volume, and the wider impacts of butterfly gardens (Daniels) and butterfly houses (Veltman) in ecotourism extend to many aspects of conservation through marketing and education adjuncts. A third, related, component is appreciation of butterflies in more natural environments: 'in the wild', as parts of broadening more traditional experiences – thus, studies in South Africa (Huntly *et al.* 2005) suggested that many people would embrace chances to expand their experiences beyond the more conventionally sought large mammals, and learn about other taxa – including invertebrates. Increased awareness of environmental changes and changes in biodiversity has led to photography becoming a predominant focus, with collecting of specimens restricted in many places as a putative threat increasing vulnerability of species already affected by habitat change. In Britain, and some other places in which the fauna is well-documented and largely easily identifiable, collecting for other than scientific purposes is viewed widely as unnecessary.

Buchmann and Nabhan (1996) designated 'butterfliers' as 'a new breed of butterfly hunters who do their hunting with binoculars, notebook, and pencil'. Wider benefits from such exercises include acceptance of the needs to preserve remnant habitats such as fragments of native forests in largely altered landscapes, as foci for remaining insect diversity: again, a context in which butterflies can be important flagships and benefit from local cultural heritage – as in the remnant 'totem forests' of West Africa (Larsen 1995). Particular species may drive these efforts – the brilliant blue Ulysses swallowtail (*Papilio ulysses*) is a major tourism icon for tropical Queensland, and has helped in publicizing the wider plight of butterflies in the region. Almost by definition, nocturnal moths and small, less conspicuous species, are not as amenable to demonstrating biodiversity to visitors – although a well-attended light sheet at night is an impressive picture of insect variety to any onlooker!

Notwithstanding any stigma associated with collecting, particularly in parts of the northern temperate zones, many scientists have emphasized that collecting is indeed necessary for fundamental advances in basic documentation that involve systematic and biological study, and clarification of conservation needs and level of threat. Entomological and recreational aspects of interacting with Lepidoptera thus intergrade continually, and collecting is a major avenue for recruitment of interest and expertise – many of the world's leading lepidopterists had their interests initiated or enhanced by childhood collecting and mentorship from older enthusiasts (Cheesman and Key 2007). Many lepidopterists initiate and develop their interests through recreational pursuits, and those pursuits are a core contribution to entomological understanding. Indeed, the major institutional reference collections of butterflies and larger moths are founded in bequests and donations from hobbyists. In Australia, in an entomological milieu that lags a century or more behind the natural history fervour of Britain, Moulds (1999) noted that in five major museum collections, totalling some 249 000 butterflies, more than 80% of these were collected by hobbyists, non-professional lepidopterists. Their enthusiasms have thereby accumulated, and their collective legacy to later generations furnished both documentation and the foundation for further education and investigation. Recently, however, some concerned scientists have remarked the diminished recruitment into interests in Lepidoptera and many other aspects of natural history in the face of competing interests for young people, the progressive complications of undertaking organized field work within school or higher level curricula, and the stigmas associated with collecting specimens. Lack of informed and enthusiastic leadership and mentorship is also important. In contrast, the lures of competitive sports and computer games and related technologies provide strong incentives for limited leisure time activity within peer groups, whilst natural history is not seen as an equally valid pastime amongst urban populations, in particular. Recruitment of young people into naturalists clubs and the like is reported widely to be low. Sociological impediments to specimen collecting are believed widely to affect one of the major avenues by which lepidopterists have traditionally gained their lifelong interest, through childhood butterfly (and, more rarely, moth) collecting. One modern complication has been the satisfaction of the 'collector urge' by artificial objects. Pokémon trading cards in Japan were invented as a substitute for insects that had been lost from the inventor's childhood collecting localities, for example (Cheesman and Key 2007). The positive values of hobbyist collecting of Lepidoptera must be distinguished carefully from the ravages of commercial collecting for black market trade in protected species, and remain important in education. But, in many advanced societies, collecting of insects is viewed with disdain and in many places is discouraged strongly by formal prohibitions and social pressures.

Particular well-led and effectively coordinated school-based exercises can do much to redress this, with conservation a central inducement that can garner support and publicity, and provide 'purpose' accompanied by a sense of excitement and value. Scott (2002) summarized the important roles of schools in developing a broad community conservation programme for the Richmond birdwing butterfly (*Ornithoptera richmondia*) in Australia. Organized and sponsored activities led to several hundred schools participating by planting larval food plant vines in school grounds, and monitoring these for caterpillar presence, and several schools participated also in a novel 'Adopt-a-Caterpillar' scheme whereby caterpillars supplied under licence (as a legally protected species) were observed throughout their development, leading to comments on this as a valuable learning experience for students and the school community. This exercise combined fundamental natural history and ecology education based on a conspicuous flagship species with intricate biology together with meaningful conservation measures and, importantly, was not undertaken in isolation but as a well-coordinated larger project supported by good publicity and effective coordination and feedback that provided participants with a sense of ownership and individual significance within the conservation plan. Such sense of ownership is reflected in the network of more than 400 people actively concerned about conservation of the birdwing and providing critical support such as fostering larval foodplant vines and otherwise managing habitats throughout its range, with the operations coordinated effectively at a landscape level through group field days, workshops and meetings, and a well-produced newsletter.

Community participation is a key element of insect species conservation (New 2009, 2010), and demands this sense of value and importance rather than of simple direction *ex cathedra* by scientists or 'authority' who are seen sometimes as remote from reality and to take volunteer or community help for granted. Such support is, rather, an important resource that must be cultivated carefully. It follows that 'citizen scientists' are critical in insect conservation, with their interests initiated commonly through Lepidoptera projects (see Johansen and Auger, this volume). Their involvement at early stages of planning, together with other interested parties and 'stakeholders', may help ensure that all points-of-view can be incorporated effectively into conservation planning, rather than the constituencies being presented with a firm but possibly more controversial plan. The central importance of informed lepidopterists has relevance in another important context – using Australia as an example, the staff responsible for threatened species evaluations and management in most State or Territory conservation agencies (and the Commonwealth agency) are rarely entomologists, and may lack confidence in initiating novel programmes for butterflies and other invertebrates. If, as is almost always the case, a choice of focal taxa is available it is entirely natural that staff will gravitate toward taxa with which they are more familiar and work within their individual 'comfort zone'. Vertebrates and vascular plants benefit accordingly. The guidance provided by hobbyist entomologists and others is also a critical resource that may determine the effectiveness of the management undertaken. Ensuing community participation in Lepidoptera conservation, with due regard to necessary health and safety issues, can be diverse and individual experiences enhanced by variety, responsibility and clearly defined worth of individual or team exercises – whether providing advice, detecting species on field surveys, monitoring abundance or distribution, actively managing sites, captive rearing, record keeping, fundraising or other activities that may arise. However, there are normally two phases in any such programme. The initial excitement of discovery of a 'notable' species or population in need of conservation, often attended by a feeling of novelty and the reality of strong media interest, may be followed by the long-term need for extended conservation management involving repetitive routine measures over many years. Such sustained endeavour, however vital, is rarely deemed newsworthy, and interest can lapse easily. Small isolated urban remnant sites, for example, are key habitats for many insects, and continued site maintenance may be needed to counter successional changes and external intrusions. Maintaining community interest may be difficult, and establishment of a 'Friends' Group' with a sense of proprietorship may be important. Such groups inevitably undergo turnover of key enthusiasts and, as for the Richmond birdwing noted earlier, a periodical newsletter and effective liaison across interest groups can do much to sustain interest especially if accompanied by group activities – both scientific and social.

This scenario parallels the difficulties of initiating and sustaining interest in natural history in an era when the traditional 'sense of wonder' at natural phenomena has been dulled into bland acceptance by series of magnificent documentary films and photographic books, giving the false impression that such remarkable access is commonplace. The enormous efforts and costs taken to produce these are usually not obvious to the spectator. Nevertheless, many such productions also convey an air of 'remoteness' simply because they deal with places or organisms unlikely to be directly familiar. The growth of interest in the real world, based in a familiar local environment and observing its Lepidoptera inhabitants, has helped to overcome one of the main barriers to understanding – as emphasized earlier, the foundation that 'people like butterflies' is a critical aspect for conservation as demonstrating willingness to participate directly, and encouraging others to do so. It follows that relevant information must be made available and easily comprehensible. One of the most worthy aspects of interaction between people and Lepidoptera is assuredly that of undertaking practical conservation in the face of the changes wrought through human activity. Dedicated butterfly reserves for rare or threatened species have appeared in many parts of the world, most of them for ecologically specialized species whose intricate biology alone can capture public imagination and interest. Ant-associated Lycaenidae, such as the Brenton blue and Karkloof blue (*Orachrysops niobe, O. ariadne*) in South Africa and the Eltham copper and Bathurst copper (*Paralucia pyrodiscus lucida, P. spinifera*) in Australia, are important awakeners of conservation interest in those regions and exemplify numerous cases in which butterflies have become important 'flagship species' with wider influences in fostering insect conservation (New 2011). All of these listed examples have common names referring to places where they occur. Earlier efforts to conserve the large blue (*Maculinea arion*) in Europe stimulated wider interest there, drawing also on traditional memory of extinction of the large copper (*Lycaena dispar dispar*) in Britain in the mid-nineteenth century. Many such butterflies have strong local community associations, their welfare accepted readily as important by public and officialdom alike. Organized targeted surveys using standard counting methods (such as the 'transect walks' pioneered in Britain; Pollard and Yates 1993) are often well-attended, and participation can be encouraged in many ways. Likewise, surveys such as the 4th of July Butterfly Count in North America, now ongoing for more than 30 years combine scientific data-gathering with 'giving butterfliers a chance to socialise and have fun' (Ries 2008) and raising public awareness of butterflies. Larger Lepidoptera, such as brightly coloured morphos, monarchs and swallowtail butterflies, progressively augmented by lesser-known members of the order, are a major counter to entomophobia, and have been of critical significance in raising awareness of the importance of insects and their conservation on both ethical and practical grounds. The umbrella influences of butterflies are important – thus, whilst calls for butterfly gardens are widespread and received sympathetically, the ecological roles of Hymenoptera in gardens (as emphasized by Daniels this volume and Grissell 2010) are almost certainly more significant, even though the public perception of 'wasps and ants' is far less welcoming, and calls for 'wasp gardens' indeed unusual.

CONCLUSION

Major organizations such as the Xerces Society and the North American Butterfly Association (NABA) in North America and Butterfly Conservation in Britain are vital 'rallying points' for coordinating and fostering interest, and their publications have inestimable value in documenting information and conservation news in ways that create sympathetic interest and support.

Thus, NABA's mission is 'to increase public enjoyment and conservation of butterflies' and its recent initiative (October 2010) of opening the National Butterfly Center in Mission, Texas provides a hub for wide-ranging informative displays and access to adjacent butterfly-rich field sites that demonstrate habitat enrichment. The Center is 'dedicated to education, conservation and scientific research on wild butterflies', so encouraging the major themes discussed in this chapter. That recreational activities such as butterfly watching can remain fun and be acknowledged as such, as well as contribute meaningfully to knowledge and conservation, is a lesson worthy of the widest possible dissemination.

REFERENCES

Asher, J., Warren, M., Fox, R., Harding, P. *et al.* (2001) *The Millennium Atlas of Butterflies in Britain and Ireland.* Oxford: Oxford University Press.

Buchmann, S. L. and Nabhan, G. P. (1996) *The Forgotten Pollinators.* Washington DC: Island Press.

Cheesman, O. D. and Key, R. S. (2007) The extinction of experience: a threat to insect conservation? In *Insect Conservation Biology*, ed. A. J. A. Stewart, T. R. New and O. T. Lewis. Wallingford, UK: CABI. pp. 322–348.

Clark, J. F. McD. (2009) *Bugs and the Victorians.* New Haven, CT: Yale University Press.

Grissell, E. (2010) *Bees, Wasps, and Ants. The Indispensable Role of Hymenoptera in Gardens.* London: Timber Press.

Hogue, C. L. (1987) Cultural entomology. *Annual Review of Entomology*, **32**, 181–199.

Huntly, P. M., Van Noort, S. and Hamer, M. (2005) Giving increased value to invertebrates through ecotourism. *South African Journal of Wildlife Research*, **35**, 53–62.

Kristensen, N. P., Scoble, M. and Karsholt, O. (2007) Lepidoptera phylogeny and systematics: the state of inventorying moth and butterfly diversity. *Zootaxa*, **1668**, 699–747.

Larsen, T. B. (1995) Butterfly biodiversity and conservation in the Afrotropical region. In *Ecology and Conservation of Butterflies,* ed. A. S. Pullin. London: Chapman & Hall, pp. 290–313.

Lemelin, R. H. (2007) Finding beauty in the dragon: the role of dragonflies in recreation and tourism. *Journal of Ecotourism*, **6**, 139–145.

Lemelin, R. H. (2009) Goodwill hunting: dragon hunters, dragonflies and leisure. *Current Issues in Tourism*, **12**, 235–253.

Montgomery, S. L. (1983) Carnivorous caterpillars: the behavior, biogeography and conservation of *Eupithecia* (Lepidoptera: Geometridae) in the Hawaiian islands. *GeoJournal*, **7**, 549–556.

Moulds, M. S. (1999) The history of Australian butterfly research and collecting. In *Biology of Australian Butterflies*, ed. R. L. Kitching, E. Scheermeyer, R. E. Jones and N. E. Pierce. Collingwood, Australia: CSIRO Publishing. pp. 1–24.

New, T. R. (2009) *Insect Species Conservation.* Cambridge: Cambridge University Press.

New, T. R. (2010) Butterfly conservation in Australia: the importance of community participation. *Journal of Insect Conservation*, **14**, 305–311.

New, T. R. (2011) Launching and steering flagship Lepidoptera for conservation. *Journal of Threatened Taxa*, **3**(6), 1805–1817.

Pollard, E. and Yates, T. J. (1993) *Monitoring Butterflies for Ecology and Conservation.* London: Chapman and Hall.

Ries, L. (2008) What can we learn from the 4th of July counts? *American Butterflies* (Summer 2008), 34–36.

Robinson, G. S., Tuck, K. R. and Shaffer, M. (1994) *A Field Guide to the Smaller Moths of South-East Asia.* London: The Natural History Museum

Salmon, M. A. (2000) *The Aurelian Legacy: British Butterflies and their Collectors.* Berkeley, CA: University of California Press..

Scott, S. (2002) School and community participation in the Richmond Birdwing conservation project. In *Conservation of Birdwing Butterflies*, ed. D. Sands and S. Scott. Chapel Hill, Australia: SciComEd Pty & THECA, pp. 185–193.

Young, M. (1997) *The Natural History of Moths.* London: T. and A. D. Poyser.

7

Dragonflies: their lives, our lives, from ponds to reserves

MICHAEL J. SAMWAYS

INTRODUCTION

We as humans are becoming increasingly detached from, and at variance with, nature (Stokes 2006). This is the extinction of experience (Miller 2005). Human reality on most days is being insulated from the vagaries of weather in our sheltered dwellings and offices and simply forgetting just how much we rely on a healthy planet to sustain us. Those who are exposed to nature on a daily basis, as may be the case for a farmer in the tropics, are, in turn, simply engaged in survival. We see biodiversity around us, but we do not fully appreciate that this biodiversity, and its emergent properties of ecosystem services, are being eroded at an alarming rate (Butchart *et al.* 2010). It is almost as if we are losing touch with nature, its variety, beauty and its function. Yet we know too, that we cannot be complacent, as nature can always deal a blow, whether physical events from tornados to tsunamis, or biological ones, from parasites to pests.

But all is far from lost, so long as we begin to rescue the extinction of experience based on firstly appreciating nature, then respecting it (Samways 2007). There is an old adage that there are three times we go to a zoo: as a child, as a parent, as a grandparent. And these visits stay in our memory probably more than having spent an equivalent time on a computer. As there are now more people living in the urban environment than the rural one, any meaningful contact with raw nature is a very important life-line for future appreciation of the value of nature in our lives. Urban parks, as well as zoos, provide some experience of this, but both share the disadvantage that, whether wild animals in pens or trees on a lawn, they are negentropic entities requiring our continued maintenance. They are not part of open-ended communities functioning as they would in their natural environment.

Dragonflies are highly visible, regularly present, aesthetic creatures that cannot be missed by even the most unobservant of human individuals. Besides, everyone knows what a dragonfly is: a 'jigamanzi' or 'water dancer' in the Zulu language. Furthermore, they feature strongly in most cultures, from subjects in ancient Chinese art to decorations on bed linen on sale in our local shopping mall. But what is really interesting is that the dragonflies seen in an urban environment are living, breathing individuals living a natural life, both in the water as larvae, and in the air as adults. This is brought home by the fact that they do not lend themselves to captivity, with captive populations a near impossibility. They must have their normal, open environment to rest, feed and breed. This means that observation of a dragonfly, even in the local park, is truly to observe a wild animal, right in front of us. And they ask nothing of us. While they are there for their own sake, they are there nevertheless for us to be able to appreciate and enjoy. This is rescuing the extinction of experience in an urban environment. We are experiencing wild nature, with the added benefit of not generally being eaten, bitten or stung as we often would in remote and truly wild locations. The downside, at least for the aficionados, is that the species we do see in the local park are those species which can tolerate an urban environment, and this may exclude those species that require special habitats and which may be listed as threatened or endangered. But even so, certain botanical gardens, as in the case of those in the Cape of South Africa, may be home and sanctuary to certain globally threatened species (Willis and Samways 2011).

DRAGONFLIES IN CULTURE

Dragonflies through the centuries have been considered as highly symbolic. They grow up as brown, rather grubby larvae living inconspicuous lives in water. Then they emerge as glistening free-flying adults that are free to largely dominate their air space. True dragonflies even have larvae which breathe through their rectum! The adults at least represent elegance, simplicity and freedom, cleanliness out of murkiness. It is beauty out of the beast. It is not surprising therefore that they have featured strongly in art and poetry through the ages

(Lemelin 2009). They are part of the fabric of life that is highly visible and comfortable for us. They create an aura of pastoralism.

This interest in dragonflies is worldwide, and they are the first group of invertebrates to receive a global assessment of threats facing them in today's world (Clausnitzer *et al.* 2009). This was the outcome of many dragonfly specialists around the world coming together and pooling their knowledge. There are even detailed assessments of their conservation status across southern Africa (Suhling *et al.* 2009) and even the whole of Africa (Dijkstra *et al.* 2011). Then there are many field guides available from North America (Dunkle 2000) and Europe (Djkstra and Lewington 2006) to Namibia (Suhling and Martens 2007), South Africa (Samways 2008), Hong Kong (Wilson 2004) and Australia (Theischinger and Hawking 2006) to name but a few. Furthermore, they are highly regarded as specific service providers.

DRAGONFLIES AND PONDS

While dragonflies occur in both running and still water, it is at ponds (water bodies less than 10 hectares) where these insects are mostly seen, and where they interface most with human culture. The Pond Manifesto (2008) emphasizes the importance of maintaining global freshwater biodiversity and ensuring its sustainable use. Amazingly, ponds as a whole represent 30% of the global surface area of standing water, and are recognized as important functional and conservation features of the landscape (Céréghino *et al.* 2008; Nicolet *et al.* 2004). Except for the permanently frozen ones, all these ponds across the world are very likely to have at least one species of dragonfly, with the warmer areas of the globe being home to as many as three dozen species. Some species will inhabit temporary pools after rain, with the Globe Skimmer *Pantala flavescens* even inhabiting small pools along vehicle tracks across the tropics, where temperatures reach that of tepid bath water. This remarkably mobile species is the only one to inhabit the small volcanic lakes on Easter Island (Samways and Osborn 1998).

Small ponds are widespread in ornamental gardens, and so are easily accessible, and where adult dragonflies are regularly seen, at least at the warmest time of the year (Steytler and Samways 1995). The important point is that ponds hold a fascination for children and adults alike, and it is beside these water bodies where there is the strongest possibility for rescuing the extinction of experience. Open ponds, such as drinking troughs for livestock, with a concrete surround and no vegetation, will attract certain species as tourists, and in the case of certain libellulid dragonflies, they may perch on the warm masonry. In Africa, the colourful red *Trithemis kirbyi* is one of those species (Suhling *et al.* 2006). But generally, such highly artificial ponds do not sustain dragonfly populations because vegetation is lacking. Plants, both submerged and emergent, are necessary not just for creating the physically appropriate, and therefore attractant, conditions, but are also vital for several behavioural activities from hawking and perching to egg laying (Chovanec 1994). These conditions can be critical, and when changed, either by natural seral succession or by invasion by alien plants both in and beside the water, there will be a change in the dragonfly assemblage. Similarly, cutting back or mowing of vegetation can remove essential resting and oviposition sites, with the red-listed damselfly species *Pseudagrion coeleste umsingaziense* being threatened by excessive mowing of the long grass on which it perches beside ponds (Samways 2006). For some species, essential vegetation may be some distance from the water, and in the case of *P. c. umsingaziense*, this includes trees within easy flying distance, no more than 50 m from the water's edge, on which to perch in the shade on very hot days.

Conditions in the pond are as important as those at its edges. Macrophytes, such as water lilies and emergent grasses, are essential for many species on which to perch, find mates, and into which to lay eggs. In some cases, invasive alien plants actually benefit some species (Stewart and Samways 1998). For example, *Acisoma panorpoides* increases in local abundance when such water plants fill a pond, while *Pseudagrion massaicum* can benefit from ponds seemingly clogged with green algae. This illustrates how forgiving some species are of human-modified pondscapes, so long as the right habitat conditions prevail. This may involve some shading of the pond for it to be suitable, with *Pseudagrion hageni* only favouring ponds with a margin of trees providing deep shade. On Mayotte Island in the Comoro archipelago, several narrow range endemic species are tolerant of a tree canopy of alien mango trees because they are essentially shade loving (Samways 2003). This contrasts with the situation in South Africa, where most of the narrow range endemic dragonfly species are highly intolerant of local invasion by alien trees, and they can become locally extinct. Fortunately, the situation is reversible, with their return coming about when the alien trees are removed (Samways and Sharratt 2010). A similar situation occurred in the Seychelles, where, on Cousine Island, ponds were established as water refugia for giant tortoises. Dragonflies colonized these ponds in accordance with Island Biogeography Theory (Samways 1998), but later disappeared as the island was restored to its historic condition and indigenous trees began to shade out the ponds (Samways *et al.* 2010).

Other conditions in a pond can also make a difference. Certain fish species are highly predatory on some dragonfly larvae, and ponds designed to attract these insects should ideally be free of fish predators. If the presence of fish is unavoidable, then there is likely to be a change in relative dragonfly abundance (Johansson *et al.* 2006). Ponds need not be too deep, with most species within the first 60 cm, where water weed is most abundant (Samways *et al.* 1996a). In areas where freezing occurs, greater depth may be required, and a muddy bottom necessary to provide a suitable habitat and shelter in the depths of winter.

Dragonflies can often be observed alongside other organisms in protected areas and game reserves. Pools in both rest camps and in the reserve itself are important focal points for dragonfly observers. Where large game animals are present, there is considerable physical impact on the local dragonfly assemblage, with some species being excluded by the impact while others are highly tolerant. In Botswana, for example, the impact of large game animals, especially elephants, tends to drive out species such as the common pond damsel (*Ceriagrion glabrum*), the common bluetail (*Ischnura senegalensis*), the black percher (*Diplacodes lefebvrii*) and the grizzled pintail (*Acisoma panorpoides*), while species such as the banded groundling (*Brachythemis leucosticta*) and the long skimmer (*Orthetrum trinacria*) are highly tolerant of such disturbance (Samways and Grant 2008).

Ponds (can) play an important role in the conservation landscape of Europe (Chovanec 1994), while in Africa, artificial ponds <2 hectares in extent, mostly used for water storage and irrigation, can play a very important role in increasing the area of occupancy of various species in any one area, with these ponds also acting as stepping stone habitats by the stronger fliers, as well as habitats per se for a whole host of species (Samways 1989). However, at least in Africa, these species are widespread generalists, with none of the red-listed species being recorded from these habitats. A negative point regarding encouragement of dragonflies, as top predators in an area, is that they can have a considerable impact on populations of other insects, even smaller members of their own order. This predator impact can have, for example, a marked effect on butterfly (Sang and Teder 2011) and bee (Knight *et al.* 2005) populations.

DRAGONFLY RESERVES

There are essentially two ways of viewing human cultural inter-action with dragonfly conservation landscapes. Firstly, species may be recorded from various sites that are not set aside specifically for dragonflies but are sites of special interest, high biodiversity and/or aesthetic appeal. Guides have been written, highlighting sites where particular species may be found, at the right time of year of course (Samways and Grant 2006). Such guides are available for various protected sites in Britain (Hill and Twist 1998) and for national botanical gardens in South Africa (Willis and Samways 2011). The second approach is to set up reserves specifically for dragonflies, which may be for a specific species, such as the Norfolk hawker (*Aeshna isosceles*) in the UK (Merritt *et al.* 1996), or for dragonflies in general, irrespective of their threat status. In the former case, these reserves are usually for locally threatened species which benefit from an increase in the extent of good quality habitat. In the case of dragonfly reserves, these tend to have strong local cultural value and are of wide appeal to the general public. Japan leads the way with such reserves, with Nakamura Dragonfly Reserve being the first of its kind and very popular. It was developed from former rice paddies, and while attracting a wide range of species they are generally widespread generalists. The value of such reserves is cultural and not primarily aimed at conservation. However, they do contribute towards rescuing the extinction of experience, and the children visiting such reserves may well be inspired to conserve dragonflies and other biota later in their lives, either professionally or as citizen scientists (Primack *et al.* 2000; Figure 7.1).

Nothing disappoints more than an unfulfilled promise, and so it is important that sites and dragonfly reserves point out to the public which species can be seen, and the chances of seeing a particular species (Suh and Samways 2001), as well as at what time of year (Suh and Samways 2005). In the KwaZulu-Natal Botanical Garden, South Africa, a dragonfly awareness trail has been developed, with specific viewing points and information boards illustrating the species likely to be seen at particular locations with certain species combined with easy access. The concept of 'core resident species' was developed to include those species, so as not to disappoint through not being present (Niba and Samways 2006). However, these 'certain species', while being reliable, do not add that little extra excitement of seeing a rare species. So, as with bird watching, dragonfly watching can be spiced with mention

Figure 7.1   Dragonfly pond in Japan. Photo by Michael Samways.

of possible rarities that might be observed, and even encouragement given to record any species that may never have been recorded at the site before, and may appear following adverse conditions such as a particularly warm and wet spell of weather. Indeed, there is strong likelihood that new arrivals will be recorded, which are not necessarily new residents, with the passing of the years (Oertli 2008).

RECOMMENDATIONS FOR CONSTRUCTING A GOOD DRAGONFLY POND

Generally, the aim of a pond from a cultural point of view is to attract a wide range of species in an aesthetic setting. Where children and the elderly are concerned, safety and accessibility is also important. Ponds ideally should have a range of microhabitats from edges with marshy, flooded areas and tall grasses and sedges, some bare soil areas, some rocks (on which certain species like to sun themselves and gain a thermal advantage; McGeoch and Samways 1991), some reeds, and parts of the pond which are shaded by bushes or trees. The margin should drop away gradually, and while a depth of more than a half metre or so is not necessary, allowance must be made for siltation over time. Lilies and reeds can be planted in the soft bottom, but where there is a plastic or concrete lining, pots can be provided to accommodate these plants, especially those that have vigorous growth and may need to be contained. Alternatively, if feasible, earthen banks can be created (Primack *et al.* 2000). Choice aquatic plants can influence which species will survive, with the threatened European species *Aeshna viridis* requiring the water plant *Stratiotes aloides* for optimal survival, with the plant supplying some protection from fish predators (Rantala *et al.* 2004). A greater species richness of macrophytes will lead to more opportunities for a range of species (Butler and deMaynadier 2008). An abundance of macrophytes also helps decrease turbidity, which also encourages high dragonfly species richness (D'Amico *et al.* 2004). Some sticks or branches, protruding out of the water are important additions as they provide perching points for certain species. Very important is that the water level must be kept constant (Samways *et al.* 1996b), unless the aim is to have a seasonal pond, but it will not generally attract the same set of species in comparison with a permanent pond with a constant water level (Suhling *et al.* 2005). Whether or not there is an inflow and outflow can affect the species present (Samways and Stewart 1997), as some species prefer the moving water at these areas, with *Zygonyx natalensis* on a dragonfly trail in South Africa, seen only at the waterfall at the outflow of a dammed pond (Willis and Samways 2011). If there is an option for ponds in various settings, it adds variety to have them of different sizes and with different degrees of shade (Biggs *et al.* 2004). Indeed, a set of small-sized ponds will have more species and higher conservation value than a single, large pond of equivalent surface area, but it is worth bearing in mind that large ponds will attract species that the small ponds do not (Oertli *et al.* 2002). In turn, ponds that are more isolated will have a greater difference in species present than ponds close together owing to more limited dispersal than would seem to be the case for such good fliers as dragonflies, as well as various levels of breeding success (McCauley 2006). Occasionally, it may be necessary to clean out the pond where vegetation clogs it, so leading to a reduced dragonfly assemblage (De Marco *et al.* 1999). (See Figure 7.2.)

CONCLUSION

It is always interesting to keep records of the coming and going, and phenology, of the various species over time (van Strien *et al.* 2010). Species' arrivals at new ponds do not depend on the presence of already established species, with the presence of suitable microhabitats being more important than issues associated with restrictions on species packing (Osborn and Samways 1996). Nevertheless, there can

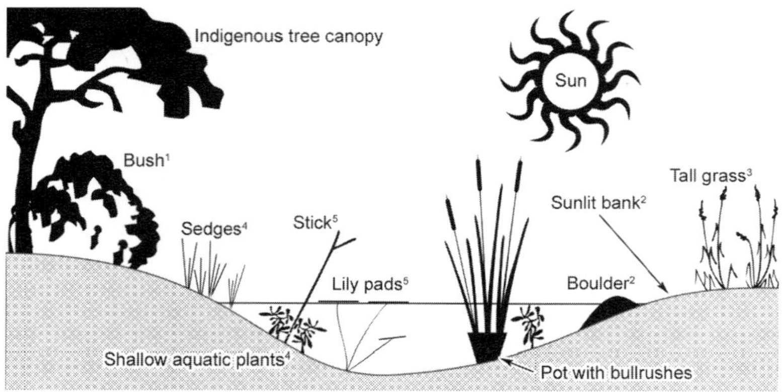

Figure 7.2   The design of a dragonfly pond, with five key features (1–5). See text for details. From Samways (2008).

be competition, especially among larvae, leading to changes in local abundance (Suutari *et al.* 2004). In small ponds, created as bird baths on the ecotourism island of Cousine, Seychelles, conspecific and allospecific predation between individuals of *Tramea limbata* and *P. flavescens* often results in just a single individual of *P. flavescens* surviving through to adulthood (Samways *et al.* 2010). Moore (1991) recorded species turnover and population levels of dragonflies at a pond he created in Cambridgeshire, UK, over a 27-year period, and showed among other things the accuracy of recording adults, which has proved to be such a quoted metric in many papers since then (Oertli 2008). Observations like this can be very valuable to determining long-term changes in species distributions and population levels, as has been done for dragonflies in Germany and elsewhere in Europe (Ott 2010). Ott has shown just how dynamic the dragonfly fauna of ponds can be, and has recorded the northward shift in geographical range of species in response to global climate change. So, overall, ponds and dragonflies can have considerable scientific merit as well as cultural and recreational values.

REFERENCES

Biggs, J., Bilton, D., Williams, P. *et al.* (2004) Temporary ponds of eastern Poland: an initial assessment of their importance for nature conservation. *Archives des Sciences*, **57**, 73–84.
Butchart, S. H. M., Walpole, M., Collen, B. *et al.* (2010) Global biodiversity: indicators of recent declines. *Science*, **328**, 1164–1168.
Butler, R. G. and deMaynadier, P. G. (2008) The significance of littoral and shoreline integrity to the conservation of lacustrine damselflies (Odonata). *Journal of Insect Conservation*, **12**, 23–36.

Céréghino, R., Biggs, J., Oertli, B. and Declerck, S. (2008) The ecology of European ponds: defining the characteristics of a neglected freshwater habitat. *Hydrobiologia*, **597**, 1–6.

Chovanec, A. (1994) Man-made wetlands in urban recreational areas: a habitat for endangered species? *Landscape and Urban Planning*, **29**, 43–54.

Clausnitzer, V., Kalkman, V. J., Ram, M. *et al.* (2009) Odonata enter the biodiversity debate: the first global assessment of an insect group. *Biological Conservation*, **142**, 1864–1869.

D'Amico, F., Darblade, S., Avignon, S., Blanc-Manel, S. and Ormerod, S. J. (2004) Odonates as indicators of shallow lake restoration by liming: comparing adult and larval responses. *Restoration Ecology*, **12**, 439–446.

De Marco, P., Latini, A. O. and Reis, A. P. (1999) Environmental determination of dragonfly assemblage in aquaculture ponds. *Aquaculture Research*, **30**, 357–364.

Djkstra, K. D.-B. and Lewington, R. (2006) *Field Guide to the Dragonflies of Britain and Europe.* Gillingham, UK: British Wildlife Publishing.

Dijkstra, K.-D. B., Boudot, J.-P., Clausnitzer, V. *et al.* (2011) Dragonflies and damselflies of Africa (Odonata): history, diversity, distribution, and conservation. In *The Diversity of Life in African Freshwaters: Underwater, Under Threat*, ed. W. Darwall, K. Smith, D. Allen *et al.* Gland, Switzerland: IUCN, pp. 126–173.

Dunkle, S. W. (2000) *Dragonflies Through Binoculars: A Field Guide to Dragonflies of North America.* Oxford: Oxford University Press.

Hill, P. and Twist, C. (1998) *Butterflies and Dragonflies: A Site Guide.* Chelmsford, UK: Arlequin.

Johansson, F., Englund, G., Brodin, T. and Gardfjell, H. (2006) Species abundance models and patterns in dragonfly communities: effects of fish predators. *Oikos*, **114**, 27–36.

Knight, T. M, McCoy, M. W., Chase, J. M., McCoy, K. A. and Holt, R. D. (2005) Trophic cascades across ecosystems. *Nature*, **473**, 880–883.

Lemelin, H. (2009) Goodwill hunting: dragon hunters, dragonflies and leisure. *Current Issues in Tourism*, **12**, 235–253.

McCauley, S. J. (2006) The effects of dispersal and recruitment limitation on community structure of odonates in artificial ponds. *Ecography*, **29**, 585–595.

McGeoch, M. A. and Samways, M. J. (1991) Dragonflies (Odonata: Anisoptera) and the thermal landscape: implications for their conservation. *Odonatologica*, **20**, 303–320.

Merritt, R., Moore, N. W. and Eversham, B. C. (1996) *Atlas of the Dragonflies of Britain and Ireland.* Huntingdon, UK: Natural Environment Research Council.

Miller, J. R. (2005) Biodiversity conservation and the extinction of experience. *Trends in Ecology and Evolution*, **20**, 430–434.

Moore, N. W. (1991) The development of dragonfly communities and the consequences of territorial behaviour: a 27-year study on small ponds at Woodwalton Fen, Cambridgeshire, United Kingdom. *Odonatologica*, **20**, 203–231.

Niba, A. S. and Samways, M. J. (2006) Development of the concept of 'core resident species' for quality assurance of an insect reserve. *Biodiversity and Conservation*, **15**, 4181–4196.

Nicolet, P., Biggs, J., Fox, G. *et al.* (2004) The wetland plant and macroinvertebrate assemblages of temporary ponds in England and Wales. *Biological Conservation*, **120**, 261–278.

Oertli, B. (2008) The use of dragonflies in the assessment and monitoring of aquatic habitats. In *Dragonflies and Damselflies: Model Organisms for Ecological and Evolutionary Research*, ed. Alex Córdoba-Aguilar. Oxford: Oxford University Press, pp. 79–95.

Oertli, B., Joye, D. A., Castella, E. *et al.* (2002) Does size matter? The relationship between pond area and biodiversity. *Biological Conservation*, **104**, 59–70.

Osborn, R. and Samways, M. J. (1996) Determinants of adult dragonfly assemblage patterns at new ponds in South Africa. *Odonatologica*, **25**, 49–58.

Ott, J. (2010) The big trek northwards: recent changes in the European dragonfly fauna. In *Atlas of Biodiversity Risk*, ed. J. Settele, L. Penev, T. Georgiev, *et al.* Sophia, Bulgaria: Pensoft, pp. 82–83.

Primack, R., Kori, H. and Mori, S. (2000) Dragonfly pond restoration promotes conservation awareness in Japan. *Conservation Biology*, **14**, 1553–1554.

Rantala, M. J., Ilmonen, J., Koskimäki, J., Suhonen, J. and Tynkkynen, K. (2004) The macrophyte, *Stratiotes aloides*, protects larvae of dragonfly *Aeshna viridis* against fish predation. *Aquatic Ecology*, **38**, 77–82.

Samways, M. J. (1989) Farm dams as nature reserves for dragonflies (Odonata) at various altitudes in the Natal Drakensberg mountains, South Africa. *Biological Conservation*, **48**, 181–187.

Samways, M. J. (1998) Establishment of resident Odonata populations on the formerly waterless Cousine island, Seychelles: an Island Biogeography Theory (IBT) perspective. *Odonatologica*, **27**, 253–258.

Samways, M. J. (2003) Threats to the tropical island dragonfly fauna (Odonata) of Mayotte, Comoro Archipelago. *Biodiversity and Conservation*, **12**, 1785–1792.

Samways, M. J. (2006) National Red List of South African dragonflies (Odonata). *Odonatologica*, **35**, 341–368.

Samways, M. J. (2007) Rescuing the extinction of experience. *Biodiversity and Conservation*, **16**, 1995–1997.

Samways, M. J. (2008) *Dragonflies and Damselflies of South Africa*. Sophia, Bulagaria: Pensoft.

Samways, M. J. and Grant, P. B. C. (2006) Honing Red List assessments of lesser-known taxa in biodiversity hotspots. *Biodiversity and Conservation*, **16**, 2575–2586.

Samways, M. J. and Grant, P. B. C. (2008) Elephant impact on dragonflies. *Journal of Insect Conservation*, **12**, 493–498.

Samways, M. J. and Osborn, R. (1998) Divergence in a transoceanic circumtropical dragonfly on a remote island. *Journal of Biogeography*, **25**, 935–946.

Samways, M. J. and Sharratt, N. J. (2010) Recovery of endemic dragonflies after removal of invasive alien trees. *Conservation Biology*, **24**, 267–277.

Samways, M. J. and Stewart, D. A. B. (1997) An aquatic ecotone and its significance in conservation. *Biodiversity and Conservation*, **6**, 1429–1444.

Samways, M. J., Caldwell, P. M. and Osborn, R. (1996a) Spatial patterns of dragonflies (Odonata) as indicators for design of a conservation pond. *Odonatologica*, **25**, 157–166.

Samways, M. J., Osborn, R. and Van Heerden, I. (1996b) Distribution of benthic invertebrates at different depths in a shallow reservoir in the KwaZulu-Natal Midlands. *Koedoe*, **39**, 69–76.

Samways, M. J., Hitchins, P. M., Bourquin, O. and Henwood, J. (2010) *Tropical Island Recovery: Cousine Island, Seychelles*. Oxford: Wiley-Blackwell.

Sang, A. and Teder, T. (2011) Dragonflies cause spatial and temporal heterogeneity in habitat quality for butterflies. *Insect Conservation and Diversity*, **4**(4), 257–264.

Stewart, D. A. B. and Samways, M. J. (1998) Conserving dragonfly (Odonata) assemblages relative to river dynamics in a major African savanna game reserve. *Conservation Biology*, **12**, 683–692.

Steytler, N. S. and Samways, M. J. (1995) Biotope selection by adult male dragon-flies (Odonata) at an artificial lake created for insect conservation in South Africa. *Biological Conservation*, **72**, 381–386.

Stokes, D. L. (2006) Conservators of experience. *BioScience*, **56**, 6–7.

Suh, A. and Samways, M. J. (2001) Development of a dragonfly awareness trail in an African botanical garden. *Biological Conservation*, **100**, 345–353.

Suh, A. and Samways, M. J. (2005) Significance of temporal changes when designing a reservoir for conservation of dragonfly diversity. *Biodiversity and Conservation*, **14**, 165–178.

Suhling, F. and Martens, A. (2007) *Dragonflies and Damselflies of Namibia*. Windhoek, Namibia: Gamsberg Macmillan.

Suhling, F., Sahlén, G., Kasperski, J. and Gaedecke, D. (2005) Behavioural and life history traits in temporary and perennial waters: comparisons among three pairs of sibling dragonfly species. *Oikos*, **108**, 609–617.

Suhling, F., Sahlén, G., Martens, A., Marais, E. and Schütte, C. (2006) Dragonfly assemblages in arid tropical environments: a case study from western Namibia. *Biodiversity and Conservation*, **15**, 311–332.

Suhling, F., Samways, M. J., Simaika, J. P. and Kipping, J. (2009) The status and distribution of dragonflies (Odonata). In *The Status and Distribution of Freshwater Biodiversity in Southern Africa*, ed. W. Darwall, K. G. Smith, D. Tweddle and P. Skelton. Gland, Switzerland: IUCN, pp. 48–65.

Suutari, E., Rantala, M. J., Salmela, J. and Suhonen, J. (2004) Intraguild predation and interference competition on the endangered dragonfly *Aeshna viridis*. *Oecologia*, **140**, 135–139.

The Pond Manifesto (2008) European Pond Conservation Network, Europe. Available at: www.europeanponds.org.

Theischinger, G. and Hawking, J. (2006) *Complete Guide to the Dragonflies of Australia*. Canberra, Australia: CSIRO.

van Strien, A. J., Termaat, T., Groenendijk, D., Mensing, V. and Kéry, M. (2010) Site-occupancy models may offer new opportunities for dragonfly monitoring based on daily species lists. *Basic and Applied Ecology*, **11**, 495–503.

Willis, C. and Samways, M. J. (2011) *Water Dancers of South Africa's National Botanical Gardens*. Pretoria, South Africa: South African National Biodiversity Institute.

Wilson, K. D. P. (2004) *Field Guide to the Dragonflies of Hong Kong*, 2nd edn. Hong Kong: Agriculture, Fisheries and Conservation Department.

Part II   Insects and Leisure

8

Relating to aquatic insects: becoming English fly fishers

ADRIAN FRANKLIN

ADRIAN FRANKLIN

INTRODUCTION

In this chapter I analyse why and how, through the development of a fly-fishing recreation culture, English anglers have developed a deep association with, and understanding of, the aquatic insects that provide the focus of their enthusiasm. Today, that understanding combined with the substantial economy of recreational fly-fishing and its socially powerful 'disciples' means that the environmental needs of the insects figure in the conservation and management of fisheries just as much as the needs of the trout themselves. This extraordinarily intense association with insects that mainstream society eschews (but which spread rapidly to the United States, Canada, Australia, Chile, Scandinavia and to Japan) is extremely unusual and unique so its development is all the more important to understand (see Franklin 1996; 2002). This essay develops a textual analysis of the history of this association with insects and the nexus with trout through a reading of the rich literature of fly-fishing, beginning in the medieval period and stretching through to the contemporary period. It can be seen that although it has the appearance of an ancient practice, its essential properties and knowledge are actually a very recent development and relate more to modern than traditional practices with nature.

Fly-fishing is a quintessential English pastime. So self-assuredly at home on the banks of beautiful southern chalk streams or wading the rocky runs of upland northern becks, the culture and fraternity of English fly-fishing seems as old as the hills. Its adherents know the

The Management of Insects in Recreation and Tourism, ed. Raynald Harvey Lemelin. Published by Cambridge University Press. © Cambridge University Press 2013.

123

specific aquatic flies of their home streams and tie imitations of them with the same dedication and attention to detail as any accomplished hunter-gatherer in the anthropological record. Fly-fishing for trout features frequently on TV shows such as 'Midsomer Murders', 'Morse', 'Harry Enfield and Chums', 'A Bit of Fry and Laurie' and 'Foyle's War', as well as works of serious literature: it features in the work of William Shakespeare (*Henry IV Pts 1 and 2*), W. B Yeats (*The Fisherman*), Jane Austen (*Pride and Prejudice,* 1813), Thomas Hughes (*Tom Brown's School Days,* 1857) Rudyard Kipling ('On dry-cow fishing as a fine art', 1890), Arthur Ransome (*Rod and Line,* 1929), Henry Williamson (*Salmo the Salmon,* 1936), John Hilaby (*Within the Streams,* 1949) and Agatha Christie (e.g., *After the Funeral,* 1953). It is perennially used to establish the way of life of the English countryside and its gentlemanly codes. It was surely significant, for example, that in the first nervous conversation between Mr Darcy and Jane Bennett's uncle, Mr Gardiner in Jane Austen's *Pride and Prejudice,* fly-fishing was chosen as their opening topic of conversation. Jane Austen knew full well that this was something that the landed gentleman and a merchant would share in common: from the last quarter of the eighteenth century it had become increasingly attractive to wealthy merchants and shopkeepers. The fishing tackle trade enjoyed a boom during the early to mid-nineteenth century and the rapid expansion of the railways took the now urban-based 'leisure classes' back into the countryside to almost every stream that held trout (Franklin 1996, 2002).

Such a long tradition, seemingly surviving the urbanization and modernization of England, gave the impression of a primordial impulse and culture at work and precisely the kind of historical continuity that appealed to the romantic sensibility of Victorian and early twentieth-century England (Franklin 1999). These impressions have some justification. One of the earliest and most popular books in the English language was Izaak Walton's *Compleat Angler* of 1653 (it has been in print ever since) and in this, fly-fishing for trout is extolled as the most esteemed form of the gentle sport of angling. Indeed, angling is compared with hunting and falconry and found to be superior, as the most civilising and healthy pastime.

However, fly-fishing in England has an even older published record in *The Treatise of Fishing with an Angle* (attributed to Dame Juliana Berners), printed in the second Book of St Albans, in 1496 – with an earlier written manuscript surviving from 1450 (McDonald 1963). It is very obvious that the style if not the content of *The Compleat Angler* derived from this earlier text, which was the first to compare angling to the most noble of sports, hunting, fowling and hawking. Until then, angling had not been mentioned in medieval writing, let alone claimed kinship with these chivalric sports (McDonald 1963: 4), but clearly, angling had a special appeal in the mid-fifteenth century at a time when the idea of chivalry was coming to be seen as decadent and old fashioned. McDonald suggests that this relates to profound changes that had transformed the social relations of the medieval world.

This world, comprising the twin estates of aristocrat and serf had been challenged by a third estate, of middling ranks, freeborn and respectable, with aspirations to make their way in the world through industry and commerce. As McDonald (1963: 4) says, 'The Treatise of Fishing with an Angle is not addressed to noblemen but to all who are "virtuous, gentle and freeborn"'. In such a world where social mobility could accompany careers in industry, business and the professions, new forms of association came into being that looked to the codified practices of chivalry (if *not* their sports) as a way of ordering their lives and comportment. According to Elias (1994), it was especially with the lengthening chains of social dependency that accompanied their more complex division of labour, that new forms of civilized behaviour became written down. After all, the familiar worlds of village and court were giving way to the more fluid, mobile and culturally diverse social exchanges of commerce and trade – and such relations with diverse strangers required new forms of social glue. Such a world required first and foremost a more quiet and orderly sensibility, one less prone to excitement and violence (a functional feature of the war-torn medieval period) and one more given to tolerance, patience and cooperation. The new sensibility was codified following the manner in which chivalry was formalized, through written books, and these guides, ostensibly about sports and leisure, were scarcely concealed metaphorical works aiming to establish values and social norms. These books were structured in a very formal way, normally having two essential components. The first, and opening section always involved an *argument*, a reasoning behind the need to be recruited into a practising fraternity. In *The Treatise* and the *Compleat Angler*, the chivalric sport of hunting was not dismissed so much as criticized for having serious flaws. Against the flaws of hunting, angling was favourably compared, particularly because it taught quietude, modesty in the presence of nature, it was the basis of hearty companionship between fishing friendships and because it bestowed a healthy (and holy) lifestyle on its practitioners. Angling then was both metaphor and practice. As a metaphor for life, angling clearly related to a more cooperative, attentive and peaceful citizen, capable of slotting into the emerging complexities of early modern life. Here, idealized was the seriously engaged *individual*, capable of self-directed deployment while at the same time an amiable part of a larger fraternity. Here the fraternity included not only humans but also their non-human components in nature: the insects, the riverside vegetation and the river itself. Fly-fishing, for trout in particular, concentrated these cooperative virtues and sentiments by encouraging a studied and mannered connection between the insect, human and piscatorial world, between the air, the land and the water. It was a perfect metaphor.

In other words, far from having primordial roots, it seems that angling had been a relatively obscure and insignificant activity until this time. As a less aggressive and destructive pastime, and one that established close bonds between people and the somewhat remote worlds of fish and insect, angling also chimed well with an emerging early modern England with its emphasis on building stronger links between the historically separate cultures of its nation. It was one of the first signs that nature was going to become a strong vehicle for nation building.

THE TREATISE OF FISHING WITH AN ANGLE (1496)

In *The Treatise of Fishing with an Angle* of 1496 the angler who fails to catch fish is rewarded by the benefits of his immersion in nature, particularly the rich aquatic landscape close to every village and town: by 'his wholesome and merry walk at his ease, and a sweet breathe of the sweet smell of the meadow flowers, that makes him hungry. He hears the melodious harmony of birds. He sees the young swans, herons, ducks, coots, and many other birds with their broods, which seem to be better than all the noise of the hounds, the blasts of horns, and the clamour of birds that hunters, falconers, and fowlers can produce' (McDonald 1963: 46). Here is the origin of what became, by 1653 in Walton's *Compleat Angler*, the figure of the contemplative, angler-philosopher. He rises early, keeps a quiet and simple profile in nature while at the same time making it the object of his gaze and thoughts.

In *The Treatise*, it becomes apparent in the section on instruction that angling is in part about making working connections between the lives of fish and the species that they eat, for the fifteenth century angler had to procure and make everything for himself, including the baits he used. In the section describing trout, the period April to October is described as 'leaping time', a season when trout rise to take aquatic and other insects from the surface of the stream. At this time, one must 'angle for him with an artificial fly appropriate to the month. These flies you will find at the end of this treatise, and the months with them' (McDonald 1963: 57). There are earlier records of angling for trout using flies that go back to the third century AD in a poem in Greek by Aelian. Writing about what was described then as an 'unusual mode of fishing', Macedonians on the River Astreus made imitations of flies which they used to catch trout (Braekman 1980: 11). While McDonald (1963: 8) also suggests a potential predecessor in Oppian's Halieutica (AD 169) which included a comparison with hunting and instruction, he feels it is unlikely that these Greek works were known to late medieval writers. Besides, it seems more likely, given that the fly patterns are clearly specifically *English* aquatic flies, that accurately match their month of appearance in the book, that this practice/art was one that had originated or at least matured separately in England. In his book *Fly Patterns: An International Guide*, Taff Price (1986) listed the flies as described in *The Treatise* and was able to identify the real fly they imitated even from their brief list of raw materials. Thus for April and the fly listed as *The Roddyd Fly* the instruction was to make: 'A reddish wool body, with a black silk rib; wings of a drake and a red capon's hackle.' According to Price, this is thought to be the red spinner or medium olive spinner (Baëtis Vernus); when made up it certainly looks exactly like all modern patterns of this fly.

Altogether there are 12 flies listed and these strike any modern fly fisher as a representative sequence of flies that still form the basis of most anglers' fly boxes. It begins with a fly recognizably like the early season March Brown; it has a stone fly imitation for April and the up-wing mayflies for May. By mid- to late summer, sedge flies feature mostly as they do in reality today with some terrestrials making an appearance too – the easy to recognize cowdung fly and a wasp or hoverfly. In fact, the instruction in *The Treatise*, though brief, is based on a great deal of observation and experience. Besides using artificial flies during 'leaping time', it also advises the angler to 'take the [aquatic] stone fly and the grubworm under the cow turd, and the silk worm, and the bait that grows on a fern leaf' (Price 1986: 57). In August it also advises the use of 'a flesh fly' (possibly a bluebottle).

The emergence of the angler, particularly the trout angler as an innovative student of the insect world is confirmed in a later book of 1614, *The Pleasures of Princes*, by Gervase Markham. In addition to the 12 moral qualities required to be an angler as set out in *The Treatise*, Markham also held that the angler must be 'a general scholar, knowing of the liberal sciences, a grammarian, a writer....he should know sun, moon and stars from which to guess the weather...he should know geometrical angles so as to describe the channels and windings of rivers and "the art of numbering" so as to be able to take soundings' (McDonald 1963: 23). In other words, angling was becoming something of an empirical activity; one had to have empirical skills to deploy in the practice of the art. This is borne out in the appearance of another book, Henry Peacham's *Compleat Gentleman* of 1634. In it he avers the practice of *exact imitation*: 'For the making of these flyes the best way is to take the natural flye and make one so like it that you may have sport: for you must observe that flyes haunt the water for seasons of the year, and to make their like with Cottons, Woole, Silke or feathers to resemble the like' (Wilson 1989: 23). The significance of this, while evident in the natural qualities of Dame Juliana's flies, was in making the link between observation and fly imitation because prior to the emergence of the first published entomology for fly anglers (in the nineteenth century), the huge variety of flies, even within the same family genera of insects meant that there was great variation in the form, colour and size of any given type of fly (sedge, mayfly, chironomid, stonefly, etc.) from one river/locality to another. In other words, fly-fishing then took insect recognition from the generic to the specific.

Hence, it is perhaps no surprise that by 1651, in *The Art of Angling* a few years before Walton's *Compleat Angler,* Thomas Barker was the first to invent and publish *new* flies. It may be that Barker was a one-off inventor but, it is more likely that others were also routinely inventing new patterns and/or improving older ones. After all, most anglers would have been making their own flies and had the means and the skill to do so. But it is also the built-in attentiveness to their natural environment that would have given them incentive to extend the basic stock of flies handed down from *The Treatise*.

This is perhaps confirmed by the fifth Edition of *The Compleat Angler* of 1675 in which Izaak Walton included a Part Two, a special section on angling for trout and grayling, by his friend Charles Cotton. Whereas in *The Treatise* there were mostly two fly patterns suggested for each month, by 1675, an extra 16 flies were added for the months of March–May and another 12 for June. A further 12 flies were added for the remaining months of the season resulting in 40 all told.

While the seventeenth century produced many classics of angling literature, during the eighteenth century there were very few and there does not seem to be any evidence that further progress was made into refining the knowledge and practice of imitating aquatic flies, at least in print. By the time a new flurry of books appeared on the back of a new enthusiasm in the nineteenth century, there were fewer than 100 titles published.

During the nineteenth century and especially in the last quarter, the empirical impulse that had always been present in the observant and nature-orientated angler began to benefit from the great advances in the empirical sciences, education, the huge popularity of amateur field naturalism and the presence within the angling fraternity of professional men inspired to experiment and improve techniques. In their writing, but particularly in the writing of F. E. Halford and G. E. M. Skues (to continue the British example) we can see that the centuries of fly-fishing practice and experience had actually *not* resulted in a very acute appreciation of the insects that participated in their sport. While attentive perhaps to the outward appearance of the fly, it appears from these later writings that their techniques were crude and barely based on the observable behaviour of insects or the trout that fed on them. Indeed, G. E. M. Skues writing in 1921 (p. 83) compared the intelligence of the angling fraternity (as displayed by their net stock of knowledge and techniques to date) unfavourably with that of the trout themselves! Again, the illusion of the modern fly fisherman practising an intimate and timeless proximity to the natural world is revealed. Much of what we take to be so old, traditional and timeless is actually nothing short of a modern revolution.

FREDERIC M. HALFORD AND THE RISE OF THE DRY FLY

Frederic Halford (b. 1844) was the son of a very wealthy textile industrialist who began to take an interest in fly-fishing when it had attained an almost cult-like enthusiasm among the English middle classes in the first half of the nineteenth century. After a training in fly-fishing on the River Wandle he joined the famous Houghton Fishing Club on the River Test in Hampshire in 1877, where he also met his like-minded research and writing collaborator, George Selwyn Marryat. Although he had a career in the family business he retired from it at the age of 45 to dedicate himself to fly-fishing and its improvement. Such was the intellectual and social standing of fly-fishing that this was not considered to be an eccentric career decision in any way. Rather, this was considered a mix of natural history and the higher arts, and an entirely appropriate object of dedication. Halford was a much admired man throughout his career it seems.

Halford had the means, the inclination and the necessary background to make a big difference, but the huge advantage he enjoyed over previous angling writers is that for almost all of his life an aquatic entomology had been produced and published for the sole purpose of the fly angler. This enlarged and elaborated the insect world available for the angling imagination. It was written by Alfred Ronalds, a printer and a very keen fly fisherman, who was not scientifically trained but who, together with the Reverend Brown of Gratwich, were typical of amateur naturalists of their day. He constructed a sophisticated observatory *over* his local River Blyth with bird's eye views over the fast flowing section of the river (the 'scour') and two prominent eddies.

Over several years they built up an incredibly detailed knowledge of the insect–trout interface of this river and wrote it up in a very modern and scientifically rigorous manner, based as it was on a very skilfully observed and experimental methodology. According to Marinaro's (1976) *In The Ring of the Rise*, Ronalds was something more than just an excellent entomologist: he was the 'originator of a new race of angler-entomologists, reproducing (for example) his mathematically precise diagram of how refraction affects the view that fish and fly-fisher have of each other' (Herd 1993). It is clear that Halford became not only just such an 'angler-entomologist (he published *Dry Fly Entomologist* in 1887) but also an experimental/empirical angler, after Ronalds' example.

The key observation that Halford made was that while attempts had been made to create and use (dry) flies that floated as imitations of the adult flies (there were several prior champions including Mascall in 1590; Venables in 1662, Ogden in 1879 and Pulman in 1851) in fact, mostly anglers fished their adult flies (somewhat absurdly) *underwater*, as so-called 'wet flies'. In addition, while sometimes anglers did cast their flies upstream to rising fish, most fished a team of two wet flies *downstream*, even on the exalted stretch of the River Test owned by the Houghton Club at the time Halford joined it (Wilson 1989: 15). Even during the mayfly season anglers would dapple their flies on the surface *downstream*. It seems incredible, *after Halford*, that so many could make such profound mistakes. After all, the wet flies that were used were imitations of the adult flies that were never to be *seen* or eaten by trout *under water*, but always in one of four states *on the surface*: in the act of hatching on the surface, in the act of drying their wings after hatching on the surface (sitting on the nymphal shucks); in the act of returning to the surface of the water to lay their eggs or in a 'spent' or dead and floating state after having laid their eggs.

In *every* one of these states, the flies are not only on the surface, they are *floating towards a trout* that will in almost every case take up a relatively fixed station in the stream in order to pick them off, one by one. Hence, the angler who is upstream of a feeding trout (casting downstream to it) is not imitating this natural movement when he retrieves his flies, but the very opposite. Likewise the dappling mayfly fisher could not present their flies as they drifted downstream but only in an unnatural bobbing manner. The trout to whom they presented their flies were mostly quite serenely and effortlessly sipping a steady stream of flies floating towards them rather than chasing flies that were taking off, so this was also at variance with nature, or was at least a movement that was 'less typical'.

Of course, anglers carried on in this erroneous style precisely because it did work; occasionally, they would entice a trout to take their flies. Success confirmed the theory of their practice and did not lead them to see it as deficient or less successful than it might otherwise be. Nor did they quite see the necessity of exactly copying the insects' behaviour (if they even noticed it) exactly. Their visual observations of what was happening on the stream only lead them to the conclusion that when specific flies were on the water and trout were rising to them, it was sufficient to cast flies of the right imitation to them, as close as possible. That this had been done for a very long time, by some truly excellent anglers and was in all the guides, was incontrovertibly true and it was how most had been trained to fish. Nobody, it seems had thought to observe exactly what was happening in reality.

While their less than perfect theory was a supreme act of mis-observation (their flies were not where they should be or moving correctly relative to a feeding trout) those who *had* begun to believe in the potency of fishing with dry flies were hampered anyway by technical problems. It had proved too difficult to create that impossible illusion with the equipment they had. This is why the Halford revolution was technical as well as zoological/behavioural.

Once convinced of the need for an effective floating fly that could be cast ahead of a feeding trout which would then float correctly and at the pace of the river flowing towards it, it was a simple matter of making the technical innovations required. So, for example, Halford and his partner Marryat used water resistant, buoyant quill for bodies, they deployed stiffer hackle feathers that floated a dressed hook efficiently and this was topped off by 'double split wings' made from feathers that would make the fly float in a realistic upright position.

By making many false casts the air would dry this configuration so that it would float once again after each attempt on the water. Perhaps Halford's experience in the textile industry had given him an ability to think about these fibre-based fly making materials in an experimental manner.

Even so, Wilson shows that further technical innovations had to be made in order to exactly imitate the natural aquatic flies. First, constant false casting frayed the whipping that secured the hook to the leader meaning the fly (very time consuming to make or expensive to buy) had to be thrown away, but the invention of eyed hooks meant that the angler could just retie a new one. In addition, flies were normally saturated and distorted after catching a trout and this quick-change technology made a big difference. Second, the need to throw an accurate line always upstream of the trout meant that anglers needed a more effective line that would cast into the wind, should it be blowing downstream. New, heavier oil-dressed silk lines fitted the bill and because they floated too, they never sunk the flies as lines had done before. And finally, a new rod, much shorter and with more power was required that could cast such heavier lines. This new technique made possible by this new tackle certainly worked more efficiently and soon dry-fly fishermen were out-catching others *very substantially*. Even so, the technique was extremely difficult to master making it intriguingly attractive: it introduced the notion of dry-fly fishing as a 'higher or fine art'.

None of Halford's ideas were entirely new but he brought them all together and improved them as never before. He was able to do this because as an entomologist he realised that improvements to angling techniques would emanate only from a closer understanding of insects and their interactions with trout but also, as an independently wealthy, early retired professional angler he could dedicate the necessary time to the work required for such improvements. Most previous angler-writers wrote their one book and that was that. But with Halford we see the origins of 'the professional angler' (there were many from the mid-twentieth century onwards) whose life was dedicated to the study of these human–insect–fish-stream relationships and creating ever more refined and informed ways of angling. But not just the angler: Halford was an early conservationist and he wrote about trout and fishery management too, but his legacy would also be measured by later concerns to understand and protect the habitats of the aquatic insects themselves. Thus, Halford left a legacy of seven major books published between 1886 and 1913, all of which extended the field he entered immeasurably including two books on making and managing fisheries.

As a result of Halford's interventions fly-fishing became a more sophisticated, technical and biological enterprise/recreation that would require years of study and practice to perfect. He introduced substantial degrees of difficulty and gentility into what had been a more basic craft and made it an almost unique encounter with the natural world. The social world of the mobile and leisure-rich English upper and middle classes adopted it as almost the perfect form of country sport, with the result that huge pressure was placed on existing fisheries. On the basis of its success it also spread and revolutionized the trout fishing cultures of America, New Zealand, Chile, Scandinavia, Australia and Europe (see Washabaugh and Washabaugh 2000 for its subsequent development and naturalization in the United States, and French 1994, 2002 for its application in Australia).

G. E. M. SKUES AND THE EMERGENCE OF THE NYMPH

Frederic Halford left one very significant loose end. In many of his own experiences, as recounted in *Dry-Fly-Fishing in Theory and Practice* 1889 (see for example p. 117), trout often *appeared* to be rising to the surface but would not rise to an angler's dry fly. At other times when no adult flies were on the river trout were *bulging* (creating a disturbance on the surface), but were not apparently taking food from the surface. In these circumstances, Halford's autopsies showed him that they were actually taking the insect in its nymphal stage, often in vast numbers, typically as they were swimming up to the surface to hatch (their previous life-cycle stage was spent buried in the mud bed of the river or under stones). He even thought of the nymphal insect as their 'beef and mutton' while the adult was a special course, their caviar. Halford argued (1889: 115) that 'while such a nymph could be imitated by the fly tier there was no way that the swimming action of the nymph could be imitated, and so the angler was left with only the dry-fly option'. Halfordonian purists would thus ignore the trout feeding in this way.

Perhaps here, Halford was applying his principle of imitation too literally and ignoring his otherwise reliable principle of *experimentation*. It was, after all, a fairly serious lapse in his ability to make important connections between trout and insect behaviour. Before any nymph could make it to the surface to hatch it *had* to swim through the water column, past the positions trout always occupy. And, as Taff Price

(1986) argues, many trout are quite content to intercept them here rather than take them off the surface. Further, G. E. M. Skues showed that often they became so satiated with nymphs that their appetites declined markedly by the time adult flies had emerged and were 'on the water'. Equally, Halford's objection that swimming nymphs could not be imitated is not born out by the movements of the trout feeding on them: they are not forced to chase a fast swimming creature in any way and at most shift only slightly from their resting position. Here then, there was room for another sequence of interventions in the elaboration of fly-fishing theory.

Some recent writers [they include Paul Schullery (2007), John Goddard (1998) and Andrew Herd (1993)] consider George Edward MacKenzie Skues to be the greatest innovator. He was not convinced that Halford had followed through on his various observations about the nymph and set out to explore its potential. If it was the most important insect element in the trout's diet then it deserved at least more research. He took very seriously Halford's observation that the traditional alternative to his revolutionary dry fly, the wet fly, was an absurdity. Here was an insect imitation of the adult flies that could be seen flying in the vicinity of the river that was fished *under the water*: the one place in their environment that they never entered. His thesis was that Halford's revolutionary dry fly-fishing was sophisticated and perfect for occasions when flies were on the surface of the water but it did not amount to a method that could be used on most occasions when trout could be caught. Meanwhile the traditional wet fly was simply an aberration. For Skues, the modern fly fisher must be able to fish flies that accurately imitated every lifecycle stage of the aquatic flies, namely nymphs, hatching duns, spinners and spent flies. This was a tall order and it was not going to be easy to persuade an angling fraternity that had become enamoured of the dry-fly.

Skues, a London-based lawyer who fished the Abbotts Barton fishery on the River Itchen, embarked on a lifetime of experimentation that culminated in six remarkable books. He argued that the fly fisher required not only an entomology of the adult stages of the insect but their nymphal stages also – it was an extension of the need to observe even that which was not normally visible to the angler on the bank. This knowledge would then form the basis of an entirely new set of fly patterns based on the imitation of the underwater insect forms which would displace the colourful absurdities of the traditional wet fly box. In time, these were relegated to *lures*, a theoretically dubious, if occasionally useful tool since a new theory held that they worked by triggering aggression in non-feeding trout and were thus useful when all else failed.

As with Halford, Skue's revolutionary idea for fly presentation and their patterns also spawned new ways to fish the flies, and 'nymphing' became, in the hands of its adherents, as addictive and as refined a method as fishing with the dry fly. This is partly because the nymph angler has a major disadvantage relative to the dry fly angler (and thus the degree of difficulty and by extension the reward was even higher): whereas the latter can always see or hear the rise as the trout takes the fly, the signs that a trout has taken a nymph are more subtle and not at all easy to detect. Nymphing trout are often taking large numbers of nymphs very casually. They are mostly easy pickings and so the angler's nymph travelling at the speed of the river flow will also be taken gently and quietly. There will be no great movement, no splashing, no pulling or sensation imparted to the hands of the angler. If the angler fails to spot a fly being taken in a trout's mouth, the weight of the fly line will soon drag on the fly and the trout will sense it is wrong and release it, often with no panic or fuss. Thus the angler has first of all to find a trout he can see and then cast the nymph to a point just upstream of his lie and no further to either side of him than he has been willing to travel to take nymphs previously (trout are *very* habitual). Having done this, he must watch the trout for all the tell-tale signs that it has taken the fly (a deviation to take a fly that corresponds to where the angler fly must be; a showing of white as the trout opens and shuts its mouth, a bulge of water that indicates a sudden move to take the fly above him) and at exactly the right moment strike to set the hook. This entire operation required such skill, dexterity, eyesight and a quiet approach that it soon overtook the exclusive use of the dry fly as the technique of choice. It was a logical conclusion to a developmental pathway begun in fits and starts in previous times but which came together relatively quickly in a matter of some 50 years.

CONCLUSION

After the Skues revolution his techniques eventually became part of the well-rounded river fly fisher but they were particularly influential over the massive acceleration of lake and still water fisheries that expanded into every water reservoir in the UK and to an increasing number of public and private lakes. In these the aquatic life was less well known and had not been studied by anglers until the 1960s. However, as this became by far the biggest form of fly-fishing in the country it was the imitation of nymphal and hatching flies that became the single biggest focus of their sport. Here the key insect family that was most important to understand was not the upwing flies that dominated rivers but midges, the *Chironomidae* family. But that is another story.

REFERENCES

Austen, J. (1813) *Pride and Prejudice*. Whitehall, London: T. Egerton.
Braekman, W. L. (1980) *The Treatise of Angling in The Boke of St. Albans (1496)*. Brussels: Scripta.
Christie, A. (1953) *After the Funeral*. London: Collins Crime Club.
Elias, N. (1994) *The Civilising Process*. Oxford: Blackwell.
Franklin, A. S. (1996) On fox hunting and angling: Norbert Elias and the sportisation process. *Journal of Historical Sociology*, **9**(4), 432–456.
Franklin, A. S. (1999) *Animals and Modern Cultures*. London: Sage.
Franklin, A. S. (2002) *Nature and Social Theory*. London: Sage.
French, G. (1994) *Tasmanian Trout Waters*. Hobart, Australia: New Holland Publishers.
French, G. (2002) *Frog Call*. Hobart, Australia: New Holland Publishers.
Goddard, J. (1998) *The Essential G. E. M. Skues*. London: A&C Black Publishers Ltd.
Halford, F. M. (1887 [1921]) *Dry Fly Fisherman's Entomologist*. London: George Routledge and Son.
Halford, F. M. (1889) *Dry-fly-fishing in Theory and Practice*. London: H.F and G. Witherby Ltd.
Herd, A. (1993) A fly-fishing history. Available at: http://www.flyfishinghistory.com.
Hilaby, J. (1949) *Within the Streams*. London: Harvey & Blythe.
Hughes, T. (1857) *Tom Brown's School Days*. London: Macmillan.
Kipling, R. (1890) On dry-cow fishing as a fine art. *Fishing Gazette*, **13** December 1890.
Marinaro, V. (1976) *In The Ring of the Rise*. New York: Lyons & Burford.
Mascall, L. (1590) *A Booke of Fishing With Hooke and Line*. London: J. Wolfe.
McDonald, J. (1963) *The Origins of Angling*. Garden City, New York: Doubleday and Co.
Ogden, J. (1879) *Ogden on Fly Tying*. Cheltenham, UK: Low Marston Searle and Rivington.
Peacham, H. (1634 [1906]) *Compleat Gentleman*. New York: The Clarendon Press.
Price, T. (1986) *Fly Patterns: An International Guide*. London: Ward Lock.
Pulman, G. P. R. (1851) *The Vade Meccum of Fly-fishing for Trout*. London: Green and Longmans.
Ransome, A. (1929) *Rod and Line*. London: Jonathon Cape.
Schullery, P. (2007) *Skues on Trout*. Mechanicsburg, PA: Stackpole Books
Shakespeare, W. (1994) *Henry IV Pts 1 and 2*, ed. B. Mowat and P. Werstine. New York: Washington Square Press.
Skues, G. E. M. (1921) *The Way of a Trout with a Fly*. London: Adam and Charles Black.
Venables, R. (1662) *The Experienced Angler, or Angling Improved, Being a General Discourse of Angling, Imparting Many of the Aptest Ways and Choicest Experiments for the Taking of Most Sorts of Fish in Pond or River*. London: Richard Marriott.
Walton, I. (1663 [1990]) *The Compleat Angler*. London: Studio Editions.

Washabaugh, W. and Washabaugh, C. (2000) *Deep Trout: Angling in Popular Culture.* Oxford: Berg.

Williamson, H. (1936) *Salmo the Salmon.* London: Little, Brown and Company.

Wilson, D. (1989) Preface. In *Dry-Fly-fishing in Theory and Practice,* ed. F. M. Halford. London: H. F. and G. Witherby Ltd.

Yeats, W. B. (1991) The Fisherman. In *W.B Yeats Selected Poems,* ed. T. Webb. Harmondsworth, UK: Penguin.

9

An appreciation for the natural world through collecting, owning and observing insects

AKITO Y. KAWAHARA AND ROBERT M. PYLE

INTRODUCTION

Insect collecting has served as a central part of a naturalist's lifestyle for centuries. Some of the greatest scientists engaged themselves in the joys of insect collecting and observation, which often began at a young age. Charles Darwin regularly went to the coast of Wales, and to Wicken Fen near Cambridge, to hunt for beetles that he prized (Barlow 1958). Gregor Mendel, pioneer of modern genetics, enjoyed beekeeping (Večerek 1965). Theodosius Dobzhansky, a central figure in evolutionary biology, became a biologist after collecting butterflies during grade school (Alaya 1985). Alexander von Humboldt, considered the father of biogeography, collected insects during his formative years in Germany (De Terra 1955). Edward O. Wilson, one of the strongest voices in the conservation of biological diversity, credits his career to the ants he observed as a boy in Florida (Wilson 1994). One wonders whether these renowned biologists would have developed their ideas if it weren't for their interest in collecting, owning, and observing insects during their childhood.

Insect collecting can be remarkably enjoyable and brings together the excitement of exploration with learning. Every entomologist, both professional and amateur, knows the joy of capturing a prized bug in a butterfly net. This act often captivates the netters themselves, leading them into the study of insects for the rest of their lives. Charles Darwin wrote in his autobiography, 'No pursuit at

Cambridge was followed with nearly so much eagerness or gave me so much pleasure as collecting beetles. It was the mere passion for collecting...' (Barlow 1958: 20). Another famous Cantabrian, Vladimir Nabokov, later put it this way: 'Few things have I known in the way of emotion or appetite, ambition or achievement, that could surpass in richness and strength the excitement of entomological exploration' (Nabokov 1966: 126).

The joy of collecting, owning observing insects also often leads to discoveries that can serve as important contributions to science. Insects constitute approximately 58–67% of the total diversity of all eukaryotic species in the world (Alder and Foottit 2009), and the number of individual insects on the planet at any given moment is thought to be a quintillion (Williams 1964). Human benefits from invertebrates include nutrient cycling, plant pollination and propagation, maintenance of plant community composition and structure, food for insectivorous vertebrates and maintenance of animal community structure (Gullan and Cranston 2010). However, the number of insect taxonomists does not correlate with the number of insects on the planet. Only ~19% of US taxonomists work on insects while 26% work on vertebrates and 30% study plants (Gaston and May 1992). Herbert Ross once wrote, 'One thing [that should] never worry the entomologist[:] ... running out of work. There is so much to be done and so many fields in which the surface has merely been scratched' (Ross 1945: 248) – which remains the case to this day.

Unfortunately, humans are becoming increasingly distanced from nature and the outdoors. Children, who are innately curious about nature, are now frequently confined to their homes and use smart phones or sit in front of a flat screen. Parents are also going outdoors less than they used to, and it is not rare to meet adults in urban settings that have never gone hiking (Louv 2005, 2011). Cities continue to grow, and the open lot that was once a sanctuary for migratory birds may now be replaced by a shopping mall, the stream nearby that was once a haven for fish, frogs and dragonflies may now just be a concrete trench. While we can still find plants breaking through pavement, dragonflies defending territory over parking lots, butterfly gardens in people's yards, the natural environment that once existed is rapidly diminishing. How can we inspire natural history education when the global landscape is changing so rapidly? Not only do we need to advance natural history education in schools, but we also need to recapture the excitement for nature that is innately within us, according to E. O. Wilson's hypothesis of Biophilia (Wilson 1984). We will need to explore the outdoors, and an obvious place to begin is to provide butterfly nets so that children can collect, rear and observe insects.

THE BUTTERFLY NET

The butterfly net is an extremely valuable and versatile tool that can capture one's interest in natural history and has the potential to harness great interest for the natural world. It is one of the simplest and most important tools a child or adult can have, and has been used since Victorian naturalists travelled across the world. As the second author put it in an essay on the subject, 'Kids love nets because chasing insects is fun. It also brings the chaser face-to-face with exciting, novel, always-surprising life. Talk to any number of biologists, doctors, wildlife managers, and other life-science professionals, and the preponderance of them will tell you that catching bugs was a vital early stimulus for their engagement with nature' (Pyle 2009: 17). Insect nets also have the virtue of being inexpensive, or even easily constructed by kids from available materials.

Many of us who do not frequently interact with nature can immediately become interested in wildlife through the simple joy of collecting. Two of the first author's personal experiences immediately come to mind. The following are told in the first person, from the primary author's perspective.

In high school, I went on a 4-day holiday trip to Jamaica with a close neighbour, Owen, and his father, neither of whom had collected insects with a butterfly net. Following my habit of carrying my butterfly net everywhere, I packed two collapsible nets for the trip. We stayed in a large resort hotel in Ocho Rios, Jamaica, and on the morning after our arrival, we decided to go to the beach to swim. Owen immediately jumped in the water but I decided not to go in because I never was a great swimmer. As I pondered whether to go into the water, I noticed several white butterflies flying around bushes along the shoreline and contemplated returning to my room to get my net. But as an adolescent 17-year-old, I did not want to be seen with a butterfly net on a beach full of young women. I pretended to walk on the beach as a tourist, but my eyes were fixed on the butterflies. They flew from side to side across patches of vegetation, occasionally landing and spreading their wings as they rested on leaves. The species was the white peacock, *Anartia jatrophae* L., a common Neotropical lowland butterfly that I had never seen before. I couldn't hold back anymore. Like a programmed robot with a chip inserted during childhood, I ran back to the hotel room to get two rusty collapsible butterfly nets that were in the bottom of my suitcase.

Sunbathers by the hotel pool lifted their sunglasses as they watched a skinny, shirtless teenager run along the edge with a butterfly net in each hand. When I reached the beach, Owen was nowhere to be seen, but his dad was sitting on a tree stump smoking a cigarette. I rested a spare net by the stump and started chasing the butterflies with the other. Without even asking, Owen's father joined soon after I began. He too began running on the sand barefooted, and together we chased the butterflies, the weight of our bodies sinking into the sand as we ran. He caught one, then two, and many followed. Because I hadn't taught him anything about what to do after netting a butterfly, he took one net, caught one, then switched nets with me, and caught another. While I was examining each butterfly, I had a few moments to glance up to see how he was doing. What I saw was a middle-aged man running up and down the beach with a butterfly net without the time to flick the ashes off his cigarette in his mouth. He still mentions the excitement he felt chasing those butterflies that day.

Years later, I had a similar experience with another friend who had never previously collected insects. In 1999, I took a year off from college and lived in Japan. During that year, I had the opportunity to search for *Luehdorfia japonica* Leech, or 'Gifu-cho' in Japanese, one of the most spectacular swallowtail butterflies in Japan. The butterfly is somewhat challenging to find as it only flies for about a week in the spring and is restricted to pristine lowland forests. I knew of a good locality in Toyama, about 8 hours away from my home by car. Because it was so far, I decided to ask my 35-year-old businessman friend, Sada, to help with driving. We left around 9 pm, and arrived the following morning around 5 am in Toyama at a hill with a small public park on the top. Sada had brought his son, Oh, who had insisted to be taken along on a trip to collect bugs. We parked the car along the road and began walking down one of the trails together, Sada and his son behind me. As soon as we started walking, I saw what appeared to be a yellow object fluttering to the ground. Instinctively, I shouted 'There!' and pointed. Before I could begin chasing, Sada had already run ahead of me to attempt to catch it. He approached the butterfly quickly and the butterfly began to fly. He swung the net haphazardly, and the butterfly flew up into the canopy. 'Kuso!' he yelled, 'Mochotto datta no ni!' (meaning 'Darn, I almost had it!'). When the next one appeared, he swung his net and caught it in mid-flight. He shouted with joy, 'Yatta!' Because the butterfly is quite rare, we released them. Sada still speaks of how he felt when he caught that insect.

At one point during our walk, Oh pointed to the ground and said, in Japanese, 'there is one here that is bending its body.' We moved slowly to the butterfly and saw that it was a female laying an egg on wild ginger, *Asarum* sp., its host plant. I explained the natural history of the butterfly and we watched for a few minutes before it flew away. Oh carefully collected the leaves on which the eggs were laid and put them in a Tupperware container so that he could observe them. The eggs turned into caterpillars and were kept in his home and reared until they turned into butterflies. Oh is now 13, and that experience, among others, triggered an interest in the natural world, and has led him to consider a career in entomology.

CHILDREN AND NATURE

Educating children about the natural sciences is critically important at a time when global diversity is declining at an alarming rate and humans are becoming increasingly distanced from nature (Louv 2005, 2011). Regardless of ethnicity or background, children are inherently curious about the natural world (Bjerke *et al.* 1998a; Kellert 2002). Professional entomologists and insect enthusiasts reading this book may recall how playing with living things in the outdoors during their formative years led to a deep interest and respect for natural history. But does this interest peak at a particular time during their development?

While children are often perceived as being interested in animals at all ages, there is a time when their interest will heighten. From an entomological perspective, we could almost conclude that all kids are born entomologists until they learn not to be. Indeed, studies have shown that effective and emotional relationships with animals are greatest between the ages of 6 to 12 years, and gradually decline thereafter (Kellert 2002). This general trend has been described in several other independent studies (e.g., Bjerke *et al.* 1998b; Bunting and Cousins 1985). These findings imply a narrow window for reaching kids to instil a deep concern for nature, and thus an optimal period for getting nets and other tools to promote natural history education into their hands.

Kellert and Westervelt (1983) classified children's appreciation of animals into nine groups, and discovered that the most common responses can be categorized into *humanistic* (strong affection towards animals, such as pets), *naturalistic* (appreciation for wildlife and the outdoors) and *negativistic* (avoidance, dislike and fear) attitudes. The humanistic and naturalistic attitudes were highest among children who owned pets or frequently interacted with farm animals, regardless of gender or race. Children who took part in activities such as hunting and birdwatching also showed higher appreciation for natural history and lower negativistic scores than those who did not. Numerous other studies confirm this trend (Bjerke *et al.* 1998a; Kellert 1985; Paul and Serpell 1993). Unfortunately, among mammals, birds, amphibians, fish and arthropods, the latter receive the lowest humanistic and naturalistic scores and the highest negativistic score (Kellert 1993). Why do insects and other arthropods score lowest?

Numerous reasons may account for why the public perceives insects in this way. Kellert (1993) summarized some of these: it may be difficult for humans to relate to insects because they differ in morphology, behaviour and abundance from our own species, and because they often respond unpredictably in our home. Popular culture can also create inaccurate myths, which include the presumed connection between arthropods and human disease (e.g. McNeill 1976). Moreover, the English language contains terms such as 'bug-eyed', 'spidery' and 'worm', which indicate human traits that lack warmth (Hillman 1991). Parents, who are largely influenced by the media, may be at fault because they can have the greatest influence on a child's thoughts.

Surprisingly, most children know far more about insects than do their parents. Comments from a parent to a child should not be, 'don't touch that spider, it is disgusting', but might be replaced with 'spiders are very interesting animals, they produce silk, one of the strongest materials on the planet'. Kellert (1985) demonstrated that eleventh graders were significantly more knowledgeable than adults on questions concerning invertebrates. For example, 78% of all children (86% of all eleventh graders) knew that spiders do not have ten legs, while only 50% of adults knew this. Only 23% of adults, compared to 48% of 16-year-olds knew that inchworms and earthworms were not in the same phylum. Children living in rural areas scored much higher than kids in urban areas (Kellert 1985). While the outlook does not appear very positive, negativistic views can be changed by proper education and exposure to natural history at the right time during childhood. Children who form the habit of outdoor exploration may, in time, overcome their negative feelings toward certain aspects of nature. In his book *Wintergreen*, the second author describes his long conflict between a passion for butterflies and deep, crippling arachnophobia, and how he eventually bested it through intentional practice: getting to know spiders first hand, beginning with small, unthreatening species and building up to large, bulbous orb-weavers and tarantulas (Pyle 2006). A cultural change will need to take place to promote a greater appreciation of insects, much like what we find in Japan.

FOSTERING A WORLD THAT CHERISHES INSECTS: ENTOMOLOGY IN JAPAN

Insects have a central role in many Asian countries, and their presence is especially pronounced in modern Japan. Children are frequently seen carrying nets and cages in city parks (Figure 9.1). Japanese boys and their fathers frequently keep horned rhinoceros and stag beetles in their homes as pets. Beetles are sold in vending machines and department stores, and are accessible via petting zoos. Part of the pet section in a Japanese department store is converted during the summer into an 'insect corner' where different species of beetles are sold in cages, along with basic insect collecting and rearing equipment, general entomology books and plastic cages. Pet beetles are displayed alongside cat toys, dog food and goldfish. When people visit an established collecting locality for insects, they will frequently encounter other amateur collectors frantically seeking their prized insects (Kawahara 2007).

Insects are also found elsewhere in modern Japanese culture, including firefly festivals, documentaries, popular television shows, candies, toys and videogames. Children's books are also translated into Japanese directly from old Victorian entomological literature. The most popular of such books is the ten-volume series written by Jean-Henri Fabre (1823–1915), the famous French entomologist who wrote *Souvenirs Entomologiques* (Fabre 1879–1907) a monumental work that included detailed observations on insect behaviour. Fabre's books have become so popular that in the summer of 2005, the 7-Eleven convenience store chain gave away eight plastic renditions of various insects that Fabre studied as a promotion to help boost sales (Kawahara 2007). One of the most notable was *Scarabaeus typhoon* (Fischer-Waldheim) rolling a large ball of dung. Can you imagine the reaction of a 7-Eleven customer in the United States if given a ball of plastic dung with a beetle figurine as a gift for purchasing a soda?

The insect theme also appears frequently in Japanese videogames. Satoshi Tajiri, the creator of the successful animation 'Pokémon,'

Figure 9.1   Japanese children collecting insects in a city park. Photo by Kenji Nishida.

enjoyed collecting insects as a child and his experiences inspired him to develop the animation that would 'capture the excitement and fun of bug hunting' (IGN 2011). In a recent count, there were up to 15 different insect collecting games and insect battle games in Japan, available for multiple gaming platforms. Some of these were explained in Kawahara (2007) and appear in the documentary film 'Beetle Queen Conquers Tokyo' (Oreck 2010).

According to the developers of the very successful children's videogame 'Mushi-King', the game was developed not only to target kids, but also their parents. 'The parents ... can relate to [beetles] as well. Fathers remember catching and keeping beetles when *they* were kids' (Wallace 2005: A9). The first author doubts that he would be writing this article today if it were not for his father's childhood experiences collecting butterflies and the widespread availability of Japanese entomological supplies. The day that helped direct his career path as an entomologist came when his father walked into a Japanese department store and accidentally came upon insect collecting equipment. He bought a net for him, as the equipment in the store brought back memories of his childhood experience rearing insects. His father's gift led him to be immediately captivated by the net, and become fascinated with collecting anything that moved.

Thus, the combination of parents' entomological experiences during childhood and availability of entomological resources fosters proper education about insects. When these children become parents themselves, they too are likely to provide positive feedback about insects. This education process can continue over generations, and helps strengthen the society's overall appreciation for entomology and natural history. The memory of collecting a rare insect and learning about it seems never to leave one's mind. In his autobiography, Charles Darwin wrote, 'I am surprised what an indelible impression many of the beetles which I caught at Cambridge have left on my mind. I can remember the exact appearance of certain posts, old trees and banks where I made a good capture. ...I had never seen in those old days *Licinus* alive, which to an uneducated eye hardly differs from many of the black Carabidous beetles; but my sons found here a specimen, and I instantly recognised that it was new to me; yet I had not looked at a British beetle for the last twenty years' (Barlow 1958: 61).

WHERE DO WE STAND, AND WHAT CAN WE DO?

The state of natural history education in the United States is of great concern. A study showed that of American children between second and eleventh grade, only 52% knew that penguins are birds, and only 29% knew that Koala bears are not really bears (Kellert 1985). Orr (1994: 126) stated, 'Even in this time of ecological concern, high schools, colleges, and universities continue to turn out a large percentage of graduates who have no clue how their personal prospects are intertwined with the vital signs of the earth.' The second author attributed this problem largely to the lack of accessible habitat near the child's home. He wrote,

> Most kids used to wander freely and catch fireflies in a jar – or crawdads, or polliwogs – and, through those encounters, learned to connect with the land on which we all depend. These days, their attachment to electronica almost from birth, combined with parents' fears for their child's safety and the loss of accessible habitats close to home, means that this fundamental experience of roaming freely is increasingly rare. Where will our future conservationists and biologists come from, when kids no longer chase grasshoppers in real life? Well, there is no more effective defense against nature deficit disorder than the butterfly net!' (Pyle 2009: 57–58).

Thus it is of primary importance to engender greater appreciation among the general public toward insects and other arthropods. Ambitious education programmes that touch on the positive values of insects are increasingly necessary.

In order to counter the general trend in the United States, several organizations have pushed forward to change this perception. For example, the Lepidopterists' Society has developed the Outernet Project (http://www.lepsoc.org/education.php), which helps place butterfly nets into the hands of children so that they can explore, sample and learn about entomology. The society is teaming with the entomological supply company BioQuip Products (Gardena, California) to provide youngsters with a free insect net, spreading board, insect box, insect pins and insect collection booklet, when kids join The Lepidopterists' Society; or at a small expense if they choose not to join. The Outernet Project started a partnership with many local insect clubs in the United States, and some, such as the Utah Bug Club, teach children how to collect, mount, and curate butterflies (http://www.utahbugclub.org). There is also 'Firefly Watch', a project that tracks the distribution of fireflies in the Eastern United States during the summer. Participants can log onto a website and input firefly data that they observed in their backyard. The organization then utilizes these data to construct real-time distribution maps for many species and determine their population fluctuations. These data become available to the participants and educate them about fireflies.

Another method to enhance natural history education is to bring often negatively perceived 'creepy-crawly' arthropods, directly to students in the classroom (See Ernst *et al.* and Rykken and Farrell, this volume). The 'Insect Zoo' on Maryland Day, held annually at the University of Maryland College Park campus, introduces a diversity of arthropods, both live and pinned, to the public. A long line often forms outside the event room each year, and participants are encouraged not just to look, but to touch the tarantulas, stick insects and hissing cockroaches. Terrified parents watch their children hold the insects, and then later decide they should try holding them too. Instructors educate students about the natural history of each arthropod (Figure 9.2).

One of the most notable examples of bringing arthropods to students is Dr Linda Rayor's 'Spider Biology' course at Cornell University, which she has taught for more than 15 years. This very popular course teaches students about every aspect of the biology of spiders, and includes a take-home extra credit assignment that allows students to learn about the natural history of their pet spiders. We followed the

Figure 9.2   Children learn from an instructor about arthropods at University of Maryland's 'Insect Zoo', part of annual the 'Maryland Day' event. Photo by Paula Shrewsbury.

example set forth by Dr Rayor and proposed a similar project at the University of Maryland in David Hawthorne's BSCI 120 Insect Biology course in 2006, which had an enrolment of 100 undergraduate students, most of whom were taking the course simply to fulfil their science requirement. Like Dr Rayor's course, we offered the students the option of extra credit – 15% of their total grade – for caretaking pet tarantulas. At first, the students were very hesitant. But 92 of 100 decided they needed the extra points enough to push their fears aside and participate in tarantula care and observation. During the course of the semester, some became so attached to their baby tarantulas that they named them, and some even went to the extent of carrying them to every class they attended. Approximately a third of the students kept their spiders after the course ended, saying that they could not part with their pet.

In the BSCI 120 course, we also took students to a small field near the football stadium on campus. We handed each student a net and said, 'Go catch whatever you can!' Students at first were reluctant to venture into the field, but once they started searching, they couldn't stop. After the 30-minute outing was over, 2 of 15 had returned. One of the most memorable participants was a 300-pound linebacker on the division 1 college football team. After running around the field with the net that looked like a plastic spoon next to his huge body, he returned and said, 'This was the best experience I ever had in a class, I want to do it again'. The same student wrote to me several years later and told me that he hadn't forgotten that experience and had bought his son a bug collecting kit.

Some people and organizations view insect collecting as an activity best avoided, since it involves the killing of its subjects. However, the great preponderance of informed fact and opinion favours the conclusion that collecting very rarely offers a threat to the conservation of insect populations (Pyle *et al.* 1981). As fecund invertebrates, most insects replace their numbers rapidly; and people with butterfly nets are not, for the most part, very effective predators. Also, collectors commonly target the more colourful and dramatic males, which further buffers potential losses, since females are more expensive to the population. Almost all serious danger to rare insects comes from habitat (and now climate) alteration. And it can certainly be said with confidence that most insect conservationists were originally beguiled and inspired by collecting.

Furthermore, one must take into account what a child can gain from placing an insect in his or her hand. Direct contact, through making an insect collection, teaches the child about the characters that are used to distinguish taxa in ways that a photograph or illustration in a book cannot accomplish. Such close observation frequently leads to curiosity about the animals' lifeways and environmental adaptations; young entomologists become young botanists, geologists and climatologists; and all of the life sciences benefit, along with the environment.

Some prominent butterfly clubs promote optical encounters with insects to the exclusion of nets. This has led to a divisive 'watching versus catching' debate (Pyle 1992). However, such a schism in the entomological world is unnecessary and diverting from the very real need for all bug-lovers to cooperate on behalf of habitat conservation. For one thing, optics simply do not replace nets in a child's experience. Give a boy a pair of binoculars or a camera, and he will be engaged momentarily; but give a girl a net, and watch her go! Besides, if the captor is concerned about killing the insect, it can be let go after being placed in one's hand, or on a plant. Both authors frequently teach through catch-and-release. While conservation should certainly be kept in mind when collecting insects, it is generally easy to avoid species with conservation concerns mainly because they tend to be rare, and thus, coming in contact with them is not very common. If one does encounter a species of concern, it may be released without harm, or merely observed. Children taking part in the Outernet Project are provided with a copy of the Lepidopterists' Society's conservative Collecting Guidelines, to encourage good, careful practice. We believe on the whole that the social value of nets vastly outweighs their liability.

CONCLUSION

Collecting, rearing and observation can be readily done with appropriate, simple equipment. It can be conducted practically anywhere, such as in one's backyard or nearby park. Easy yet challenging, one of the most enjoyable and least expensive of outdoor activities, insect collecting can greatly improve one's appreciation for the natural world. In Japan, the availability of insect collecting equipment and supplies allows many Japanese insect hobbyists to make personal collections and further entomological research and promote interest across the natural sciences. These Japanese collections are the result of an appreciation for natural history in Japanese culture, which begins at childhood.

Japanese children are educated about insects mainly through collecting, but their interest in learning about insects is supported by Japanese culture, which produces an outstanding number of children's books, television programmes, and insect-related toys and videogames. When these children become adults, they help teach their children about entomology, and an interest in entomology continues for generations. This transfers into a concern for habitat to support the treasured insect resource. The portrayal of insects as being scary, gross and creepy is often misleading, scientifically inaccurate and does not help teach the general public about natural history. We must not harm the children's innate interest to learn about natural history by brainwashing them with incorrect facts. We should let them decide for themselves, by providing them with nets and jars, and then letting them explore the fascinating world outdoors on their own.

ACKNOWLEDGEMENTS

Harvey Lemelin kindly invited the first author to write this chapter. This chapter stems from ideas that were developed recently through interactions with Amauri Betini Bartoszeck, Julie Byrd, Jessica Oreck, Richard Louv, Thomas Simonsen, Todd Stout, Lisa Taylor and Susan Tunnicliffe. Stacey Bealmear, Julie Byrd, Robert Dirig, Shuya Fan, Amy

Hester, Luc LeBlanc, Alan Leslie and Carolina Puente assisted with obtaining literature or provided helpful comments on the text. Kenji Nishida, Mike Raup and Paula Shrewsbury provided the photograph of children collecting in Japan. We are also thankful to our families for willingly exposing us to the wide outdoors.

REFERENCES

Alaya, F. (1985) *Theodosius Dobzhansky 1900–1975*. Washington DC: National Academy of Sciences.

Alder, P. M. and Foottit, R. G. (2009) Introduction. In *Insect Biodiversity: Science and Society*, ed. R. G. Foottit and P. M. Alder. Oxford: Wiley-Blackwell, pp. 1–6.

Barlow, N. (1958) *The Autobiography of Charles Darwin, 1809–1882: With Original Omissions Restored, Edited with Appendix and Notes by his Grand Daughter, Nora Barlow*. London: Collins.

Bjerke, T., Ødegårdstuen, T. S. and Kaltenborn, B. P. (1998a) Attitudes toward animals among Norwegian adolescents. *Anthrozoös*, **11**(2), 79–86.

Bjerke, T., Ødegårdstuen, T. S. and Kaltenborn, B. P. (1998b) Attitudes toward animals among Norwegian children and adolescents: species preferences. *Anthrozoös*, **11**(4), 227–235.

Bowd, A. D. (1984) Fears and understanding of animals in middle childhood. *Journal of Genetic Psychology*, **145**, 143–144.

Bunting, T. E. and Cousins, L. R. (1985) Environmental depositions among school-age children. *Environment and Behavior*, **17**, 725–768.

De Terra, H. (1955) *Humboldt: The Life and Times of Alexander Von Humboldt 1769–1859*. New York: Alfred A. Knopf.

Eagles, P. F. J. and Muffitt, S. (1990) An analysis of children's attitudes toward animals. *Journal of Environmental Education*, **21**, 41–44.

Fabre, J.-H. (1879–1907) *Souvenirs entomologiques*, 10 volumes. Paris: C. Delagrave.

Gaston, K. J. and May, R. M. (1992) Taxonomy of taxonomists. *Nature*, **356**, 281–282.

Gullan, P. J. and Cranston, P. S. (2010) *The Insects: An Outline of Entomology*. Oxford: Wiley-Blackwell.

Hillman, J. (1991) *Going Bugs*. New York: Spring Audio, Gracie Station.

IGN (2011) Top 100 game creators of all time. Available at: http://games.ign.com/top-100-game-creators/69.html.

Kawahara, A. Y. (2007) Thirty-foot telescopic nets, bug-collecting videogames, and beetle pets: entomology in modern Japan. *American Entomologist*, **53**(3), 160–72.

Kellert, S. R. (1985) Attitudes toward animals: age-related development among children. *Journal of Environmental Education*, **16**, 29–39.

Kellert, S. R. (1993) Values and perceptions of invertebrates. *Conservation Biology*, **7**(4), 845–855.

Kellert, S. R. (2002) Experiencing nature: affective, cognitive, and evaluative development in children. *Children and Nature, Psychological, Sociocultural, and Evolutionary Investigations*, ed. P. H. Kahn and S. R. Kellert. Cambridge, MA: The MIT Press, pp. 117–151.

Kellert, S. R. and Westervelt, M. O. (1983) Children's attitudes, knowledge, and behavior toward animals. Government Printing Office Report No. 024-010-00641-2.

Louv, R. (2005) *Last Child in the Woods: Saving Our Children From Nature Deficit Disorder*. Chapel Hill, NC: Algonquin Books.

Louv, R. (2011) *The Nature Principle: Human Restoration and the End of Nature Deficit Disorder*. Chapel Hill, NC: Algonquin Books.

McNeill, W. H. (1976) *Plagues and Peoples*. Garden City, NY: Anchor Press.

Nabokov, V. (1966) *Speak Memory. An Autobiography Revisited*, New York: Putnam.

Oreck, J. (2010) *Beetle Queen Conquers Tokyo*. Argot Pictures.

Orr, D. W. (1994) *Earth in Mind: On Education, Environment, and the Human Prospect*. Washington DC: Island Press.

Paul, E. S. and Serpell, J. A. (1993) Childhood pet keeping and humane attitudes in young adulthood. *Animal Welfare*, **2**, 321–337.

Pyle, R. M. (1992) *Handbook for Butterfly Watchers*. Boston, MA: Houghton Mifflin.

Pyle, R. M. (2006) *Wintergreen: Rambles in a Ravaged Land*. Seattle, WA: Sasquatch Books.

Pyle, R. M. (2009) The beauty of butterfly nets. *Wings: Essays on Invertebrate Conservation*, **21**(2), 15–18.

Pyle, R. M., Bentzien, M. and Opler, P. A. (1981) Insect conservation. *Annual Review of Entomology*, **26**, 233–258.

Ross, H. (1945) Entomology as a career. *Bios*, **16**(4), 245–248.

Večerek, O. (1965) Johann Gregor Mendel as a beekeeper. *Bee Wild* **43**(3), 86–96.

Wallace, B. (2005) Look out, Japan is in grips of animated beetles. *The Los Angeles Times*, October **9**: A9.

Williams, C. B. (1964) *Patterns in the Balance of Nature and Related Problems in Quantitive Biology*. London: Academic Press.

Wilson, E. O. (1984) *Biophilia*. Cambridge and London: Harvard University Press.

Wilson, E. O. (1994) *Naturalist*. Washington DC: Island Press.

10

Gardening and landscape modification: butterfly gardens

JARET DANIELS

INTRODUCTION

Gardening consistently ranks as one of the most popular American pastimes. More than 100 million households in the United States have a yard or garden, and Americans collectively spend more than US $57 billion annually to maintain and enhance their landscaping (National Gardening Association 2008). This trend of consumption is only expected to escalate in the future with a rapidly growing population. While property beautification and local food production are often some of the primary reasons for participating in lawn, garden and landscape activities, a growing number of gardeners are becoming environmentally conscious. In fact, some 90% of all US households express the importance of maintaining residential, commercial and municipal landscapes in an environmentally friendly way. Gardeners and homeowners are progressively more concerned about the impact of various products and chemicals used, and recognize the role landscaping for wildlife can potentially have on biodiversity conservation and ecosystem service sustainability, especially in urban and suburban settings.

Over the last several decades, interest in attracting and watching wildlife has rocketed tremendously in popularity. Bestselling publications such as *Bringing Nature Home* (Tallamy 2007) and *Last Child in the Woods* (Louv 2005) helped solidify backyard wildlife landscaping as a mainstream endeavour and sparked a national call to arms urging increased and meaningful contact with nature. More importantly,

The Management of Insects in Recreation and Tourism, ed. Raynald Harvey Lemelin. Published by Cambridge University Press. © Cambridge University Press 2013.

they further solidified the growing marriage of avocation and conservation. Organizations such as the National Wildlife Federation have long strived to educate property owners about the benefit of landscaping for wildlife and foster direct participation through their Certified Wildlife Habitat programme. Designed with simplicity in mind, the programme provides a template that encourages the addition of key resources and wildlife-friendly practices such as providing food, water, cover, nesting sites and reduced chemical application. If basic criteria are met, participants can register their landscapes and join a growing nationwide network of certified sites. The National Wildlife Federation's efforts have expanded to include Community Wildlife Habitat, uniting individual backyards, schoolyards and public spaces, such as parks and businesses, thereby enhancing overall community sense of place, leveraging habitat conservation impact, creating new wildlife learning sites. The resulting programmes have grown significantly in popularity to include nearly 140 000 certified wildlife habitat sites across the US in 2011 (National Wildlife Federation 2011).

Not surprisingly there is often a synergistic relationship between landscaping for wildlife and observing wildlife, particularly due to the continued strong appeal of organisms such as songbirds and butterflies. According to the 2006 National Survey of Fishing, Hunting, and Wildlife-Associated Recreation, nearly one-third of the US population enjoyed observing wildlife (US Fish and Wildlife Service 2006). Of the more than 71 million recreationists that participated in activities such as watching, feeding and photographing wildlife, some 95% stayed within 1 mile of their home. More than 14 million of these wildlife-watchers enhanced or maintained landscapes for the benefit of wildlife. Together they spent over US $1.5 billion on plants that provide food, cover or other resources. While wild birds garnered the attention of a large majority of the wildlife-watchers, interest in other charismatic organisms continued to be strong. Despite the traditional negative public perception of insects and spiders, they attracted the attention of some 16 million observers, gardeners and photographers. While the bulk of this attention is likely attributable to butterflies, dragonflies and damselflies, interest in these invertebrates has risen through the last decade, regularly surpassing that of some traditionally more charismatic organisms including reptiles, amphibians and fish. This increase mirrors the steady upward trend of interest associated with around-the-home wildlife watching and landscaping.

Access to an ever expanding quantity of print and digital information resources in recent decades has additionally fuelled overall public engagement. Robert Michael Pyle's *Audubon Society Field Guide to North American Butterflies* (1981) was arguably the first comprehensive reference to feature full colour photographs of living butterflies as seen in nature, providing users of all skill levels with a straightforward way to quickly identify species. It was followed a few years later by the *Audubon Society Handbook for Butterfly Watchers* (Pyle 1984). A sound natural history primer, it emphasized conservation along with gardening and detailed the basics of how to find and photograph butterflies. Together the books helped jumpstart the butterfly watching revolution and paved the way for other popular publications to come.

Launched in 1993, the *Butterflies through Binoculars* series (Glassberg 1993, 2001; Glassberg and Minno 2000) enhanced the user experience by including distinctive field marks and side-by-side comparisons of similar species. Catering to the increasing public distaste or misunderstanding of collecting, the books emphasized butterfly watching in its purest form and provided the foundation for a new army of North American Butterfly Association members and other emerging wildlife enthusiasts across the country. The series has continued to expand to include focal organisms such as dragonflies and butterfly larvae.

Today, hundreds of additional print-based publications, websites and even smartphone applications are available to facilitate organism identification, track sightings and share observations or photographs virtually anywhere, anytime. The relatively recent explosion of wildlife-watching resources is similar to that for related wildlife gardening publications. Google Books Ngram viewer statistics confirm this relationship (Google Labs 2010). Prior to 1980 few books highlighted wildlife or butterfly gardening. By the mid 1980s, however, both subjects witnessed a significant explosion and continue to regularly generate new titles.

GROWING POPULARITY OF BUTTERFLY GARDENS

Domestic gardens cover millions of hectares of land in the United States. Collectively they represent important potential remnant green spaces for many organisms in an increasingly fragmented and urbanized national landscape. For much of the last decade, however, insects traditionally were unwelcome guests in the public landscape. Viewed predominately as vegetative pests, gardeners regularly waged war against these primarily herbivorous foes with an assortment of pesticides. Emerging awareness of the non-target impacts chemicals can have on other organisms combined with growing interest in organic gardening and an overall increased environmental literacy has gradually led to changing landscape practices.

The tremendous public interest in butterfly gardening has helped drive this movement. In part, this popularity is due to the accessibility of butterflies in urban, suburban and rural areas, and in sites that range in size from very small to very large. A majority of the nearly 800 species of butterflies found in North America north of Mexico have been recorded in gardens, yards or public green spaces from time to time. A butterfly garden is a planted or augmented landscape established specifically to attract a diversity of butterfly species from the surrounding area. There exist no size or design constraints for butterfly gardens. They can encompass many hectares or be limited to a small container or window box. Similarly, they may represent naturalized country meadows, highly ornamental suburban plantings, sections of abandoned urban lots or even entire green roofs. No matter what the final configuration, the implicit goal of any butterfly garden, beyond attracting organisms for observation or photography, is to provide a resource-rich habitat in a larger landscape matrix that would otherwise be limiting due to the impacts of human activity (Mauro *et al.* 2007).

Butterflies primarily visit gardens to feed. As most adult butterflies are generalists, they will pause at a wide variety of flowers in search of nectar. This offers numerous opportunities for gardeners or wildlife-watchers to easily view and enjoy these colourful insects. Beyond basic floral resources which simply attract adult butterflies to a particular area for a time limited period, most butterfly gardens strive to incorporate other elements including shelter and larval host plants that help support organism reproduction. For many gardeners, incorporating flowering plants in the landscape is an easy sell that can yield many benefits from increased property value to greater enjoyment of the outdoor space. Flowering plants also provide almost instantaneous results often attracting numerous butterflies soon after they are in the ground.

Larval host plants typically require a longer public learning curve. The idea of inviting herbivorous insects into the landscape specifically to devour plant material remains a challenging idea for many gardeners or property owners. The specificity of host selection additionally provides limitations. While little background research is required to select basic nectar plants, more detailed insight is necessary to properly incorporate appropriate larval host plants. The growing list of available publications and other resources on butterfly gardening has significantly facilitated this task, as has the ability to quickly and accurately identify the variety of butterfly species observed. This enables gardeners to more effectively tailor plantings to accommodate local species. The resulting opportunity to experience multiple life history stages is particularly appealing for many gardeners and wildlife enthusiasts. It further emphasizes the importance of certain critical habitat resource components, provides new family natural history learning opportunities, and helps reinforce the idea that domestic gardens can potentially aid in the maintenance of local populations.

The overall growth of butterfly gardening has produced a corresponding niche boom in the nursery industry. Many species previously valued exclusively for their ornamental value could now be promoted and sold as butterfly-attracting plants. Demand generated from this new generation of gardeners has equally advanced the market for and appreciation of native plants, often considered untidy, weedy or undesirable in traditional ornamental landscapes. While many cultivated flowers attract adult butterflies, the vast majority of larval host plants are exclusively native. Florida is one of the leading states in the US nursery and greenhouse industry, ranked second only to California with total nursery sales in excess of US $3 billion annually (Hodges and Haydu, 2003). Native plants comprised some 10.5% of this total market or over US $315 million. As a result of this growing availability, many gardeners and landscapers have come to appreciate the many practical, financial and environmental benefits of natives beyond their utility for attracting wildlife.

BUTTERFLIES IN AN INCREASINGLY BUILT ENVIRONMENT

The global landscape is becoming increasingly urbanized (United Nations Population Division 2006). Habitat loss resulting from unmanaged urban sprawl is one of the most serious threats to species survival and indeed biodiversity in general. The impacts of expanding human development are spatially broad and have proceeded at an alarming speed, particularly in recent decades. Between 1954 and 1997, the pace of urban sprawl nearly quadrupled resulting in the conversion of some 1.2 million hectares of natural or agricultural land to urbanized landscapes on an annual basis (US Department of Agriculture 1997; US Environmental Protection Agency 2000). This built environment covers some 5.6% of the total surface area of the United States (US Department of Agriculture 2009) and exceeds all lands held in conservation by the National Park Service, the Nature Conservancy and state parks combined (McKinney 2002).

A rapidly growing body of literature has focused on the impact of expanding sprawl on biodiversity along the urban–rural gradient. The resulting alterations are predominantly negative. They can fragment habitat, disrupt ecological processes and modify the composition and distribution of species, including insects (Fahrig 2001, 2003; McKinney 2008). For many taxa, species richness declines toward the highly urbanized core. Driving much of this decline are changes in the density, diversity and complexity of vegetation along the gradient with overall community shifting from being primarily composed of native vegetation at various stages of succession in rural areas to an intensely managed urban landscape dominated by nonnative plants. The gradual replacement of native vegetation with more widespread nonnative and often invasive plant species causes increased biotic homogenization, a process that increases compositional similarity of sites and threatens to dramatically reshape local environments.

Numerous studies have tried to assess the effects of increased landscape urbanization on butterfly or other focal insect communities. Kocher and Williams (2000) examined data from Fourth of July Butterfly Counts conducted at some 514 locations across North America. They found that species richness and diversity indices decreased with greater habitat disturbance. Conversely, the abundance of exotic species such as cabbage worms (*Pieris rapae*) tended to be higher in more disturbed habitats. This result is consistent with Clark *et al.* (2007) who noted that butterfly species richness decreased with measures of urbanization, while sites associated with less development and increased green space supported a greater number of species. Moreover, butterfly species with limited generations and specialized host preferences along with regionally rare species were particularly impacted. They tended to be most vulnerable to local and landscape-level changes associated with disturbance and decreased at a significantly faster rate compared to more common butterflies and less restricted specialists. Similar findings indicating disproportional sensitivity of specialist and rare species to human disturbance and habitat fragmentation have been reported (Kitahara *et al.* 2000; Rabinowitz *et al.* 1986; Reed 1992; Summerville and Crist 2001; Tscharntke *et al.* 2002). In most cases, generalist species tended to be much less significantly impacted by expanding human disturbance.

The presence of multiple anthropogenic stressors raises the concern that changes in butterfly communities will be magnified, resulting in widespread biotic homogenization (McKinney and Lockwood 1999) of butterfly communities through time. Biotic homogenization is a long-term process that results in increased genetic, taxonomic or functional similarity among communities (Olden and Rooney 2006). While taxonomic homogenization is often used to describe replacement of native species with exotic species (Olden and Poff 2003), it also refers to the replacement of specialist species with those that thrive in human-dominated systems (McKinney and Lockwood 1999). Taxonomic homogenization has been documented for a variety of organism groups, geographic locations, and time scales (reviewed in Olden 2006) and is believed to be contributing to the current 'Homogocene' time period (Rosenzweig 2001).

Because biotic homogenization results from uneven turnover of species across sites, the study of homogenization puts an emphasis on patterns of beta diversity rather than alpha diversity. This is important because community similarity may increase in a given landscape even if species richness remains the same (Olden and Poff 2003). A focus on biotic homogenization may thus provide a broader framework for conservation across landscapes rather than on individual sites.

While the specific mechanisms of biotic homogenization are not always clear, it is generally considered to result from human-induced changes such as urbanization (McKinney 2008) that promote widespread colonization of a few common species, local extinction of many rare species, or both. For instance, Ekroos *et al.* (2010) observed an increase in habitat generalist but decrease in habitat specialist species with increasing agricultural intensity, while Blair (2001) documented increased similarity of butterfly communities in more urbanized habitat types. Disentangling the interactive effects of anthropogenic changes, as well as the potential benefits of local management actions including landscaping for wildlife, are thus critical steps towards effective long-term conservation of butterfly communities.

By contrast, some studies suggest that the relative abundance of organisms does not always drop off as urbanization increases. In certain situations, species richness may peak in areas such as suburban landscapes that experience low or intermediate amounts of human disturbance and often eclipse the levels found in natural settings. Blair and Launer (1997) found that the butterfly species richness was highest in sites with moderate disturbance but decreased in both directions from this peak along the continuum toward urban and natural environments. The intermediate disturbance hypothesis predicts this trend in diversity and is consistent in many biological communities subjected to anthropogenic changes. The moderate impacts of urban sprawl may be somewhat minor and spotty in the larger landscape matrix whereby suburban yards and gardens can be found adjacent to more natural areas or larger contiguous landscapes. Species diversity is thus enhanced by spatio-temporal heterogeneity which provides an increased assemblage of available habitats and helps maintain the community in a non-equilibrium state.

Moreover, suburban landscaping practices (e.g., gardening) commonly tend to subsidize a variety of traditionally scarce environmental resources such as water or nutrients through regular irrigation and fertilizer applications. As a result, such sites often have higher net productivity than both natural systems and more disturbed urban areas. This typically translates into increased access to or availability and quality of food resources for wildlife such as butterflies, and may help promote increased reproductive success, more rapid development times and prolonged breeding opportunities.

Beyond individual sites, disturbance may benefit populations at the larger landscape level. Kun *et al.* (2009) showed that an intermediate frequency of localized disturbance helps maximize equilibrium metapopulation density and organism invasion speed. While widespread habitat fragmentation may significantly hinder metapopulation viability or cause extinction, smaller dynamic changes to the landscape may produce new habitable sites that can act as stepping stones, thereby helping connect previously isolated population groups. In this scenario, landscape disturbance can serve to increase metapopulation density and ultimately metapopulation persistence. Although, habitat fragmentation has traditionally been the dominant issue in butterfly conservation, many studies have tended to focus on the paradigm of islands of appropriate habitat embedded in a broader matrix, despite considerable variation in the terminology used to describe this (New 1991; Panzer *et al.* 1995). Although the limitations of fragmentation as a model have been recognized for some time, they appear to be less important for arthropods than for larger organisms (Debinski and Holt 2000).

GARDENS AND HABITAT

In recent years there has been a growing emphasis on the creation of green space and maintenance of remnant natural areas, particularly in urban environments. Besides providing community value for aesthetic enjoyment and recreation, such sites are becoming an increasingly important component of mainstream conservation thinking. Public and private gardens in particular constitute a substantial proportion of remnant available suburban and urban green space. As a result, ever more attention is being paid to the potential role such gardens can play to help support biodiversity and the associated basic ecosystem services.

Few studies have systematically evaluated butterfly and wildlife gardens for their overall conservation value or to help provide broad recommendations of best practices to attract and potentially support or enhance local or regional organism diversity. Vickery (1995, 2007) reviewed data from the Butterfly Conservation national garden survey collected from more than 1000 gardens throughout the UK over some two decades of monitoring. Results indicated that larger gardens attracted a greater number of butterfly species as did gardens nearer natural areas and habitats. In essence, larger gardens with appropriate habitat or which were nearby source populations could potently support organism breeding. Additionally, gardens landscaped with a greater diversity of flowering nectar plants and larval host plants attracted a broader range of butterflies when compared to gardens containing lower floral diversity.

Work by Mauro *et al.* (2007) further support these findings. Their results suggest that in order to maximize the diversity of butterflies attracted to a garden the design should strive to maximize overall size, include a variety of suitable flowering nectar plants and larval hosts, and sit adjacent to available corridors to help link areas of suitable habitat. They also found that while the landscape matrix surrounding habitat areas influenced organism composition a variety of other specific habitat characteristics including patch size, vegetative composition and permanent environmental resources were important.

Burghardt *et al.* (2009, 2010) equally showed that vegetative composition was a driver of butterfly diversity. They compared suburban yards landscaped using conventional practices with those planted entirely with native plants. The diversity and abundance of herbivorous butterfly larvae responded positively to a greater percentage of native grasses, forbs and shrubs in residential landscapes, being significantly higher on properties with native vegetation. Breeding birds responded in concert, suggesting a further trophic level effect and benefit. These results are consistent with Tallamy (2004) who suggested that native plant enhancement tends to increase herbivorous insect species richness and abundance and therefore helps provide a larger food resource base for birds and other insect feeding organisms. Suggested reasons behind the observed trends surround plant diversity. Conventional landscaping tends to create numerous pockets of green space with similar vegetative composition thereby resulting in more homogenous habitats across the suburban matrix. Yards including a greater diversity of native plant species help form a more complex and heterogeneous mosaic of habitats, often with many unique ecological features, and are thus capable of supporting a substantially larger diversity of organisms. This heterogeneity is further enhanced by the plant selection and landscape maintenance decisions of individual property owners.

Information regarding the degrees to which gardens contribute to the regional abundance or aid in the persistence of butterflies through time, especially when compared to natural habitats, is scarce. Levy and Connor (2004) examined the use of natural sites and gardens in the San Francisco Bay area of California by the pipevine swallowtail, *Battus philenor*. They found that natural areas were more effective at attracting adult butterflies and had higher rates of oviposition compared to gardens. Of the gardens surveyed, butterflies were more likely to be present in landscapes with more mature and well established populations of larval host plants. Such sites also were more capable of supporting larval development. Despite this, however, the egg densities and survival rates were overall lower in gardens compared to natural populations supporting host plant populations, thereby suggesting that even high quality gardens with extensive host resources did not represent optimal habitats.

Such results are not necessarily surprising. Gardens or other managed green space in human-dominated landscapes are not intended to replicate or replace pristine natural habitats. In the larger picture though, such landscapes with appropriate vegetative and environmental resources can attract and support butterflies and other wildlife. They may potentially act as stepping stones (MacArthur and Wilson 1967; Schultz 1998) to promote small scale biodiversity enhancement and dispersal among fragmented populations. With appropriate vegetative augmentation, they may further positively influence the carrying capacity of suburban landscapes and help provide a valuable tool to mitigate biodiversity loss in areas subjected to increasing anthropogenic impacts.

GARDENS FOR EDUCATION AND RESEARCH

About 80% of the US population lives in urban or suburban areas (US Census Bureau 2001). Despite often being disconnected from nature on a daily basis, various surveys indicate that public attitudes toward wildlife tend to be more positive among urban residents than those living in rural locations (Kellert 1987, 1989, 1991). These results suggest that wildlife plays an important role in shaping urban residents' daily experiences, and that both animals and their habitats should have a higher priority in urban planning, management and legislation (Bjerke and Østdahl 2004). Parks and other municipal green spaces have long been recognized as major contributors to the physical and aesthetic quality of urban neighbourhoods (Walker 2004). In recent years though, a more contemporary view of these diverse public resources has been emerging, one that calls attention to the broader ecological contributions they can make to help transform communities into vibrant, healthy places where residents are better connected to the environment and each other.

Saunders (2003) eloquently emphasizes that while the loss of biodiversity is typically considered one of the most detrimental outcomes of anthropogenic disturbance, society is also sadly loosing certain basic experiences and relationships with nature. Clearly, one of the most important roles parks and other natural areas can serve is as informal settings for environmental engagement and learning.

The same can be said for public and private gardens. Even small scale landscape enhancements are important. They can help encourage awareness of biodiversity and its conservation while providing a venue for interaction. These are places where people young and old may develop increased appreciation for nature through direct contact with natural environments or other related landscapes (Kola-Olusanya 2005). They can provide numerous opportunities to explore new environments, discover a variety of wildlife from birds to insects and share meaningful experiences with family and friends. These positive learning experiences can strongly influence individual environmental attitudes, preferences for nature, and the development of conservation values.

The growing popularity of butterflies and butterfly gardening offers numerous opportunities to promote engagement in nature, engender public support and acknowledgement for broader environmental issues, and potentially make a significant contribution to biodiversity conservation. Through interest in this one insect group, individuals and families can be introduced to basic ecological concepts, learn about the importance of habitat and better understand how organisms interact with each other and the environment. For similar reasons, butterflies are particularly effective in helping promote science literacy and participatory learning. Such engagement tools are especially important as the majority of the general public does not understand the process of science and has limited science knowledge (Miller 2004; National Science Foundation 2006). Repair of this gap requires increasing the quantity and quality of education opportunities that foster conservation awareness, appreciation, and knowledge among youth and adults. Today, many schools, zoos, nature centres and botanical gardens are creating demonstration butterfly and pollinator gardens or larger learning landscapes to help promote science learning and promote environmental stewardship.

Effective environmental management and organism conservation requires a comprehensive understanding of what species are present, where and when they occur, and the indirect influence of human activities on them. Traditionally, the majority of targeted conservation efforts have focused on more pristine wild areas with little attention being paid to the increasingly expanding built environment, particularly suburban landscapes, despite their potential support for biodiversity conservation and the sustainability of ecosystem functions. As a result, many scientists (e.g., Cooper *et al.* 2007) have suggested that new methods be developed to study and improve the impacts of management strategies for residential landscapes. This is particularly critical as many monitoring and research programmes frequently do not capture information from private lands (Lepczyk 2005). Observations captured from private landowners can therefore be of significant benefit to science and help inform larger conservation efforts and strategies. Such efforts that involve volunteers from the general public to collect scientific data are referred to as citizen science projects. While often considered educational tools to help increase science literacy, the ability of projects to collect high quality, long-term data has increasingly been recognized (McCaffrey 2005). In fact, Cooper *et al.* (2007) proposed that 'citizen science, because it operates over such large scales by drawing on spatially dispersed participants, can be used to create a new frontier to advance the theory and practice of conservation in residential ecosystems.'

Today, citizen science projects collect data on a wide range of ecological issues from invasive species, climate change, ecological restoration, population ecology and organism monitoring, and cover many taxonomic groups from flowering plants and butterflies to frogs and butterflies (Silvertown 2009). Residential landscapes, public and private gardens and municipal green space can provide a rich living laboratory for combining scientific research, public education and community development. The bourgeoning arena of citizen science (see Johansen and Auger, this volume) can also be particularly valuable for increasing public awareness and appreciation for butterflies and other insects, increasing general entomological knowledge and gathering information with clear conservation relevance.

REFERENCES

Bjerke, T. and Østdahl, T. (2004) Animal-related attitudes and activities in an urban population. *Anthrozoös*, **17**, 109–129.

Blair, R. B. (2001) Birds and butterflies along urban gradients in two ecoregions of the US In *Biotic Homogenization*, ed. J. L. Lockwood and M. L. McKinney Norwell, MA: Kluwer, pp. 33–56.

Blair, R. B., and Launer, A. E. (1997) Butterfly diversity and human land use: species assemblages along an urban gradient. *Biological Conservation*, **80**, 113–125.

Burghardt, K. T., Tallamy, D. W. and Shriver, W. G. (2009) Impact of native plants on bird and butterfly biodiversity in suburban landscapes. *Conservation Biology*, **23**, 219–224.

Burghardt, K. T., Tallamy, D. W., Philips, C. and K. J. Shropshire (2010) Non-native plants reduce abundance, richness, and host specialization in lepidopteran communities. *Ecosphere*, **1**, 1–22.

Clark, P. J., Reed, J. M. and Chew, F. S. (2007) Effects of urbanization on butterfly species richness, guild structure, and rarity. *Urban Ecosystems*, **10**, 321–337.

Cooper, C. B., Dickinson, J., Phillips, T. and Bonney, R. (2007) Citizen science as a tool for conservation in residential ecosystems. *Ecology and Society*, **12**, 11.

Debinski, D. M. and Holt, R. D. (2000) A survey and overview of habitat fragmentation experiments. *Conservation Biology*, **14**, 342–355.

Ekroos J., Heliölä, J. and Kuussaari, M. (2010) Homogenization of lepidopteran communities in intensively cultivated agricultural landscapes. *Journal of Applied Ecology*, **47**, 459–467.

Fahrig, L. (2001) How much habitat is enough? *Biological Conservation*, **100**, 65–74.

Fahrig, L. (2003) Effects of habitat fragmentation on biodiversity. *Annual Review of Ecology, Evolution, and Systematics*, **34**, 487–515.

Glassberg, J. (1993) *Butterflies Through Binoculars: A Field Guide to Butterflies in the Boston, New York, Washington Region*. New York: Oxford University Press.

Glassberg, J. (2001) *Butterflies Through Binoculars: The West*. New York: Oxford University Press.

Glassberg, J. and Minno, M. C. (2000) *Butterflies Through Binoculars: Florida*. New York: Oxford University Press.

Google Labs (2010) Google Books Ngram Viewer. Available at: http://books.google.com/ngrams, accessed 16 July 2011.

Hodges, A. W. and Haydu, J. J. (2003) *Commodity Outlook 2003: US and Florida Ornamental Plant Markets*. Gainesville, FL: University of Florida, available at http://edis.ifas.ufl.edu/FE374.

Kellert, S. R. (1987) Social and perceptual factors in the preservation of animal species. In *The Preservation of Species*. ed. B. G. Norton. Princeton, NJ: Princeton University Press, pp. 50–73.

Kellert, S. R. (1989) Human-animal interactions: a review of American attitudes to wild and domestic animals. In *Animals and People Sharing the World*, ed. A. Rowan. Hanover, NH: University of New England Press, pp. 137–175.

Kellert, S. R. (1991) Public views of wolf restoration in Michigan. In *Transactions of the 56th North American Wildlife and Natural Resources Conference*. Washington DC: Wildlife Management Institute, pp. 152–161.

Kitahara, M., Sei, K. and Fujii, K. (2000) Patterns in the structure of grassland butterfly communities along a gradient of human disturbance: further analysis based on the generalist/specialist concept. *Population Ecology*, **42**, 135–144.

Kocher, S. D. and Williams, E. H. (2000) The diversity and abundance of North American butterflies vary with habitat disturbance and geography. *Journal of Biogeography*, **27**, 785–794.

Kola-Olusanya, A. (2005) Free-choice environmental education: Understanding where children learn outside of school. *Environmental Education Research*, **11**, 297–307.

Kun, A., Oborny, B. and Dieckmann, U. (2009) Intermediate landscape disturbance maximizes metapopulation density. *Landscape Ecology*, **24**, 1341–1350.

Lepczyk, C. A. (2005) Integrating published data and citizen science to describe bird diversity across a landscape. *Journal of Applied Ecology*, **42**, 672–677.

Levy, J. M. and Connor, E. F. (2004) Are gardens effective in butterfly conservation? A case study with the pipevine swallowtail, *Battus philenor. Journal of Insect Conservation*, **8**, 323–330.

Louv, R. (2005) *Last Child in the Woods: Saving Our Children from Nature-deficit Disorder*. Chapel Hill, NC: Algonquin Books.

MacArthur, R. H. and Wilson, E. O. (1967) *The Theory of Island Biogeography*. Princeton, NJ: Princeton University Press.

Mauro, D., Dietz, T. and Rockwood, L. (2007) Determining the effect of urbanization on generalist butterfly species diversity in butterfly gardens. *Urban Ecosystems*, **10**, 427–439.

McCaffrey, R. E. (2005) Using citizen science in urban bird studies. *Urban Habitats*, **3**, 70–86.

McKinney, M. L. (2002) Urbanization, biodiversity, and conservation. *BioScience*, **52**, 883–890.

McKinney, M. L. (2008) Effects of urbanization on species richness: a review of plants and animals. *Urban Ecosystems*, **11**, 161–176.

McKinney, M. L. and Lockwood, J. L. (1999) Biotic homogenization: a few winners replacing many losers in the next mass extinction. *Trends in Ecology & Evolution*, **14**, 450–453.

Miller, J. (2004) Public understanding of and attitudes toward scientific research: what we know and what we need to know. *Public Understanding of Science*, **13**, 273–294.

National Gardening Association (2008) Environmental lawn and garden survey. South Burlington, Vermont.

National Science Foundation (2006) *Science and Engineering Indicators 2006*, 2 vols. Arlington, VA: National Science Foundation.

National Wildlife Federation (2011) Garden for wildlife. Available at: http://www.nwf.org/Get-Outside/Outdoor-Activities/Garden-for-Wildlife.aspx, accessed 23 August 2011.

New, T. R. (1991) *Butterfly Conservation*. Melbourne: Oxford University Press.

Olden, J. D. (2006) Biotic homogenization: a new research agenda for conservation biogeography. *Journal of Biogeography*, **33**, 2027–2039.

Olden, J. D. and Poff, N. L. (2003) Toward a mechanistic understanding and prediction of biotic homogenization. *American Naturalist*, **162**, 442–460.

Olden, J. D. and Rooney, T. P. (2006) On defining and quantifying biotic homogenization. *Global Ecology & Biogeography*, **15**, 113–120.

Panzer, R., Stillwaugh, D., Gnaedinger R. and Derkovitz, G. (1995) Prevalence of remnant-dependence among the prairie inhabiting insects of the Chicago region. *Natural Areas Journal*, **15**, 101–116.

Pyle, R. M. (1981) *The Audubon Society Field Guide to North American Butterflies*. New York: Alfred A. Knopf.

Pyle, R. M. (1984) *The Audubon Society Handbook for Butterfly Watchers*. New York: Charles Scribner's Sons.

Rabinowitz, D., Cairns, S. and Dillon, T. (1986) Seven forms of rarity and their frequency in the flora of the British Isles. In *Conservation Biology. The Science of Scarcity and Diversity*. ed. M. E. Soulé. Sunderland: Sinauer Associates, Inc., pp. 182–204.

Reed, J. M. (1992) A system for ranking conservation priorities for Neotropical migrant birds based on relative susceptibility to extinction. In *Ecology and Conservation of Neotropical Migrant Landbirds*, ed. J. M. Hagan III, D. W. Johnson. Washington DC: Smithsonian Institution Press.

Rosenzweig, M. L. (2001) The four questions: what does the introduction of exotic species do to diversity? *Evolutionary Ecology Research*, **3**, 361–367.

Saunders, C. (2003) The emerging field of conservation psychology. *Human Ecology Review*, **10**, 137–153.

Schultz, C. B. (1998) Dispersal behavior and its implications for reserve design in a rare Oregon butterfly. *Conservation Biology*, **12**, 284–292.

Silvertown, J. (2009) A new dawn for citizen science. *Trends in Ecology and Evolution*, **24**, 467–471.

Summerville, K. S. and Crist, T. O. (2001) Effects of experimental habitat fragmentation on patch use by butterflies and skippers (Lepidoptera). *Ecology*, **82**, 1360–1370.

Tallamy, D. (2004) Do alien plants reduce insect biomass? *Conservation Biology*, **18**, 1689–1692.

Tallamy, D. W. (2007) *Bringing Nature Home: How Native Plants Sustain Wildlife in our Gardens*. Portland, OR: Timber Press.

Tscharntke T., Steffan-Dewenter, I., Kruess, A. and Thies, C. (2002) Characteristics of insect populations on habitat fragments: a mini review. *Ecological Research*, **17**, 229–239.

US Census Bureau (2001) US Census of Population and Housing. Census 2000 Summary File 1. Washington DC: US Census Bureau.

US Department of Agriculture (1997) Census of agriculture, geographic area series, AC97-CD-Volume 1. US Department of Agriculture, National Agricultural Statistics Service, Washington DC, USA.

US Department of Agriculture (2009) Summary Report: 2007 National Resources Inventory, Natural Resources Conservation Service, Washington DC, and Center for Survey Statistics and Methodology, Iowa State University, Ames, Iowa. 123 pages.

US Department of the Interior, Fish and Wildlife Service and US Department of Commerce, US Census Bureau (2006) National survey of fishing, hunting, and wildlife-associated recreation. Washington DC: US Census Bureau

US Environmental Protection Agency (2000) Projecting land-use change: a summary of models for assessing the effects of community growth and change on land-use patterns. USEPA Report EPA/600/R-00/098, Washington DC.

United Nations Population Division (2006) *World Population Prospects, 2006 Revision*. New York: United Nations.

Vickery, M. L. (1995) Gardens: the neglected habitat. In *Ecology and Conservation of Butterflies*. ed. A. S. Pullin. London: Chapman and Hall, pp. 123–134

Vickery, M. L. (2007) Gardens as an aid to the conservation of some butterfly species. *Science Progress*, **90**, 223–244.

Walker, C. (2004) The public value of urban parks. In *Beyond Recreation: a Broader View of Urban Parks*. Washington (DC): The Urban Institute.

11

The role of edible insects in human recreation and tourism

ALAN L. YEN, YUPA HANBOONSONG AND ARNOLD VAN HUIS

BACKGROUND TO HUMAN ENTOMOPHAGY

There are many ways in which insects are associated with recreational activities such as tourism. Most involve observation, study or collecting them in the wild or observation in controlled environments such as insect zoos. Tourists visiting insect zoos may purchase souvenirs such as insect models, insect publications, postcards and preserved specimens. Most of these visitors are likely to consider eating insects as bizarre or even abhorrent. Yet it is a very important component of many cultures where it is normal to eat insects. The reluctance to embrace human entomophagy is driven by Western society values (DeFoliart 1999). The consumption of edible insects is becoming more important as a form of recreation and in generating tourist income in many less advanced nations. If managed correctly, it could be another important tool for insect conservation.

Insects (and other invertebrates) are regularly eaten today in Africa, Asia, Central America and South America. The number of species actually used as food is uncertain, but is in excess of 1000 (MacEvilly 2000). Human entomophagy is not widespread in Westernised societies such as Europe, North America and Australia, although the traditional owners in the latter two still practise limited entomophagy. There is very much a bias against eating insects in Westernized nations because of the belief that they are dirty and harbour diseases, and also there is the misconception that they are only eaten in times of dire famine (DeFoliart 1999; Van Huis 2003).

The Management of Insects in Recreation and Tourism, ed. Raynald Harvey Lemelin. Published by Cambridge University Press. © Cambridge University Press 2013.

There is a requestioning about this attitude. Food security is a major issue that faces humanity (Godfray *et al.* 2010) with an increasing realization that there will not be enough food available to feed the exponentially growing world population in the future if we attempt to provide a Western-based diet that is heavily reliant on protein derived from vertebrate animals such as cattle, sheep, pigs, poultry and fish. Increasing meat consumption often characterizes increasing standard of living and, with an increasing global population, there will be a limit to protein production from conventional meat sources. The amount of energy required to produce meat-based protein is very high, and there are serious environmental consequences such as the clearing of more land for grazing stock (and loss of natural biodiversity) and the increased production of emissions (Yen 2009b). Farming insects produces less greenhouse gases (Oonincx *et al.* 2010), and is a more efficient way to produce protein (FAO 2010a; Vogel 2010).

There is now more exposure to the use of insects as human food due to increasing media reports and also tourists visiting countries where entomophagy is more widespread. With new technologies, more reports are widely available through the internet. Unfortunately this reporting is usually sensationalist and has a strong 'fear' or 'wow' factor; people are dared to eat insects, and this masks the serious reasons for human entomophagy. Even though there are several books that promote consumption of insects (Gordon 1998; Menzel and D'Aluisio 1998; Ramos-Elorduy 1998; Taylor and Carter 1976), they still emphasize the bizarre and 'dare to try' attitude of the popular media in varying degrees.

The terms 'recreation' and 'tourism' need to be qualified first because of possible differing interpretations on how they are applied with regard to entomophagy. Recreation can involve eating insects as a form of recreational snack food, or eating insects while on a recreational pursuit as a tourist (whether they are from within the same country or from another country coming to experience a different culture).

This chapter will assess the relationship between human entomophagy and recreation and tourism by considering three different scenarios associated with eating insects for domestic consumption, with a tourism component, and finally dining on insects in restaurants.

ENTOMOPHAGY AND DOMESTIC USE

This occurs where there is a strong and active traditional use of insects as food or medicine. The main activities are the collection (wild harvesting) or production (farming) of edible insects for domestic consumption (either by the collector/producer or for sale at markets for domestic use), and to a much smaller extent, for restaurants or for export. In some countries, the Western values have overshadowed traditional values, such as in the Northern Transvaal (South Africa) where people in Westernized towns are ashamed to admit that they like eating grasshoppers (Van der Waal 1996).

Van Huis (2003) lists about 250 species of edible insects in Africa, mainly caterpillars, locusts and grasshoppers and beetle larvae. These are primarily collected for domestic consumption, so some consideration about the management of edible insect populations in the wild is warranted.

The prime African edible insect is the mopane worm, the caterpillar of the emperor moth *Gonimbrasia belina* Westwood. An estimated annual production of 9500 million mopane worms in a 20 000 km^2 area in South Africa alone is worth US $85 million of which 40% is earned by poor rural women (Stack *et al.* 2003; Styles, 1994). The use of mopane worms in nutrition and as a means to generate income is substantial in southern African rural households.

Both adult and hopper stages of the brown, desert, migratory and red locusts are consumed when and wherever they occur (Harris 1940). Many other species are also eaten, and larger species, particularly the females, are preferred (Barreteau 1999). Harvested and marketed orthopterans may yield more revenue to farmers than the sale of their millet (Van Huis 2003), and this is one reason why farmers do not want to treat their crops with pesticides (Cerritos and Cano-Santana 2008). The insects are cooked, fried or roasted, but they can be sun-dried for storage purposes, pounded in a mortar to powder, and eaten as a flavouring with porridge. The African edible bush cricket *Ruspolia differens* Serville (or *nsenene*) is a food item in most of eastern and southern Africa. The adults may be gathered during the day from grasses (Mors 1958), but they are more often collected at night when attracted as swarms to artificial lights of professional collectors with lights shining into the sky, or by women or children from street lights (Van Huis 2003). Agea *et al.* (2008) found that retailers in the Kampala and Makaka districts of Uganda purchased three quarters of their *R. nitidula* stock from wholesalers and the rest directly from collectors. The wholesale price is about US $0.56 per kg while the retail price is about $2.80. The insect is considered a delicacy and the retail price is about 40% higher than the price of a kilogramme of beef. Traders generated an average of more than $200 per season by its sale.

In his 1735 *Systema Naturae*, Linnaeus describes the palm weevil larvae as '*larvae assate in deliciis habentur*' (larvae are delicious). Larvae of the palm weevil (*Rynchophorous* spp.) are eaten in Asia (*R. ferrugineus* Olivier), Africa (*R. phoenicis* Fabricius) and America (*R. palmarum* L.), and in general are considered to be very delicious (Cerda *et al.* 2001) possibly because of its high fat content (Fasoranti and Ajiboye 1993). Fallen palms serve as breeding sites and can support hundreds of larvae and palms are often felled in order to promote larvae. There is a taboo on children eating palm weevil larvae in Nigeria; adults tell children that larvae feed on palm wine and that they would get drunk by eating them (Fasoranti and Ajiboye 1993). However, Fasoranti and Ajiboye (1993) suspect that the reason is an economic one; adults want to prevent the children felling palm trees to increase the number of larvae in order not to deprive the community of primary palm products such as palm oil, palm kernels and palm wine.

Several lepidopteran species are utilized as food as well as for silk production. These include representatives of genera such as *Borocera*, *Gonometa*, *Mimopacha*, *Pachypasa* (Lasiocampidae), *Anaphe* (Notodontidae) and *Goodia* (Saturniidae). In Africa, there are 42 tree species with edible caterpillars (Turk 1990). Schabel (2010) has suggested a number of ways to manage forests to protect the caterpillars. The demand for land in the face of growing human population and wood exploitation has degraded forests to savannah, which is accompanied by a loss in edible caterpillar species. Forests have also been mismanaged by over-exploitation as caterpillars are an important source of food and income (Ferreira 1995). In Malawi, farmers adjacent to Kasungu National Park have been allowed to harvest at certain times in the Park to diversify their income base and to win their support for wildlife conservation programmes (Munthali and Mughogho 1992). By allowing rural people to use the parks and reserves sustainably, the preservation of the country's biodiversity is enhanced. A management programme involved the adoption of a rotation burning policy that promotes vegetation coppicing, eases harvesting and promotes high caterpillar yields. The use of traditional food is sustainable and has economic, nutritional and ecological benefits for rural communities in sub-Saharan Africa. Future studies should focus on sustainable ways of harvesting wild populations, the use of improved conservation practices, the enhancement of cottage industries for farming insects and the development of economically feasible ways of mass-rearing edible species. This will not only benefit domestic consumption, but will enable sales of edible insects to tourists and for the export market, thus generating more income for local communities. The export demand is due to interest in edible insects in overseas countries and also due to Africans who have migrated overseas and who seek some of their traditional foods.

ENTOMOPHAGY AND TOURISM

Edible insects are becoming part of the culinary experience for tourists in some countries. This can involve snacking on insects in busy tourist locations, or eating them as part of an ecotourism holiday. Ecotourism involves participants wanting an experience that reflects the traditional way of life in the country that they are visiting. One of the arguments in favour of ecotourism is that native ecosystems are maintained for use of both the locals and for tourists and to conserve native biodiversity. The economic and environmental benefits of ecotourism have long been advocated but there have been few actual analyses. Kirkby *et al.* (2010) undertook a cost–benefit analysis of land use in an ecotourism development in Peru, and found that the net value exceeded all alternative land uses such as forestry, ranching and agriculture.

Some ecotourism experiences involve traditional lifestyles, including food gathering and preparation. Edible insects are sometimes included in the context of sustaining ecosystems through use of non-wood forest products, although they generally comprise a relatively minor part of the total experience. Tourist operators can be found that have edible insects as part of their trips in the headwaters of the Amazon in Ecuador, Ariau in Manaus (Brazil), Borneo, Tado in West Flores (Indonesia) and the bushtucker experience with Australian Aborigines in either central or northern Australia. In Kenya, 80% of the 1.5 million tourists who visit each year are associated with ecotourism. The International Centre of Insect Physiology and Ecology (ICIPE) in Nairobi has pioneered a set of rural community livelihood support mechanisms based on the sustainable use of natural forest products, including 'commercial insects' (honeybees and wild silkmoths). The project worked at three globally significant, biodiversity-rich forest sites (Kakamega, Arabuko-Sokoke and Mwingi), their buffer zones and adjacent villages. It provided such incentives through the gathering, cultivating and marketing of traditional commercial insects products, such as silk and honey, but also through the introduction and development of innovative non-traditional activities such as butterfly farms or the gathering of royal jelly and stingless bee honey, which has medicinal applications.

Figure 11.1   Street vendor sells edible insect in Bangkok. Photo by Y. Hanboonsong.

Food markets in Asia (especially Thailand, Laos, Cambodia and China) often have insect vendors (Figure 11.1). They sell to either locals or foreign tourists (Figure 11.2), and some organized tourist group tours will include a stop at one of these markets (e.g., the Don Makai Market in Vientiane, Laos). Insects are sold at markets that now cater for both locals and for tourists. For example, at the Klong Toey market in Bangkok, the customers can be 50% Thai and 50% foreigners. The insects sold at the Klong Toey market are purchased at the Talat Rong Kluea market on the Cambodian border, at Mae Sot on the Burmese border, or from Cambodian and Burmese vendors at Talat Thai (Pathum Thai province). Insects are purchased live (dead insects do not get a good price) and sold cooked. The cricket *sading* is bred on farms at Lop Buri, Khon Kaen, Roi Et and Korat. The Talat Rong Kluea market has over 100 stalls; 90% of insects are caught in Cambodia because it is thought that no pesticides are used. Bombay locusts from Pailin province in Cambodia are most popular; they occur naturally from May to September, but are now stored frozen to ensure year round availability (Fernquest 2010).

The practice of eating insects is common in Thailand. Traditionally insect consumption is mostly found in the north and

Figure 11.2   Tour group at the 'Ruammal edible insect' cart at
Phitsanulok province in northern Thailand. Photo by Y. Hanboonsong.

northeast region. Over 158 species of edible insects are eaten by people in the northeast and approximately 50 species are consumed in the North (Rattanapan 2000). The different insect-eating habits in various regions may depend on cultural practices, religion or geographical area. The northeast region often suffers food shortages due to infertile soils, drought or floods, and insects, which are easy to find and can be harvested from paddy fields, natural ponds and streams or forests, have become a part of life and culture (Hanboonsong 2010). This pleasure of eating insects is increasingly popular and spreading widely across the country. Bamboo caterpillars (*Omphisa fuscidentalis* Hampson) are the most popular edible insects in Northern Thailand. Giant water bugs (*Lethocerus indicus* Lepeletier & Serville) and house crickets (*Acheta domesticus* L.) are popular edible insects in the northeast region. Wasps, bees and palm weevils (*R. ferrugineus*) are well-known edible insects in southern Thailand. Predaceous diving beetles, water scavenger beetles and immature weaver ants (*Oecophylla smaragdina* Fabricius) are also eaten widely in the country. Eating insects is no longer seen as only a food for poor or rural people; urban and city people, even high-income earners, also consume insects. The people eat insects not only for their nutritional content, but also their crunchy texture and taste

(Hanboonsong *et al.* 2000; Yhoung-Aree 2010) and are now a staple snack among urban dwellers (Yhoung-Aree and Viwatpanich 2005).

In the past all species of edible insects have been harvested in the wild. However, recent farming practices in insect rearing have been developed. Farming house crickets (*Acheta domesticus*), locally called *sading*, occurs in northeast Thailand. A farmer's income from a medium size cricket farm is an average of US $2000–2500 from each harvesting cycle (45 days). Palm weevil farming (*R. ferrugineus*) is mainly found in the southern region of Thailand. Farmers can earn US $12 per kg of weevil larvae for each 30 days breeding cycle. The production yield is dependent on the farm size. Small-scale farmers can produce 100–300 kg of palm weevil larvae for each 30-day period (Hanboonsong, unpublished data).

The edible insect market in Thailand begins with the farmer or producer who brings the insects (mainly crickets) from his own and other farms to sell to the wholesalers. Some farmers also sell to buyers who sell on to wholesalers. Wholesalers then distribute the products to retailers and/or street vendors. For other unfarmed species, collectors can sell directly to the wholesalers or at the market. Large wholesale edible insect markets are at Klong Toey in Bangkok, Talat Thai in Pathum Thani province close to Bangkok. In addition, the biggest edible insect market is at Talat Rong Kluea on the Thai–Cambodia border, in Sa Kaeo province. With increasing demand for edible insects, supplies of these insects are also brought into Thailand from neighbouring countries such as Laos and Cambodia to sell at Talat Rong Kluea market. These imported edible insects are mainly collected in the wild. It has been reported that over 700 tonnes per year of edible insects are imported into Thailand to sell at Talat Rong Kluea. The top five most popular imported species are silkworm pupae (*Bombyx mori* L.), ground crickets (*Acheta testacea* Walker), leaf eating grasshoppers (*Cyrtacanthacris tatarica* (L.)), mole crickets (*Gryllotalpa africana* Palisot de Beauvois) and large green grasshoppers (*Chondracris rosea* De Geer) (Ratanachan 2009).

Carts filled with trays containing different kinds of fried edible insects are commonly seen and available in every city in Thailand and this has attracted the attention of almost every foreigner walking past. For example the 'Ruammaland Edible Insect' cart at Phitsanulok province in northern Thailand is a well-known place for tourists to stop. It is even recommended as a 'must see' in tourist guide books. Many tour groups, whether Thai or foreign come to see the cooking process and to taste the insects. In addition the tourists can buy tinned insects to take home and can even put their own photos as labels on the tins. Edible insects can be found ready to eat, deep fried from street vendors or in a package such as for bamboo worms at airport shops or souvenir shops at Chiangmai. Uncooked insects such as bamboo caterpillars, grasshoppers and crickets are sold in frozen packs like other meats and are available at retail supermarkets throughout the country. The edible insect industry in Thailand is now emerging as a new sector of economic empowerment for the country.

There is considerable publicity about the consumption of fried spiders in Cambodia. A species of tarantula, *Haplopelma albostriatum* Simon (Thai zebra tarantula), locally named *a-ping*, is used. While widely reported and now a part of the overseas tourist agenda, there is very little information about this practice and most information is anecdotal. It is thought that the practice arose around the town of Skuon (Kampong Cham Province), where people used tarantulas as traditional medicine for heart, throat and lung problems, but it became a food item during the famine associated with the Khmer Rouge regime. Fried spiders are now sold in street stalls in Skuon, at the Kampong Thom market and served in restaurants in Phnom Penh. It is possible to purchase spiders by mail order at very high prices through companies that claim that they are only obtained from areas that are monitored against over-collecting. Other sideline industries related to the spiders include an exhibition of edible insects and other invertebrates at the Skuon Spider Sanctuary, tourist companies bringing in groups to sample the spiders and local guides who will take tourists out to find spiders. Spiders are wild-collected from forests or in cashew nut plantations during the day. Vendors buy live spiders from collectors who find the spider burrows and then dig up the spiders (Rigby 2002). Vendors can sell 100–200 spiders/day, and there are about a dozen vendors at Skuon. The sale of spiders is a very important source of income for poor farmers where the average daily wage is US $2/day. However it is not a practice with a long traditional history and is an example of how a relatively new cuisine becomes more popular through local demand and is further encouraged by tourism. There is a fear that these spiders could be collected to extinction, and vendors are reporting a sharp decline in numbers and blame farmers for the clearing and burning of forests. Population data for the spiders are lacking.

Mexico has had a long history of using edible insects with over 400 recorded species; the harvesting of insects such as ant eggs, maguey worms and grasshoppers is well documented (Miranda Román *et al.* 2011), and while domestic consumption is important, the sale of edible insects to tourists and to restaurants is becoming more important for local collectors. Miranda Román *et al.* (2011) provide a detailed description of the collection and preparation of edible insects in the tourist areas of Otumba and Teotihuacán (Mexico). They are mainly sold to tourists. One Mexican government agronomist believed that more Mexicans would be eating bugs were it not for decades of advertising campaigns by international companies pushing less nutritious foods (Anonymous 1989). With a renewed interest in Mexico of pre-Hispanic cuisine, the demand for insects amongst tourists and in restaurants has increased, and the poor farmers have an important income stream to support them. For example, insects are collected in Santa Lucia Ocotlan to sell at the market in Oxaca, and on a good day sales can bring in 800 pesos (US $86). Insects are a new form of income for villagers as well as a diet supplement (20 million Mexicans live on less than US $1/day). Sometimes availability is only for a short period. Farmers collect grasshoppers rather than spray with pesticides; some plant cheap corn to attract grasshoppers. Maguey worms can sell for US $40 a dozen at upscale restaurants (Anonymous 2011). Mezcal is a distilled alcoholic beverage made from the maguey plant (*Agave americana* L.) in Oaxaca, and the international marketing gimmick involves adding an insect larva to each bottle.

ENTOMOPHAGY AND RESTAURANT DINING

In many countries, the central location for edible insects has been the home, the village or the local market. Insects are now making an appearance in restaurants, and there are many examples of restaurants serving insect dishes, often at prices too expensive for most of the local people. Examples include mopane worms in Zimbabwean restaurants (Anonymous 1988), and ants, moth larvae and grasshoppers in some of the finest Mexican cuisine restaurants in Mexico City (Anonymous 1989, 1991, 1992). Insects have become haute cuisine in Mexico City with restaurants charging high prices for escamoles (ant eggs sautéed in butter and served in hot tortillas with guacamole, or 'Mexican caviar'), deep-fried maguey worms that resemble fried pork rind, salsa made out of crushed jumiles (*Euschictus crenator* Fabricius), and Chapulines (grasshoppers) marinated in lemon juice, salt and chilli powder. The ant eggs are only available in spring, so restaurants freeze supplies to ensure all year availability. These dishes fetch high prices in more exclusive restaurants who charge up to US $25 per dish (Ramos-Elorduy *et al.* 2011). At the Fonda Don Chon restaurant in Mexico City, the plates range in cost from US $20 for a load of 500 to 1000 *escamoles,* to $6 for a plate of 200–300 *jumiles.* The Riscal Restaurant serves 30 to 40 orders each of *escamoles* and *gusanos* every day (Anonymous 1991, 1992). Still, there are the dangers of over-exploitation and lack of regulation on collection, distribution and commercialization (Ramos-Elorduy 2006; Ramos-Elorduy *et al.* 2006). Insects are also served in restaurants in Sarawak and Kelantan (Malaysia), China, Japan and in Taiwan.

In nations where entomophagy is not a standard practice, a small number of restaurants serve insects. They are marketed as a gourmet experience. One of the early ones was The Insect Club in Washington DC during the early 1990s (DeFoliart 1993), but it has closed down. Other examples include the Brooklyn Kitchen in San Francisco as part of the Mexican feast (Gordiner 2010) and the Archipelago restaurant in London (Taylor 2010); the founder of the English Wahaca restaurant chain, Masterchef Thomasina Miers, is keen on insects and took her managers and chefs to Mexico to introduce them to entomophagy (Palmer 2011).

Will Westerners ever take to insects as food? At Wageningen University (the Netherlands), insects as food were promoted in 1995. At first, many people laughed – and cringed – but interest gradually became more serious. In 2006 the 'Wageningen – City of Insects' science festival was organized to promote the idea of eating bugs; it attracted more than 20 000 visitors. Over the past 2 years, three Dutch insect-raising companies, which normally produce feed for animals in zoos, have set up special production lines to raise locusts and mealworms for human consumption. Now those insects are sold, freeze-dried, in two dozen retail food outlets that cater to restaurants. A few restaurants in the Netherlands have already placed insects on the menu, with locusts and mealworms (beetle larvae) usually among the dishes. In 2008, insects were promoted as food during a large food fair in Amsterdam and thousands of people tasted them. Now, the Ministry of Agriculture in the Netherlands is funding a new US $1.3 million research programme to develop ways to raise edible insects on food waste, such as brewers' grain (a byproduct of beer brewing), soyhulls (the skin of the soybean) and apple pomace (the pulpy remains after the juice has been pressed out). The research focuses on how protein could be extracted from insects and used in processed foods.

THE FUTURE

The increased use of insects as a food resource for humans has important environmental and social benefits. Potentially there could be greater food security with a source of protein and other nutrients that is less environmentally damaging to produce (Yen 2009a). In some cases, it may be feasible to harvest insect pests of plants as a food source instead of treating them with environmentally damaging insecticides. It can build upon traditional knowledge of different cultures, and it is important that this information is acknowledged properly including the recognition of intellectual property rights (Yen 2009b).

The list of edible insects is probably larger than currently known, and there is the potential to discover species with different nutritional qualities. Most species of edible insects are wild harvested, and there is the danger of overharvesting, or in some cases, habitat destruction as a result of harvesting. However, sustainable harvesting protocols can be developed, as with the mopane worm (Vantomme *et al.* 2004), and this can involve sustainable collecting of insects, to controlled ranching in the natural environment, or insect farming. It would be important to determine which of these edible species could be farmed in captivity. At the local level, edible insects are an important form of additional income. There is also an increased international demand for edible insects. Much of this is due to human migration and the demand by migrants for some foods that they were fond of in their home country. Examples include Africans in Europe who buy mopane worms, and Asians in North America.

The major challenge is to get people to recognize that edible insects are not just a tourism gimmick and that in many cultures, they are preferred food items that are an important source of nutrition. Whether it is for the right reasons or not, human entomophagy is very much in the media. Most of it is driven by the media exploiting the wow or fear factor, and this approach is unlikely to engender a rational response to edible insects.

In Westernized societies, education about the future shortage of protein would be a more appropriate strategy to adopt to promote the use of food insects. The media-hype strategy is probably the wrong approach and will have little long-term benefit. One of the issues facing many people in Westernized societies is the thought of eating insects – something they consider dirty. It may be a case of having insects served in different ways. One report on the internet describes *Maeng Da* (water bug dip) creamy and delicious, and questioned whether this was because the reporter was not struggling with eating a whole grub or whether because its texture was attractive and taste enhanced by being pounded and mixed with chilli. Cerda *et al.* (2001) surveyed the responses of Venezuelan non-Amerindian tourists to *aler-ito* (larvae of the palm worm *R. palmarum*), a food of the local Indians that is encouraged both as a nutritional food and as an income gener-ating item by selling it to tourists. The majority of those surveyed (over 80%) had a positive response to the palm worm; 6% gave a negative response mainly because of 'food appearance'. In reality, most people in Westernized societies do not like having a whole animal (head, feet, skin, etc.) served up to them, and some of this dislike may extend to whole insects.

As insects fare badly in the public relations stakes when com-pared to more charismatic vertebrates such as birds and mammals, there are many institutions that promote insects through aware-ness festivals that have displays, education programmes and often live insects. Sometimes edible insects are part of the education pro-gramme. Dunkel (1997) lists several of these events in the United States and Canada: the Audubon Institute (New Orleans) organized an event called 'The Incredible Edible Insect' that attracted over 1000 people in 1997 and in 1998, Los Angeles Natural History Museum, The Bug Bowl (Purdue University, West Lafayette, Indiana), North Carolina Museum of Natural History, The Food Insect Festival at the Insectarium (Montreal), Bugfest at the Metro Park, Cleveland, Ohio, Provincial Museum of Alberta, Smithsonian Museum of Natural History and the Insectarium (Philadelphia). More recently, the Natural History Museum in London set up a travelling exhibition in shopping centres on the theme of edible insects (Fairman 2010). In the UK, a theme park in Surrey handed out free dried insect snacks as a trial to see if they should be on sale permanently in their new Asia display. Insect tast-ings have been held in tertiary institutions that have entomology sub-jects, and at professional entomology meetings such as the New York Entomological Society (The New York Bug Banquet) (DeFoliart 1992), in Australia, and at the SASI Insects in Captivity conference in 1997. More recently at the Shanghai World Expo 2010, there was an exhib-ition 'The Incredible Edible Insect: A Technical Cooperation Project in LAO PDR' that included insect tacos served in restaurants and cafes near the Expo area (FAO 2010b).

There could be negative aspects to promoting edible insects. For very popular species, it could create a demand that exceeds sup-ply. Overharvesting could result in local extinction of insect species and/or environmental damage (Ghaly 2009; Ramos-Elorduy 2006; Yen 2009b). One unforeseen consequence of the use of edible insects by a traditional culture is the collecting of wood-dwelling grubs in parts of Australia as bait by recreational fishermen; these species were a favoured food source of Australian Aborigines (Yen 2009b). On a social level, some members of various cultures may forgo traditional food sources for additional income. They sell insects that they would normally consume and purchase less healthy foods with the income.

However the positives outweigh the negatives. Overharvesting and environmental damage associated with the collecting of edible insects can be controlled by education and by regulation. No doubt this is very difficult in regions where there is significant poverty. This needs to be supplemented by programmes that ensure better long-term food security, such as insect farming technology (if appropriate), better food storage facilities (to reduce the chances of contaminated food insects) and better market access (transport, outlets). New food preservation technologies can result in seasonally abundant species being harvested and stored.

In regions where entomophagy is accepted and it provides an important component of the diet, the issue is a matter of sustainable production of edible insects (whether they are wild harvested or farmed). The issue is how tourism and export markets can be developed to avoid over-exploitation and to ensure an additional income source to the people on the land. This could involve establishing insect farms that are an additional source of income, and have the potential to be integrated into more complex food production systems such as aquaculture and poultry.

Overseas tourists could play an important role in disseminating information about the use of insects as an ecologically sustainable source of food for humans if the fear and wow factors are overcome. They can obtain a greater understanding about diets of different cultures, and also learn how food production can be linked to sustainable nature conservation, and that there may be other ways of reducing insect pests other than use of pesticides.

Edible insects are potentially another type of flagship to promote insect and habitat conservation. An example is the sustainable harvesting of mopane worms which can lead to long term conservation of mopane forests in Africa (Vantomme *et al.* 2004). A strategy involving selected edible insects as flagships for insect and habitat conservation could be developed around iconic and important edible insects such as mopane worms (*G. belina*) in Africa, bamboo worms (*O. fuscidentalis*) in Asia, and witjuti grubs (species of cossid moth larvae) in Australia. Weaver ants (*O. smaragdina* and *O. longinoda*) are multipurpose species because they are used as food, as medicine, as pet food and as a biological control agent in Africa, Asia and Australia. These species are important to the local communities in which they are found, and educational programmes for both locals and tourists would be invaluable for conservation.

REFERENCES

Agea, J. G., Biryomumaisho, D., Buyinza, M. and Nabanoga, G. N. (2008) Commercialization of *Ruspolia nitidula* (nsenene grasshoppers) in Central Uganda. *African Journal of Food Agriculture Nutrition and Development*, **8**(3), 319–332.

Anonymous (1988) Caterpillars find their way to city restaurants. *Zimbabwe Herald*, 12 April 1988, reprinted in *The Food Insects Newsletter*, **1**(1):4.

Anonymous (1989) Mexico still paradise ... Mexico City. More salsa.... Mexico City. *The Food Insects Newsletter*, **2**(2), 4.

Anonymous (1991) They ate what? *The Food Insects Newsletter*, **4**(3), 7–8.

Anonymous (1992) Mexican insect delicacies as seen through the eyes of a Campesino. *The Food Insects Newsletter*, **5**(3), 5.

Anonymous (2011) Bugs: they do grow on trees. Mexico's tradition of eating bugs becomes lucrative. Associated Press, 16 June 2005. Available at: http://www.msnbc.msn.com/id/8248519/ns/business-consumer_news/.

Barreteau, D. (1999) Les Mofu-Gudur et leurs criquets. In *L'homme et l'animal dans le bassin du lac Tchad. Actes du colloque du reseau Mega-Tchad, Orleans 15–17 Octobre 1997*, ed. C. Baroin and J. Boutrais. Paris : Editions IRD (Institut de Recherche pour le Developpement), Collection Colloques et Seminaires, no. 00/354. Université Nanterre, pp. 133–169.

Cerda, H., Martinez, R., Briceno, N. *et al.* (2001) Palm worm (*Rhynchophorus palmarum*) traditional food in Amazonas, Venezuela: nutritional composition, small scale production and tourist palatability. *Ecology of Food and Nutrition*, **39**, 13–32.

Cerritos, R. and Cano-Santana, Z. (2008) Harvesting grasshoppers *Sphenarium purpurascens* in Mexico for human consumption: a comparison with insecticidal control for managing pest outbreaks. *Crop Protection*, **27**, 473–480.

DeFoliart, G. (1992) The New York Bug Banquet: a day to remember. *The Food Insects Newsletter*, **5**(2), 1–2, 10.

DeFoliart, G. (1993) The Insect Club. *The Food Insects Newsletter*, **6**(2), 1–2, 11.

DeFoliart, G. R. (1999) Insects as food: why the western attitude is important. *Annual Review of Entomology*, **44**, 21–50.

Dunkel, F. V. (1997) Food insect festivals of North America. *The Food Insects Newsletter*, **10**(3). 1–9.

Fairman, R. J. (2010) Instigating an education in insects: the 'Eating Creepy Crawlies' exhibition. *Antenna*, **34**(4), 169–170.

FAO (2010a) *Promoting the Contribution of Edible Forest Insects in Assuring Food Security*. Rome: Forest Economy, Policy and Products Division, FAO Forestry Department programme, FAO.

FAO (2010b) FAO and SAEP LAAI LAAI! 2010 winning chef showcased insects at Shanghai World Expo 2010 from 19 to 22 September 2010. FAO Press Release, 27 September 2010.

Fasoranti, J. O. and Ajiboye, D. O. (1993) Some edible insects of Kwara State, Nigeria. *American Entomologist*, **39**, 113–116.

Fernquest, J. (2010) A guided tour through Thailand's world of edible insects. *Bangkok Post*, **12** May 2010.

Ferreira, A. (1995) Saving the mopane worm: South Africa's wiggly protein snack in danger. *The Food Insects Newsletter*, **8**(1), 6.

Ghaly, A. E. (2009) The use of insects as human food in Zambia. *OnLine Journal of Biological Sciences*, **9**(4), 93–104.

Godfray, H. C., Beddington, J. R., Crute, I. R. *et al.* (2010) Food security: the challenge of feeding 9 billion people. *Science*, **327**, 812–818.

Gordiner, J. (2010) Waiter, there's soup in my bug. *New York Times*, 21 September 2010. Available at: www.nytimes.com/2010/0922/dining/22bug.html.

Gordon, D. G. (1998) *The Eat-A-Bug Cookbook*. Berkeley, CA: Ten Speed Press.

Hanboonsong Y. (2010) Edible insects and associated food habits in Thailand. In *Edible Forest Insects as Food: Humans Bite Back Proceedings of a Workshop Focused on Asia-Pacific Resources and Their Potential for Development, 19–21 February 2008, Chiang Mai, Thailand*. RAP Publication 2010/02, ed. P. B. Durst, D. V. Johnson, R. N. Leslie and K. Shono. Bangkok: FAO, pp. 173–182.

Hanboonsong, Y., Ratanapan, A., Utsunomiya, Y. and Masumoto, K. (2000) Edible insects and insect-eating habits in Northeastern Thailand. *Elytra*, **28**(2), 355–364.

Harris, W. V. (1940) Some notes on insects as food. *Tanganyika Notes and Records*, **9**, 45–48.

Kirkby, C. A., Giudice-Granados, R., Day, B. *et al.* (2010) The market triumph of ecotourism: an economic investigation of the private and social benefits of competing land uses in the Peruvian Amazon. *PLoS ONE*, **5**(9), e13015.

Linnaeus, C. (1735) *Systema Naturae*, 1st edn. Leiden, The Netherlands: T. Haak.

MacEvilly, C. (2000) Bugs in the system. *Nutrition Bulletin*, **25**, 267–268.

Menzel, P. and D'Aluisio, F. (1998) *Man Eating Bugs*. Berkeley, CA: Ten Speed Press.

Miranda Román, G., Quintero Salazar, B., Ramos Rostrol, B. and Armando Olguín-Arredondo, H. (2011) Le recolección de insectos con fines alimenticios en la zona turística de Otumba y Teotihuacánm Estado de México [The collection of edible insects in the tourist areas of Otumba and Teotihuacan, Estado de Mexico]. *Pasos*, **9**(1), 81–100.

Mors, P. O. (1958) Grasshoppers as food in Buhaya. *Anthropology Quarterly*, **31**, 56–58.

Munthali, S. M. and Mughogho, D. E. C. (1992) Economic incentives for conservation: bee-keeping and Saturniidae caterpillar utilization by rural communities. *Biodiversity and Conservation*, **1**, 143–154.

Oonincx, D. G. A. B., van Itterbeeck, J., Heetkamp, M. J. W. *et al.* (2010) An exploration on greenhouse gas and ammonia production by insect species suitable for animal or human consumption. *PloS ONE* **5**(12), e4445.

Palmer, A. (2011) An insect banquet. *The Financial Times*, 11 Mar 2011. Available at: www.ft.com/cms/s/2/cfcal13c0-49f2-11e0-acf0-00144feab49a.html.

Ramos-Elorduy, J. (1998) *Creepy Crawly Cuisine*. Rochester, VT: Park Street Press.

Ramos-Elorduy, J. (2006) Threatened edible insects in Hidalgo, Mexico and some measures to preserve them. *Journal of Ethnobiology and Ethnomedicine*, **2**, 51.

Ramos-Elorduy, J., Pino, J. M. and Conconi, M. (2006) Ausencia de una reglamentaión y normalización de la explotación y commercialización de insectos comestibles en México [Lack of a legislation and regulation of the exploitation and commercialization of edible insects in Mexico]. *Folia Entomologica Mexicana*, **45**, 291–318.

Ramos-Elorduy, J., Moreno, J. M. P., Vázquez, A. I. *et al.* (2011) Edible Lepidoptera in Mexico: Geographic distribution, ethnicity, economic and nutritional importance for rural people. *Journal of Ethnobiology and Ethnomedicine*, **7**(2), 1–22.

Ratanachan, N. (2009) Edible insect and scorpion in Thailand–Cambodia border Rong Kluea Market town, Sa Kaeo province. *Kamphaengsean Academic Journal*, **8**(1), 20–28.

Rattanapan, A. (2000) Edible insect diversity and cytogenetic studies on short-tail crickets (genus *Brachytrupes*) in Northeastern Thailand. Unpublished MSc, thesis, Khon Kaen University, Thailand.

Rigby, R. (2002) Tuck into a tarantula. *The Sunday Telegraph*, **22** September 2002.

Schabel, H.G. (2010) Forest insects as food: a global review. In *Edible Forest Insects as Food: Humans Bite Back Proceedings of a Workshop Focused on Asia-Pacific Resources and Their Potential for Development, 19–21 February 2008,Chiang Mai, Thailand*, RAP Publication 2010/02. ed. P. B. Durst, D. V. Johnson, R. N. Leslie and K. Shono. Bangkok: FAO, pp. 37–64.

Stack, J., Dorward, A., Gondo, T. *et al.* (2003) Mopane worm utilisation and rural livelihoods in Southern Africa. Presentation for International Conference on Rural Livelihoods, Forests and Biodiversity, 19–23 May, 2003, Bonn, Germany, pp. 1–31.

Styles, C. (1994) Food for Africa? Mopane worms! Ten-GRA Publications (Commemorative edn.), pp. 1–2.

Taylor, G. (2010) Dining on insects: Anyone for crickets …? *The Independent*, 21 October 2010. Available at: www.independent.co.uk/life-style/food-amd-drink/news/.

Taylor, R. L. and Carter, B. J. (1976) *Entertaining with Insects*, Santa Barbara, CA: Woodbridge Press Publishing Company.

Turk, D. (1990) Leguminous trees as forage for edible caterpillars. *Nitrogen-Fixing-Tree-Research-Reports*, **8**, 75–77.

Van der Waal, B. C. W. (1996) 'Importance of grasshoppers as traditional food in villages in Northern Transvaal, South Africa. In *Ethnobiology in Human Welfare*, ed. S. K. Jain. New Delhi: Deep Publications, pp. 35–41.

Van Huis, A. (2003) Insects as food in Sub-Saharan Africa. *Insect Science and its Application*, **23**(3), 163–185.

Vantomme, P., Göhler, D. and N'Deckere-Ziangba, F. (2004) Contribution of forest insects to food security and forest conservation: the example of caterpillars in Central Africa. *ODI Wildlife Policy Brief*, **3**, 1–4.

Vogel, G. (2010) For more protein, filet of cricket. *Science*, **327**, 811.

Yen, A. L. (2009a) Edible insects: traditional knowledge or western phobia? *Entomological Research*, **39**, 289–298.

Yen, A. L. (2009b) Entomophagy and insect conservation: some thoughts for digestion. *Journal for Insect Conservation*, **13**, 667–670.

Yhoung-Aree, J. (2010) Edible insects in Thailand: nutritional values and health concerns. In P.B. Durst, D.V. Johnson, R.N. Leslie and K. Shono (eds.) *Edible Forest Insects as Food: Humans Bite Back Proceedings of a Workshop Focused on Asia-Pacific Resources and Their Potential for Development, 19–21 February 2008,Chiang Mai, Thailand*, RAP Publication 2010/02. ed. P. B. Durst, D. V. Johnson, R. N. Leslie and K. Shono. Bangkok: FAO, pp. 201–216.

Yhoung-Aree, J. and Viwatpanich, K. (2005) Edible insects in the Lao PDR, Myanmar, Thailand, and Vietnam. In *Ecological Implications of Minilivestock*, ed. M. G. Paoletti. Enfield, NH: Science Publishers, Inc. pp. 415–440.

Part III   Insects and Tourism

12

Butterfly conservatories, butterfly ranches and insectariums: generating income while promoting social and environmental justice

KO VELTMAN

INTRODUCTION

Butterfly and insect pavilions in museums and zoos have become popular international visitor attractions, especially in Western Europe and North America, and more recently, Asia. In this chapter the author discusses how butterfly and insect exhibits provide educational opportunities and can be used to promote invertebrate diversity and conservation. In Costa Rica, Kenya and various other locations, butterfly-breeding programmes provide an opportunity for visitors to see insects in their natural habitats. More importantly, these programmes provide a sustainable source of butterfly pupae to European and North American butterfly pavilions, help to alleviate poverty and reduce deforestation. In many cases, the appeal of these large, colourful tropical butterflies and moths has been combined with local butterfly conservation efforts. In these cases, tropical butterflies and insects have become environmental ambassadors, helping to educate visitors on not only the importance of preserving tropical ecosystems but also local ecosystems.

HISTORY OF BUTTERFLY AND INSECT EXHIBITIONS

Insect collections have been publically displayed in museums since the eighteenth century (Penang Butterfly Farm 2007). As the painting

The Management of Insects in Recreation and Tourism, ed. Raynald Harvey Lemelin. Published by Cambridge University Press. © Cambridge University Press 2013.

of 'La chasse aux papillons' by Manet depicts, viewing and collecting butterflies is a well-established leisure activity, practised by numerous cultures across the world (Russell 2003). 'In the mid-eighteenth century, English butterfly collectors began calling themselves Aurelians, from the Latin *aureolius*, a reference to the golden chrysalis of some species' (Russell 2003: 6). In 1876, Walter Rothschild, started his own natural history museum (Russell 2003). By the time of his death 63 years later, he had collected 2.25 million butterflies and moths, and was recognized as the world's preeminent butterfly enthusiast (Russell 2003). The first live insect exhibit aimed at the general public was opened in 1897 at the London Zoo. The Amsterdam Zoo started its own insect displays one year later. In 1960, in Sherborne, Dorset, the first butterfly house opened as a public attraction (Hughes and Bennett 1991). While a number of butterfly exhibits were opened in the 1960s and 1970s, a rapid expansion of these establishments occurred in the 1980s. In 1988 the first butterfly houses were opened in Coconut Creek in Florida and Callaway Gardens in Georgia, United States. Butterfly houses 'as private businesses or connected with zoos, stately homes and garden centres' (Hughes and Bennett 1991: 47) can now be found all over the world. By the end of the twentieth century there were more than 50 butterfly houses in the UK alone, and hundreds throughout the world. Observing butterflies in natural settings or as part of specimen counts continues to attract thousands of individuals every year (Pyle 2009).

In contrast to these small butterfly pavilions that tended to showcase only a few butterflies and moths, modern butterfly houses now often measure over a 1000 m² and are inhabited by more than 50 butterfly species from different parts of the world (see Figure 12.1). In these larger butterfly houses visitors are able to enjoy and interact with some of the world's most spectacular and colourful animals. Furthermore, large pavilions with naturalized landscapes, ponds and waterfalls provide an opportunity for visitors to gaze upon feeding, mating and flying butterflies and witness their amazing life cycles (eggs, caterpillars and pupae).

Consisting of breeding programmes in developing and now developed nations, transportation of specimens and displays of adult butterflies and moths, the butterfly industry has developed rapidly and has as a result become highly professionalized. The need for butterfly breeders and exhibitors to coordinate these efforts resulted in 1997 in the creation of the International Association of Butterfly Exhibitors and Suppliers (IABES, www.iabes.org). The mission statement of the organization is:

Figure 12.1   Families visiting a butterfly pavilion. Photo by Ko Veltman.

'Advancing the international butterfly industry through representation, communication, education and marketing.' Sustainability and fair trade are two of the guiding principles of the organization. Through an elected board of governors, annual meetings, a newsletter and a website, members stay in close contact and are notified of the latest developments in butterfly breeding programmes. In parallel with the expanding number of butterfly exhibitions in Europe and the United States, breeders in the tropics, also known as butterfly ranchers, have established their own breeding programmes largely aimed at supplying these facilities. These butterfly ranches are discussed next.

BUTTERFLY BREEDING

The growing popularity of live exhibits in North America, Europe and now Asia, have created a positive spin-off requiring the import of live specimens from where butterflies are found in Central and South America, Africa and Asia. Organizations such as IABES have implemented proactive guidelines and by-laws to ensure the sustainability of the industry. In some of the supplying nations, education and preservation strategies combined with economic benefits, have shifted previous extractive activities to more sustainable activities.

Figure 12.2   A butterfly ranch. Photo by Ko Veltman.

On these 'ranches' butterflies are maintained in big mesh wire enclosures together with an abundance of flowering plants as food sources (Figure 12.2). Young caterpillars and eggs are harvested and placed on shrubs within the enclosure, to protect them from predators. In some cases, young caterpillars and eggs are raised in enclosed control environments called 'laboratories'. These laboratories are clean and tidy places covered by a roof that keeps the rain and sun out. A tile floor enables easy cleaning to ensure a sterile environment. Together with clean water and fresh air, these laboratories create ideal breeding conditions for developing caterpillars.

Depending on the species, each pupa is worth between US $1.00 and $2.50. Farmers united through international agreements receive about 70% of the final selling price. In a community where households sometimes earn less than US $400 per year, income derived from the rearing of butterflies is often welcomed. On average household income has seen a 25% increase since the beginning of these butterfly farms. The main limit of further expansion in butterfly pupae sales is the small size of the live butterfly market, especially during wintertime. In this period most European and northern American butterfly exhibitions are closed.

Once caterpillars have changed into pupae, they are ready for transport. Fresh butterfly pupae are packed and shipped by postal firms to Europe and North America. After arrival the pupae are immediately hung for a short period in an emerging chamber under perfect conditions (i.e. 24°C and 70–80% humidity). Once emerged, butterflies are released into the butterfly house where they will live for a period of days or months, depending of the species. The average lifespan of these butterflies is about 3 weeks. Due to the short lifespan of most tropical butterflies, exhibits order new pupae every 2 or 3 weeks.

On-site breeding programmes of the blue morpho (*Morpho peleides*), the great owl butterfly (*Caligo memnon*) and the atlas moth (*Attacus atlas*) although more expensive, have often been carried out for educational purposes: to show the different species of caterpillars and to tell the story of metamorphosis. Increasing governmental restrictions are also making the importation of certain pupae difficult, therefore on-site breeding programmes, in this case, are indispensable. Despite these challenges, the industry is constantly seeking ways to reduce their carbon footprint and reducing the energy-use of certain butterfly pavilions. On the other hand, since visitors do not have to travel to remote sites to see butterflies, the carbon footprint is lessened, information is provided and social justice is promoted. Highlighted next is an overview of some of the most popular butterfly and moths in butterfly pavilions, followed by some of the most popular insect species found in insectariums. These observations were gathered from the author's three decades of research in the area.

BUTTERFLY PAVILIONS

The atlas moth

The atlas moth (*Attacus atlas*) is a huge and attractive moth from Asia. Since it not easily disturbed, it is often sought out by viewers and photographers. The one drawback of this species is that the adult moth is short-lived, with a lifespan of only a few days. However, since the attractive greenish caterpillars will easily grow in captivity, they can also be displayed while they are being raised.

The blue morpho

Found in the rainforest in Central and South America, the blue morpho (*Morpho peleides*, Morphidae), with its iridescent blue wings and a gracious flight, is one of the most beautiful and popular butterflies. In fact a movie entitled *The Blue Butterfly* (2004) featuring Georges Brossard, the founder of the Montreal Insectarium, was made on the blue morpho. The blue morpho is an active butterfly that is always patrolling along the sides of shrubs and bushes. Since morpho butterflies feed only on rotting fruit, display stations consisting of rotting fruits can be established throughout butterfly pavilions. Approximately 10% of the butterfly population in a butterfly exhibition should be morphos to achieve optimal attractiveness of the collection. The species that is most often available and easy to propagate is the medium-sized *Morpho peleides*. Many other species of morphos are also available and are even more beautiful, of a bigger size and larger wingspan. These species, however, are more canopy dwellers that require a bigger greenhouse and deposit their eggs higher up in the trees, and are therefore less suitable for the average exhibition.

The great owl butterfly

The great owl butterfly (*Caligo memnon*, Brassolidae) also from tropical parts of the Americas, is generally known as an owl because of the very large and prominent eye-spots on the underside of their wings. Since they are exceptional fliers and will tend to congregate in the mornings and evenings, owl butterfly should compose approximately 10% of the total butterfly collection. Owl butterflies need large flight areas, especially during the vigorous nuptial chases, with frequent stops for a short rest, often on people's heads. Owl butterflies like morphos can live for many months and only feed on rotting or very ripe fruits. These feeding patterns make them ideal for photography.

The passionflower butterfly

The various Heliconid species (*Heliconius* sp.) show a wide range of colours and patterns. The common name for the Heliconids is the 'Longwing' due to the wing shape. They live in South and Central America and are, unlike most other species, active under cloudy conditions, which make them perfect for use in butterfly houses in countries with winter-like conditions. Many species like *Heliconius erato*, *Dryas julia* and *Heliconius melopmene* are suitable for small greenhouses and can be easily bred in large numbers all the year round, as long as a supply of pollen is provided.

The swallowtail

Swallowtails (*Papilio* spp.) can be found all over the world. A short-lived species (a couple of weeks), adults are often vigorous butterflies with a strong flight, especially during sunny days. Since they require sunlight for flight and feeding, they can become inactive on cloudy days.

INSECTARIUMS

There are more than 100 insectariums found throughout the world, some of these are specialized viewing units focusing solely on insects (i.e., insectariums), while others are insect pavilions located in zoos (e.g., the Bug House at the London Zoo). Numerous types of arthropods: insects (e.g., stick insects, beetles, praying mantis, ants, bees); spiders (tarantulas, orb spiders, black widows); scorpions and myriapods (centipedes and millipedes) are displayed at these insect pavilions. The dislike of these creatures by some people (especially adults) can be offset by the ability to view these animals in controlled settings, the interest of children, and the appeal of the negative sublime (see Lockwood, this volume). Most invertebrates displayed in insect houses are from tropical countries and those species which are easily bred. While insectariums may feature arachnids and myriapods, the ensuing section focuses solely on insects.

Beetles

Beetles are the largest and most diverse groups of animals with more than 300 000 species known. Often seen in insect houses are rose beetles (*Cetoninae* and *Goliathonae*). These colourful beetles are active at daytime and easy to rear. Other species like the burrowing beetles and flower beetles *Pachnoda* and *Eudicella* are often found in insect collections.

Honeybees

Honeybees (*Apis mellifera*) are fascinating creatures. They function as a super-organism, working selflessly for the whole colony and are an essential part of our ecosystem. In addition to gathering nectar to produce honey, bees perform a vital role in pollination. Insect pollinated plants make up one-third of the human diet. An observation hive placed in an insect house can be a major attraction because bees are always active and have an amazing story to tell. Through an observation hive, visitors are able to see how work is allocated, how larvae are fed and how information is transferred through a communication process known as the 'dance of the bees'.

Leafcutter ants

Leafcutter ants belong to the genera *Atta* and *Acromyrmex*. They are distributed from northern Texas to central Argentina. Because leafcutter ants cut vegetation and carry it into their nest where they cultivate fungi, they can transform landscapes by defoliating trees and destroying agricultural crops. The worker ants are divided into three different types or castes, according to size. The 'minima ants' are the gardeners, specialized for moving around in the small spaces of the fungus garden. The 'media ants' cut leaves and bring the leaf fragments back to the nest to grow the fungi on it. The 'maxima ants' are the soldiers that take care of protecting the nest. There is just one queen. In zoos, 'displays of leafcutting ants began to appear in the United States, Europe and the UK in the mid-1970s (Powell 1991: 98). Today, these displays are found in numerous zoos and insectariums throughout the world. The ant's ability to adapt to controlled viewing environments in insectariums combined with their activities make these displays popular and educational displays for visitors.

Stick insects

Mostly nocturnal, Phasmida (stick and leaf insects or walking sticks) are not the most visually stimulating animals. However, since they are easily fed and maintained, they can become quite large. They are easily seen and handled (they will often feign death when being handled), they are quite tolerant to a range of indoor climates and their breeding abilities make them ideal for display. The most popular include the Indian Stick Insect (*Carausius morosus*) and the Giant Prickly Stick-insect (*Extatosoma tiaratum*) (Phasmid Study Group).

CONCLUSION

This chapter illustrates how butterfly and insect exhibits can generate income and promote social justice, while entertaining and educating visitors. Butterfly houses can be a centre for education. The educational value of butterflies and insects is immense, especially for future generations. (TITAG 2008). Talks, educational tours and handling sessions can be a fascinating discovery moment, particularly if the relation to our local environment is included. Education in nature conservation through butterflies and insects can be promoted successfully in a butterfly or insect conservatory. At the same time economic development in Asia, Africa and South and Central America is stimulated. In this way, new concepts of butterfly houses and insect zoos will be popularized for the good of education, promotion of nature awareness and environmental sustainability.

REFERENCES

Hughes, D. G. and Bennett, P. M. (1991) Captive breeding and the conservation of invertebrates. *International Zoo Yearbook*, **30**(1), 45–51.
Penang Butterfly Farm (2007) *Butterfly Conservation: Roles of Butterfly Farms and Breeding for Conservation of Lepidoptera*. Penang, Malaysia: Penang Butterfly Farm. Available at: http://www.butterfly-insect.com/butterfly-conservation.php.
Powell, R. (1991). The educational value of leaf-cutting ant colonies and their maintenance in captivity. *International Zoo Yearbook*, **30**(1), 97–107.
Pyle, R. M. (2009) The beauty of butterfly nets. *Wings: Essays on Invertebrate Conservation*, 15–18.
Russell, S. A. (2003) *An Obsession with Butterflies*. Cambridge, MA: Perseus Publishing.
TITAG (2008) *Why Invertebrates?* Seattle, WA: Terrestrial Invertebrate Taxon Advisory Group. Available at: http://www.titag.org/whyi.html.

13

Insect festivals: celebrating and fostering human–insect encounters

GLEN T. HVENEGAARD, THOMAS A. DELAMERE, RAYNALD HARVEY
LEMELIN, KATHLEEN BRAGER AND ALAINE AUGER

INTRODUCTION

Insect festivals, or special events celebrating insects, are growing in number and can attract large numbers of people. The Festa del grillo (the festival of crickets) in Florence, Italy, one of the world's oldest known insect festivals, is only one of many such events. Other notable festivals include le Festival international du Film de l'Insecte hosted by l'Organisme de l'Environnement du Languedoc-Roussillon de France, attended by over 20 000 people since its inception in 1995 (Pinault 2003). In 2009, the 2nd Pestival held in London, England, drew 200 000 people (Pestival 2011). In Japan, festivals such as the Dragonfly Citizen Summit provide dragonfly enthusiasts with the opportunity to practise and perfect their skills (Kadoya *et al.* 2004; Primack *et al.* 2000). Elsewhere, thousands of firefly enthusiasts in Japan and in the Great Smoky Mountains National Park congregate to watch these animals (Brown 2011), while the Annual Texas Butterfly Festival of Mission and the El Cieolo Butterfly Festival in Cd. Mante, Tamaulipas, Mexico, attest to the popularity of these insects.

From a tourism and recreation perspective, insects interest and attract people in many ways (Lemelin 2009). Large concentrations of insects attract nature lovers for viewing and photography. Rare insects attract scientists and specialists wanting to gain knowledge or observe new species. Many people just want to observe the beauty and subtle features of insects. These interactions can generate interest from tourists who may want to experience how insects contribute to human well-being (e.g., pollinators, food source, singing, or cultural significance; Yi *et al.* 2010). Even insects that negatively impact people, as Lockwood (this volume) explains, generate emotions that can translate into visitor interest. That interest is evident, for example, among many ecotourists to South Africa's game parks who responded positively to the potential for invertebrate-related activities such as the provision of indigenous knowledge and walks for children and adults (Huntly *et al.* 2005).

Many communities around the world have capitalized on human interest in insects by developing, marketing and managing insect tourism attractions. For example, many urban zoos have built butterfly gardens to diversify their attractions (see Veltman, this volume). A major mainstream tourism attraction is the famous glow-worm cave in Waitomo, New Zealand (see Hall, this volume). Other communities and organizations develop insect festivals to celebrate local insect features, but also to support other community objectives (Lemelin 2007).

In this chapter, we provide background on festivals and their role, the characteristics of insect festivals around the world, potential impacts (both positive and negative) of insect festivals and strategies for managing and/or mitigating those impacts. The goal of this chapter is to examine critically the popularity of insects for tourism and recreation, as exemplified through insect festivals.

BACKGROUND

Festivals are 'themed, public celebrations' (Getz 2005: 21) that are generally recurring in nature (Delamere 2008). Historically, festivals (also called fairs, jamborees, carnivals and a variety of other terms) regularly marked key parts of the annual calendar by celebrating local features, such as religious ceremonies, agricultural products or cultural artefacts (e.g., music and art). More recently, festivals have been rediscovered, reinvigorated, reinvented and created (Picard and Robinson 2006).

Festivals serve a variety of community functions. From the local perspective, they may be organized to enhance the image of the community (Janiskee and Drews 1998) by portraying the community in the best possible light. Festivals can help develop a local sense of community (Derrett 2003), as they build on the work of community leaders and volunteers (Delamere and Hinch 1994). Festival celebrations infuse the community with a palpable sense of pride and well-being, showing aspects of the community which make it unique and special (Delamere 2001). Indeed, many communities take their very identity from their signature, hallmark festivals (Small 2007; Delamere 2008). Individual residents of those communities may also gain personally through improved health and well-being as well as having the opportunity to learn new things and experience novel activities (Delamere 2008). Festivals provide important recreational opportunities (Mayfield and Crompton 1995) for local residents, and for their guests (domestic visitors and international tourists). Festivals can also help to maintain the activities or features being celebrated, whether they are cultural (Getz 1991) or natural (Hvenegaard 2011; Lawton 2008).

Negative impacts may also accrue to the local community. The disruption of normal routines and the intrusiveness of the festival into the lives of the local residents may be problematic (Delamere 2001; Fredline *et al.* 2003). Aspects such as noise and litter increasing to unacceptable levels may bring a level of discomfort to the community (Delamere 2001). Crowding and increased traffic, both vehicular and pedestrian (Small 2007), can create stresses for the individual and the community. Local facilities may become overused during the festival and there may also be amenity loss for the residents as their community-based facilities are repurposed for the time of the festival (Delamere 2001). Festival organizers need to fully comprehend the effects of their activities upon the local host community (Rollins and Delamere 2000).

Festivals can generate significant economic impacts (Walo *et al.* 1996), some of which stays in the local community (Hvenegaard and Manaloor 2007). The idea of the festival as tourist attraction is not a recent phenomenon. Local advocacy groups, chambers of commerce, economic development offices and tourism authorities have long seen festivals as a vehicle for expanding the tourism season or for creating a new tourism season (Delamere 2008: 131). Political interest from various levels of government (grants or sponsorship) can be translated into civic repositioning, desirable community development, or infrastructure projects (Dinaburgskaya and Ekner 2010). Additional positive economic and tourism impacts include

> the attraction of tourists into the host community/region; tourists staying longer and spending more money in the community (depending on the length of the festival); merchandise and concession sales opportunities; retention of profits in the local community; increased sales for local businesses; and local employment opportunities (Getz 2005: 392).

These positive impacts must, however, be balanced against potentially negative economic impacts. There are opportunity costs (alternative uses for grant and sponsorship monies), costs associated with leakage (dollars spent outside the host community on festival-related goods and services), or money leaving the local economy and being spent elsewhere during the festival (Getz 2005).

An important point to remember with respect to negative festival impacts is that the festival celebration must have a strong base of local support. If the residents of the host community do not support the festival initiative, it will inevitably fail regardless of the amount of tourism generated by the festival. In general, tourists will not go where the local residents do not go (Plog 2004); local involvement and support is imperative to ensure the overall success of the festival.

Festivals celebrating the natural environment focus on larger groups or specific species of birds, mammals, fish, insects, plants, landforms, water or the sky. Such festivals are growing in number and attendance (Lawton 2008). For example, the number of wildlife festivals in North America grew from ten in 1992 to over 240 in 2002 (DiGregorio 2002; Lawton 2009; National Fish and Wildlife Foundation 1999). Typically, festivals are of a short duration (usually 1–7 days). Most festivals are planned, organized and executed by volunteers and a few paid staff from community groups, special interest organizations, and/or tourism agencies. A few festivals are organized by special event planning organizations. Festivals are generally open to the public (Lawton and Weaver 2010) and typically offer activities such as guided walks, presentations, competitions, children's activities and trade shows (Hartley 2005). Many small festivals attract fewer than a hundred visitors, although a small number of festivals can attract several thousand visitors. Many festivals can be considered a form of ecotourism (Slotkin 2003), with their focus on the natural environment, environmental education and sustainability (Lawton 2008; Singh *et al.* 2007).

INSECT FESTIVAL CHARACTERISTICS

In order to assess the extent and dynamics of insect festivals, we conducted a web search of insect festivals around the world. Our team had Japanese, French and English language skills, so the results may be slightly biased toward those language regions. We did not include special events like pollinator festivals, entomological events or nature festivals that included insects as only a small part of the overall attraction. From these internet sites, we recorded information about festival location, attendance, timing, history, frequency, duration, fee structure, focus species and purpose.

Our search produced an inventory of 107 festivals with a focus on insects, all of which were from the northern hemisphere. North America hosted 82 festivals. Of 66 insect festivals in the United States, 15 were in the western states, 11 in the Midwestern states, 12 in the northeastern states and 28 in the southern states. While we can't make a direct comparison with websites from an earlier time period, the 1999 Directory of Birding and Nature Festivals listed only four insect-related festivals in the United States (National Fish and Wildlife Foundation 1999). Of 16 insect festivals in Canada, four were in the west, 11 were in the east, and one covered the whole country. Another 25 occur around the world, with 16 in Japan and three in the UK. While not included in our survey, Yi *et al.* (2010) report more than 100 traditional festivals in China related to insects.

Attendance at the inventoried insect festivals varies widely (Figure 13.1). Only one festival listed its attendance at fewer than 100, while three festivals reported 100 000 attendees or more, including the United Kingdom's Pestival, Ohio's Woollybear Festival, and Florida's Insect Festival at Fort Pierce. Saunders (2004) summarized a visitor survey of three insect festivals (with a focus on butterflies) in Ohio in 2003 and 2004. Attendees tended to be 30–49 years old and female. Most (75–86%) attended with children. Most attendees had no butterfly-related training or hobbies.

Given their location, most insect festivals occur during the northern hemisphere's spring, summer and autumn months, especially April through September. Most of the insect festivals reviewed were fairly new, having originated in the last 10 years (Figure 13.2). Nevertheless, 20 festivals have existed for more than 20 years and five festivals have been operating for 31 or more years. Almost all of the insect festivals examined occur on an annual basis. Of the 107 festivals, 56 occur on a single day, with a small number occurring over three or more days.

Of the 83 festivals that include a cost description, 51 were listed as free events (of which five were free on the condition that one pays for admission to the site). The other 32 festivals charged either a general price for every person or a variety of prices based on age or status. Twenty of these festivals allowed certain demographics in for free, generally children. Others gave special consideration to disabled persons and their attendants, university/college students, seniors, military

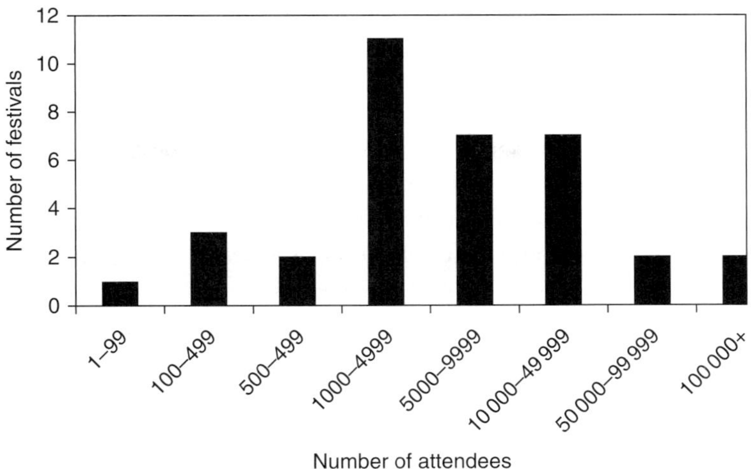

Figure 13.1   Insect festival attendance.

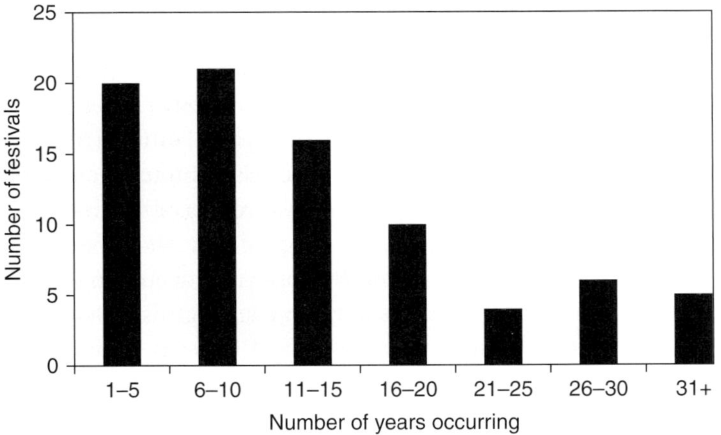

Figure 13.2   History of insect festivals.

personnel and local residents (i.e., The ButterflyFest in Jacksonville, Florida). Entrance fees ranged from US \$1–16 per person. The average entrance fee for participants aged 19–59 was US \$8.04, and was much less for children and slightly less for seniors.

These insect festivals tend to focus on the entire class of Insecta or on higher profile groups. Of the 107 insect festivals, 41 focused on all insects. These generic festivals were given a wide variety of names, including, for example, insect festival, insect fair, bug fest, bug bowl, insect-ival, bug-a-licious and pestival. The remaining insect festivals focused on particular taxonomic groups or species. There were 31 festivals focused on butterflies in general, five of which targeted the monarch butterfly, one on the Karner blue butterfly and one on woolly bear caterpillars. Another 16 festivals focused on fireflies, while eight festivals focused on dragonflies or damselflies. A variety of other taxonomic groups were the target of a few festivals each, including moths (3), fire ants (2), ladybugs (2), black flies (2), crickets (1), mountain pine beetles (1), mosquitoes (1),and cockroaches (1).

PURPOSES OF INSECT FESTIVALS

We analysed the major purposes identified on the festival websites. Many festivals had more than one purpose. Education was the most common purpose of these festivals. Thirty-one festivals stated that learning, education or awareness about insects was a primary goal. For example, the Bug City event in Garfield Heights, Ohio, seeks to help people 'to learn what insects do, why they are important, and even what they taste like' (Cleveland Metroparks 2011). Twelve festivals highlighted the importance of getting people to understand something about insects in relation to humans, such as pollination or other values. For example, the Palisade International Honeybee Festival (2011) in Colorado 'brings awareness to the honeybee crisis through education and the arts. Our goal is to increase public awareness of the impact of bees on our food sources and the wide range of benefits of bee products.' Similarly, three festivals specifically stated that debunking myths and/or decreasing phobias of insects were important goals. Celebration was the second most common purpose with 21 festivals, with a focus on the insects in general, their migration, and insect-related processes and products. These festivals often used activities related to entertainment and education to achieve this purpose (see Figure 13.3).

Many insect festivals sought to increase sustainability in some fashion. Conservation of the species itself seems to be a common theme in insect festivals; nine festivals specifically mentioned protection of the focal species, insects in general, or a specific area as a purpose. For example, three zoo-based insect festivals in Ohio work to build support for the restoration of insects and their habitats (Saunders 2004). Most conservation impacts seem to be accomplished in a secondary way by increasing visitor awareness, understanding and appreciation of insects which could lead to conservation action. Less common, but more immediate for conservation, are the few festivals

Figure 13.3   Insect enthusiasts attending an insect symposium. Photo by Glen Hvenegaard.

that raise money for insect or habitat protection, those that specifically promote protection, and that highlight the green movement. Two festivals stated that the main purpose of their festivals was to promote green practices to reduce human impact on the environment.

Nine festivals included having fun as a purpose. These festivals often focus on entertaining kids with various creative activities. For example, the Bugfest in Stockton, CA, encourages kids to 'have fun with bug-themed crafts, kids' games, live insect races, and face painting. Dig in the dirt with the arthropod petting zoos. Enjoy fly-tying demonstrations by the Delta Fly Fishers. Or take a rest and listen to fascinating presentations given by local specialists' (Oak Grove Nature Center 2011). A Dutch insect festival invites visitors on a fascinating, inspiring, versatile and wonderful introduction to the world of insects in order to experience more fun.

Developing an emotional and aesthetic connection with insects was a focus for many festivals. Five festivals highlighted the importance of close experiences with insects. This involves, for example, holding, catching and photographing insects. Four festivals, all in Japan, stated that honouring the focal species was important. Another four festivals stated that a purpose was to highlight the beauty of insects. For example, the Bugfest in Wichita Falls, Texas, helps visitors 'experience the beauty and fascination of bugs and learn about their importance in our daily lives' (River Bend Nature Center 2011). Three festivals stated that developing an appreciation for insects was important.

A variety of other purposes were listed on the insect festival websites. Six festivals were fundraisers for outputs related to insects, from protection to pest control. Although several festivals include entomophagy activities, only three listed that the promotion of the edibility of insects was a primary purpose. The focus of two festivals was on making fun of pests, while another two promoted control of pests. Two more festivals stated that the purpose of the festival was to allow people to have the opportunity to explore the area. Showcasing the area was a purpose of only one festival. Only one festival stated that increasing the economic impact of nature tourism was a purpose of the festival.

INSECT FESTIVALS AND SUSTAINABILITY

Insect festivals (and wildlife festivals in general) have the potential to promote sustainability of their target species and increase the awareness of conservation initiatives. However, few studies of wildlife festivals, let alone insect festivals, describe this direct relationship (Getz 2008; Weaver and Lawton 2007). According to Hvenegaard (1994), wildlife tourism occurs at the junction of four key components: (1) a historical relationship between wildlife and tourists; (2) a set of wildlife species and habitats, as chosen by tourists, but influenced by marketing and resource management; (3) tourists undertaking experiences with wildlife, with some visitor management restrictions; and (4) nearby communities which host these wildlife tourism activities. As for all ecotourism activities, insect festivals are at the centre of interactions among supply (venues, industry), demand (tourists), institutions, and external environments (Weaver and Lawton 2007).

Within this framework, on the one hand, activities associated with insect festivals can generate many negative impacts. These impacts depend on many factors, including the festival's scale, timing, duration and activities. For example, an annual wasp festival in Japan holds an event in which participants compete to collect the biggest nest, whether from the wild or at home (Nonaka 2010), which has the potential to deplete native wasp populations. From this perspective, festival developments can degrade habitats upon which insects depend.

On the other hand, four potential conservation benefits for species and habitats can arise from tourism-related insect festivals (Brandon 1996; Higginbottom and Tribe 2004; Hvenegaard 2011;

Weaver 2002). These benefits can be generated by providing (Diamantis 1999; Higginbottom and Booth 2003; Ross and Wall 1999; Sekercioglu 2002): (1) incentives to establish protected areas; (2) resources for wildlife and habitat management; (3) economic impact to nearby communities, thereby encouraging residents to conserve wildlife and their habitats; and (4) support for conservation by educating local residents and non-local visitors. While there is considerably more evidence for these impacts from studies on wildlife festivals and ecotourism, this section summarizes the sparse evidence for these benefits arising from insect festivals.

First, with respect to habitat protection through protected areas, an insect targeted by a festival can serve as a flagship species, which in this particular instance is a charismatic micro-fauna 'that is used as a symbol to arouse public interest in the animal and its habitat and promote broader ecological and economic values of conservation' (Smith and Sutton 2008: 127). Flagship species can capture the imagination of the public and motivate them to encourage conservation action, particularly through public support and fund raising. While large mammals are typically flagship species (Walpole and Leader-Williams 2002), Guiney and Oberhauser (2008) and Lemelin (2007, 2009, this volume) have argued that certain colourful, easy to observe and familiar insect species like butterflies, dragonflies and bees are also good candidates for flagship species. According to Smith and Sutton (2008), flagship species are especially effective if they are endemic to one area, well known beyond that area, have cultural and economic importance, can serve as an umbrella species, and have a declining population.

A few examples illustrate how butterflies can serve as flagship species. First, Hou (2000) reports how the Yellow Butterfly Festival in southern Taiwan facilitated an environmental movement that opposed the development of the proposed Meinung Dam. This dam would have flooded the so-called 'Yellow Butterfly Valley', which provides unique habitat for 110 species of butterflies. At the annual festival, environmental organizations and politicians pledged opposition to the dam. Second, migratory insect species that cross international boundaries can also attract significant public and political attention. Because the monarch butterfly uses habitats in Canada, the United States and Mexico during its life cycle, several conservation agencies from these countries now endorse the North American Monarch Conservation Plan which has resulted in a network of 13 wildlife reserves to support monarch butterfly protection (Guiney and Oberhauser 2008). The five festivals in Canada and United States focused on monarchs play an important role in raising the profile of monarch conservation and this network of protected areas. Third, the Karner Blue Butterfly Festival in Wisconsin celebrates this butterfly species and its habitat, realizing that any protection for this species will help countless other organisms as well (Guiney and Oberhauser 2008).

The Pestival is a triennial mobile festival examining insect–human interactivity in bioscience, through contemporary art, cinema, music, comedy, and education. It brings together eminent international artists and scientists, along with local and global communities, to collaborate on cutting-edge interdisciplinary art projects in order to challenge existing stereotypes about insects good and bad (vectors and pollinators), and understand the role of these animals within the biosphere. The first Pestival, held at the London Wetland Centre in Barnes in 2006, hosted 10 000 people. In 2009, the second Pestival, featuring 50 free interactive events and numerous experts from the arts and sciences and held at London's Southbank Centre, hosted 200 000 visitors. In addition, 10 million people worldwide were reached by the Pestival press campaign and the use of social media. Media coverage included the BBC, *Time Out*, *New York Times*, Nippon TV, *The Times*, *Telegraph*, *Guardian* and leading science journals and arts magazines. Of the 10% of visitors surveyed, 98% had changed their view of insects and their importance to the health of the planet and our daily lives (Pestival 2011). The next Pestival will take place in Brazil in 2012.

Second, insect festivals can help increase resources for insect protection and management of their habitats, particularly by attracting volunteers for insect monitoring and raising funds. A relatively new approach to community involvement is citizen science (see Johansen and Auger, this volume), or volunteer participation of non-professionals in data collection for ongoing research projects that can aid management (Cooper *et al.* 2007; Oberhauser and Prysby 2008). The citizen science model invites the public to participate in scientific projects (e.g., monitoring insects) using established protocols; the information gathered can be collated, analysed and published in scientific papers (Cooper *et al.* 2007; Johansen and Auger, this volume).

Insect festivals can play a key role in informing the public of citizen science opportunities and recruiting volunteers. For example, participants in the Dragonfly Festival in Manitoba learn about the important role of dragonfly monitoring for ongoing management decisions. Many insect enthusiasts participate in long-term monitoring projects such as the Monarch Larva Monitoring Project which began in 1996 (Oberhauser and Prysby 2008). In this case, volunteers are recruited (via the internet, word-of-mouth or through their various nature networks, including festivals), trained about the research protocols, choose their own monitoring sites, collect data and enter their data online. The resulting publications have addressed distribution, abundance, threats, land uses and climate change (Oberhauser and Prysby 2008). In addition, citizen science projects provide valuable learning opportunities (e.g., increase scientific knowledge and critical thinking), incentives to protect habitats (e.g., advocacy, environmentally responsible practices) and sharing of their work with others (Cooper, *et al.* 2007; Oberhauser and Prysby 2008).

Insect festivals (and insect tourism in general) can also serve to raise funds for the management of key insect habitats or protected areas targeting insects. For example, the Joseph A. Strasser Butterfly Festival in Jacksonville, Florida, is the largest fundraiser for the Tree Hill Nature Center while the Dragonfly Festival in Roswell, New Mexico, raises funds to support projects at the Bitter Lake National Wildlife Refuge. More often, the insect festivals in our survey raise funds to support the sponsoring organizations (e.g., nature centres and protected areas). Many of these organizations, in turn, manage habitats for nature protection. Habitat management is critical for maintaining healthy insect populations. In fact, habitat manipulation can quickly serve to attract diverse or targeted insect populations.

In other cases, funds raised support zoos, museums and community groups. For example, Zoobilation at the Akron Zoo and the Insect Fair at the Columbus Zoo in Ohio use the money raised to support zoo operations. The Butterfly Festival in Plano, Texas, has raised US $77 000 for various charities (Great Plano Kiwanis Club 2011).

Third, insect festivals can generate benefits for local communities through economic impact, employment, infrastructure and local pride. Only a few studies are available about economic impact. In summarizing the benefits of two Korean festivals, Kim *et al.* (2008a) note the hope that local governments have for these festivals in helping the local economies through an increase in production and employment. The total economic effect of the 2007 Firefly Festival in Muju County, involving 750 000 visitors, was US $15 million (Kim, *et al.* 2008a). In 2007, US $1.2 million was invested in the Butterfly Festival in Hampyeong County (Kim, *et al.* 2008a), with an attendance of 1.7 million in 2006 (Kim *et al.* 2008b). Festival-related expenditures resulted in direct and indirect earnings of US $11.2 million. Participation by residents in economic opportunities, such as vending booths, has increased nine fold since 1999 (Kim, *et al.* 2008b). As a result, there have been positive socio-cultural impacts such as increased community unity, congeniality and volunteerism, along with reduced negative images of past events.

Despite these positive economic impacts, some negative impacts result from festival tourism. For example, developments at the Hampyeong Butterfly Festival resulted in less local land available for other industries (e.g., agriculture), seasonal fluctuations of festival tourism, and low occupancy rates during the off-season (Kim, *et al.* 2008b). Similarly, negative socio-cultural impacts of this festival include income inequality, inappropriate business ethics and traffic congestion. The expected community benefits from butterfly tourism at the monarch butterfly reserve in west-central Mexico have not materialized due to bureaucratic constraints, lack of local control, poor planning and lack of organization (Barkin 2003). Studies in recreational ecology have demonstrated that increased human activities in recreation settings can also damage habitats, disrupt feeding and mating patterns and in some cases result in death (Lemelin 2007; Liddle 1997). There is also a potential for the decline or outright discontinuation of a festival that is dependent upon the migration of the star attraction, the scope and timing of which will have an enormous impact on the planning cycle for the festival (Rollins and Delamere 2000).

Fourth, insect festivals have the potential to deliver important educational benefits to local residents and non-local visitors. From the analysis of festival purposes earlier, it is clear that insect festival organizers seek to educate attendees about many aspects of insects. However, little is known about how those efforts result in increased knowledge, changed attitudes or increased efforts at insect-friendly and environmentally responsible actions. Saunders (2004) reported that the amount of time people spent at an insect festival was positively correlated with the likelihood of saying that the festival increased their awareness of insects and their links to habitat and human activities. In addition, many festival attendees expressed interest in attracting butterflies to their backyards, learning to identify butterflies, and learning about butterfly conservation issues. Festival attendees were more likely than zoo visitors to express interest in volunteering to help restore butterfly habitat or collect butterfly data. Last, 42–72% of festival visitors indicated interest in butterfly restoration projects (Saunders 2004). As Louv (2008) outlines, exposure to nature and insects through a variety of educational, recreational, and cultural activities helps to shape how we perceive nature. As Rykken (this volume) and Mitchell (this volume) suggest, experiential approaches to learning and modern technology can facilitate and promote these processes. Interpretation and environmental education is particularly important when poor weather conditions reduce the viewability and availability of insects at festivals.

CONCLUSION

Given our human history of ambivalence and outright hostility to most insects, it will take time to overcome past negative attitudes. For many cultural groups, insects have significant cultural values, including those related to literature, language, music, the arts, interpretive history, religion and recreation (Hogue, 1987). Most people recognize that insects have important economic value which is well documented for critical ecosystem services provided by insects. Most information is available for four such services (i.e., dung burial, pest control, pollination and 'wildlife nutrition', which is estimated to be worth about US $60 billion per year (Losey and Vaughan 2006); but we should also consider the economic value of insects as provided through insect festivals as a subset of insect tourism and recreation.

The recent increase in popularity of insect festivals represents a growing public and professional interest in insects and their connections to nature and people. Insect festivals can nurture community pride and solidarity; generate economic impacts; educate festival visitors; increase interactions among entomologists, professional amateurs (see Pearce, this volume), and the general public; and raise resources to conserve insects. Furthermore, many organizations, governments and individuals already support insect festivals. Insect festivals and programmes like NatureServe Canada's *The Status and Conservation of Butterflies in Canada* 'promote and support public butterfly education activities such as wildlife and butterfly festivals' (Hall 2009: 7; see Figure 13.4).

If, on face value, insect festivals provide a net benefit to society and to the environment, it is helpful to understand how to manage and maintain current festivals. Organizers must ensure careful planning, management, communication, and evaluation to realize a festival's benefits (Delamere 2001; Millar 2003; Delamere 2008). Like all festivals and events, insect festivals should be evaluated according to their objectives (Brunson 2002; Carlsen *et al.* 2001). Careful analyses will allow organizers to decide on how, or if, these festivals will be offered in the future. An analysis of festival failures suggest many reasons, including weather, lack of corporate sponsorship,

Figure 13.4   Identifying a dragonfly in the field. Photo by
Glen Hvenegaard.

overreliance on one source of money, inadequate marketing and
lack of advance planning (Getz 2002). In addition, more work is
needed to develop appropriate planning and management capacities
to support sustainable festivals, according to the triple bottom line
approach of economic, social/cultural, and environmental dimensions (Getz and Andersson 2008; O'Sullivan and Jackson 2002). To
the triple bottom line perspective could be added a fourth element,
the political dimension (Delamere 2008). Indeed, birding festivals,
as examples of wildlife festivals in general, do not have a clear relationship with sustainability objectives and practices (Lawton 2009).
Last, insect festivals should play a role in developing a conservation ethic that can be applied to insects, in terms of intrinsic value,
ecosystem integrity or human utility. This ethic should consider a
variety of temporal, spatial, and cultural perspectives, and address
insects from the individual, species, or habitat scales (Lockwood
1987; Samways 1990).

ACKNOWLEDGEMENTS

Financial support was provided by two different grants from the Social
Sciences and Humanities Research Council of Canada.

REFERENCES

Barkin, D. (2003) Alleviating poverty through ecotourism: promises and reality in the Monarch butterfly reserve in Mexico. *Environment, Development and Sustainability*, **5**(3–4), 371–382.

Brandon, K. E. (1996) Ecotourism and Conservation: A Review of Key Issues. Environment Department Papers, Biodiversity Series No. 33. Washington DC: The World Bank.

Brown, R. (2011) Fireflies, following their leader, become a tourist beacon. *New York Times*, **16** June 2011, p. A27.

Brunson, M. W. (2002) Evaluating a special nature-based tourism event. *Utah Recreation & Tourism Matters*. Available at: http://www.agmrc.org/media/cms/rf10_6EFE7AB486EC1.pdf.

Carlsen, J., Getz, D. and Soutar, G. (2001) Event evaluation research. *Event Management*, **6**(4), 247–527.

Cleveland Metroparks (2011) Bug City. Available at: http://www.clemetparks.com/events/bug_city.asp, accessed 30 August 2011.

Cooper, C. B., Dickinson, J., Phillips, T. and Bonney, R. (2007) Citizen science as a tool for conservation in residential ecosystems. *Ecology & Society*, **12**(2): 1–11.

Delamere, T. A. (2001) Development of a scale to measure resident attitudes toward the social impacts of community festivals, part II: verification of the scale. *Event Management*, **7**, 25–38.

Delamere, T. A. (2008) Festivals. In *Arts and Cultural Programming: A Leisure Perspective*, ed. G. Carpenter and D. Blandy. Champaign, IL: Human Kinetics, pp. 129–142.

Delamere, T. A. and Hinch, T. (1994) Community festivals: celebration or sell-out? *Recreation Canada*, **52**(1), 26–29.

Derrett, R. (2003) Making sense of how festivals demonstrate a community's sense of place. *Event Management*, **8**(1), 49–58.

Diamantis, D. (1999) The concept of ecotourism: evolution and trends. *Current Issues in Tourism*, **2**(2/3), 93–122.

DiGregorio, L. (2002) Flight path: birding festivals beckon. *Birding*, **34**(1), 77.

Dinaburgskaya, K. and Ekner, P. (2010) Social impacts of the Way Out West Festival on the residents of the city of Göteborg. Unpublished Master's Thesis, University of Gothenburg, Sweden.

Fredline, L., Jago, L. and Deery, M. (2003) The development of a generic scale to measure the social impacts of events. *Event Management*, **8**, 23–37.

Getz, D. (1991) *Festivals, Special Events, and Tourism*. New York: Van Nostrand Reinhold.

Getz, D. (2002) Why festivals fail. *Event Management*, **7**, 209–219.

Getz, D. (2005) *Event Management and Event Tourism*, 2nd edn. New York: Cognizant Communication Corp.

Getz, D. (2008) Event tourism: definition, evolution, and research. *Tourism Management*, **29**(3), 403–428.

Getz, D. and Andersson, T. D. (2008) Sustainable festivals: on becoming an institution. *Event Management*, **12**(1), 1–17.

Great Plano Kiwanis Club (2011) The festival. Available at: http://www.greater-planokiwanis.org/event_details.htm, accessed 9 Sept 2011.

Guiney, M. S. and Oberhauser, K. S. (2008) Insects as flagship conservation species. *Terrestrial Arthropod Reviews*, **1**, 111–123.

Hall, P.W. (2009). *Sentinels On the Wing: The Status and Conservation of Butterflies in Canada*. Ottawa, ON: NatureServe Canada.

Hartley, D. (2005) Birding festivals. *Birding*, **37**(1), 34–37.

Higginbottom, K. and Booth, R. (2003) Contributions of non-consumptive wildlife tourism to conservation. In *Nature-based Tourism, Environment and Land Management*, ed. R. Buckley, C. Pickering and D. B. Weaver. Wallingford, UK: CABI, pp. 181–195.

Higginbottom, K. and Tribe, A. (2004) Contributions of wildlife tourism to conservation. In *Wildlife Tourism: Impacts, Management and Planning*, ed. K. Higginbottom. Altona, Australia: Common Ground Publishing [for] Cooperative Research Centre for Sustainable Tourism, pp. 99–123.

Hogue, C. L. (1987) Cultural entomology. *Annual Review of Entomology*, **32**, 181–199.

Hou, J. (2000) Cultural production of environmental activism: two cases in southern Taiwan. Paper submitted for The Fifth Annual Conference on the History and Culture of Taiwan. Available at: http://www.international.ucla.edu/cira/paper/TW_Jeff_Hou.pdf, accessed 30 August 2011.

Huntly, P. M., Van Noort, S. and Hamer, M. (2005) Giving increased value to invertebrates through ecotourism. *South African Journal of Wildlife Research*, **35**(1), 53–62.

Hvenegaard, G. (1994) Ecotourism: a status report and conceptual framework. *The Journal of Tourism Studies*, **5**(2), 24–35.

Hvenegaard, G. (2011) 'Potential conservation benefits of wildlife festivals', *Event Management*, **15**(4), 373–386.

Hvenegaard, G. and Manaloor, V. (2007) A comparative approach to analyzing local expenditures and visitor profiles of two wildlife festivals. *Event Management*, **10**(4), 231–239.

Janiskee, R. L. and Drews, P. L. (1998) Rural festivals and community reimaging. In *Tourism and Recreation in Rural Areas*, ed. R. Butler, C. M. Hall and J. Jenkins, West Sussex, UK: John Wiley and Sons Ltd., pp. 157–175.

Kadoya, T., Suda, S. and Washitani, I. (2004) Dragonfly species richness on man-made ponds: effects of pond size and pond age on new established assemblages. *Ecological Research*, **19**(5), 461–467.

Kim, S. A., Kim, K. M. and Oh, B. J. (2008a) Current status and perspective of the insect industry in Korea. *Entomological Research*, **38**, S79-S85.

Kim, Y., Kim, S. S. and Agrusa, J. (2008b) An investigation into the procedures involved in creating the Hampyeong Butterfly Festival as an ecotourism resource, successful factors, and evaluation. *Asia Pacific Journal of Tourism Research*, **13**(4), 357–377.

Lawton, L. (2008) Are US-based birding festivals a form of ecotourism? In *Proceedings of the 18th Annual Council for Australian University Tourism and Hospitality Education (CAUTHE) Conference. Tourism and Hospitality Research, Training and Practice: 'Where the Bloody Hell are we?* ed. S. Richardson, L. Fredline, A. Patiar and M. Ternel. Gold Coast, Australia: Griffin University, pp. 203–209.

Lawton, L. (2009). Birding festivals, sustainability and ecotourism: an ambiguous relationship. *Journal of Travel Research*, **48**(2), 259–267.

Lawton, L. and Weaver, D. B. (2010) Normative and innovative sustainable resource management at birding festivals. *Tourism Management*, **31**(4), 527–536.

Lemelin, R. H. (2007) Finding beauty in the dragon: the role of dragonflies in recreation and tourism. *Journal of Ecotourism*, **6**(2), 139–145.

Lemelin, R. H. (2009) Goodwill hunting? dragon hunters, dragonflies & leisure. *Current Issues in Tourism*, **12**(3), 235–53.

Liddle, M. (1997) *Recreation Ecology*. London: Chapman.

Lockwood, J. A. (1987) The moral standing of insects and the ethics of extinction. *The Florida Entomologist*, **70**(1), 70–89.

Losey, J. E. and Vaughan, M. (2006) The economic value of ecological services provided by insects. *BioScience*, **56**(4), 311–323.

Louv, R. (2008) *Last Child in the Woods: Saving our Children from Nature Deficit Disorder*, updated and expanded edn. Chapel Hill, NC: Algonquin Books.

Mayfield, T. L. and Crompton, J. L. (1995) Development of an instrument for identifying community reasons for staging a festival. *Journal of Travel Research*, **33**(3), 37–44.

Millar, N. S. (2003) *How to Organize a Birding or Nature Festival*. Colorado Springs, CO: American Birding Association.

National Fish and Wildlife Foundation (1999) *1999 Directory of Birding and Nature Festivals*, Washington DC: National Fish and Wildlife Foundation.

Nonaka, K. (2010) Cultural and commercial roles of edible wasps in Japan. In *Proceedings of a Workshop on Asian-Pacific Resources and Their Potential for Development. Food Insects as Food: Humans Bite Back,* ed. P. B. Durst, D. V. Johnson, R. N. Leslie and K. Shono. Bangkok, Thailand: United Nations Food and Agriculture Organization, pp. 123–130.

Oak Grove Nature Center (2011) 3rd Annual 'Bug Fest. Available at: www.caravannews.com/issues/2011/April/13.htm, accessed 30 August 2011.

Oberhauser, K. S. and Prysby, M. D. (2008) Citizen science: creating a research army for conservation. *American Entomologist*, **54**(2), 103–105.

O'Sullivan, D. and Jackson, M. J. (2002) Festival tourism: a contributor to sustainable local economic development? *Journal of Sustainable Tourism*, **10**(4), 325–342.

Palisade International Honeybee Festival (2011) 3rd Annual Palisade International Honeybee Festival. Available at: http://palisadehoneybeefest. com/uploads/Welcome_Vendors_2011_PIHB-1.pdf, accessed 30 Aug 2011.

Pestival (2011) Pestival: the art of being an insect. Available at: http://pestival. org, 6 Sept 2011.

Picard, D. and Robinson, M. (2006) Remaking worlds: festivals, tourism and change. In *Festivals, Tourism and Social Change: Remaking Worlds*, ed. D. Picard and M. Robinson. Clevedon, UK: Channel View Publications, pp. 1–31.

Pinault, G. (2003) FIFI 2003, fifth international insect film festival. Available at: www.bio.net/mm/dros/2003-May/006412.html, accessed 30 Aug 2011.

Plog, S. (2004) *Leisure Travel: A Marketing Handbook*. Upper Saddle River, NJ: Pearson Prentice Hall.

Primack, R., Kobori, H. and Mori, S. (2000) Dragonfly pond restoration promotes conservation awareness in Japan. *Conservation Biology*, **14**(5), 1553–1554.

River Bend Nature Center (2011) BugFest! Available at: http://riverbendnaturecenter.wordpress.com/?s=bug&submit=Search, accessed 30 August 2011.

Rollins, R. and Delamere, T. A. (2000) A critical assessment of the role of festivals in rural communities. Paper presented at the Rural Communities and Identities in the Global Millennium Conference. Nanaimo, BC: Malaspina University-College.

Ross, S. and Wall, G. (1999) Ecotourism: towards congruence between theory and practice. *Tourism Management*, **20**(1), 123–132.

Samways, M. J. (1990) Insect conservation ethics. *Environmental Conservation*, **17**(1), 8.

Saunders, C. D. (2004) Partnering for Ohio butterflies: an evaluation of educational events and temporary exhibits about butterflies at three zoos.

Available at: http://www.butterflyrecovery.org/education/docs/EvalFestivals.pdf, accessed 30 August 2011.

Sekercioglu, C. H. (2002) Impacts of birdwatching on human and avian communities. *Environmental Conservation*, **29**(3), 282–289.

Singh, T., Slotkin, M. H. and Vamosi, A. R. (2007) Attitude towards ecotourism and environmental advocacy: profiling the dimensions of sustainability. *Journal of Vacation Marketing*, **13**(2), 119–134.

Slotkin, M. H. (2003) Ecotourism in practice: birding and wildlife festivals. In *How Science Can Support Environmental Protection: Florida Tech-BME Partnership Programme Yearbook 2003*, ed. G. Nelson and I. Hronszky. Budapest: Ariszlolelesz Publishing Co., pp. 57–69.

Small, K. (2007) Social dimensions of community festivals: an application of factor analysis in the development of the social impact perception (SIP) scale. *Event Management*, **11**, 45–55.

Smith, A. M. and Sutton, S. G. (2008) The role of a flagship species in the formation of conservation intentions. *Human Dimensions of Wildlife*, **13**(2), 127–140.

Walo, M., Bull, A. and Breen, H. (1996) Achieving economic benefits at local events: a case study of a local sports event. *Festival Management and Event Tourism*, **4**(3–4), 95–106.

Walpole, M. J. and Leader-Williams, N. (2002) Tourism and flagship species in conservation. *Biodiversity and Conservation*, **11**(3), 543–547.

Weaver, D. (2002) Asian ecotourism: patterns and themes. *Tourism Geographies*, **4**(2), 153–172.

Weaver, D. and Lawton, L. J. (2007) Twenty years on: the state of contemporary ecotourism research. *Tourism Management*, **28**(5), 1168–1179.

Yi, C., He, Q., Wang, L. and Kuang, R. (2010) The utilization of insect resources in Chinese rural area. *Journal of Agricultural Science*, **2**(3), 146–154.

14

Glow-worm tourism in Australia and New Zealand: commodifying and conserving charismatic micro-fauna

C. MICHAEL HALL

INTRODUCTION

Although more usually associated with Waitomo Caves in New Zealand, snare-forming glow-worms (*Arachnocampa* spp.) are an important element of tourism in both Australia and New Zealand. The larvae of all glow-worms construct mucus tubes from which they hang a snare or web of silk and mucus to capture prey that is attracted by its bioluminescence (Baker 2002a; Richards 1960). In Australia glow-worm tourism has been described as 'a multi-million dollar industry, thereby making glow-worms a commercially valuable organism' (Baker 2003: 13). In New Zealand Waitomo Caves attracts on average more than 400 000 tourist visits annually, and in summer visitor numbers rise to approximately 2000 people per day. However, despite their longstanding role as a tourist attraction relatively little research has been conducted into their tourism significance or their management as compared to other charismatic fauna and flora. This chapter therefore provides an overview of the different glow-worm species in Australia and New Zealand, their role in tourism and the development of appropriate visitor management strategies.

GLOW-WORMS

Australian and New Zealand glow-worms are the larval stage of a primitive fly (fungus gnats) (Diptera: Keroplatidae: *Arachnocampa* spp.)

The Management of Insects in Recreation and Tourism, ed. Raynald Harvey Lemelin. Published by Cambridge University Press. © Cambridge University Press 2013.

which, in their larval stage, are glow-worms. Glow-worms are found in a variety of habitats – such as caves, mine tunnels, railway tunnels, rainforest banks and damp overhangs – but always within areas of very high humidity (Baker 2002a). In terms of natural environments they are mostly found in caves, or sheltered locations in rainforests, particularly along riverbanks, where there is protection from wind. Eight species are present in Australia; five of them have been recently described (Baker 2010; Baker *et al.* 2008) and one in New Zealand (Table 14.1). All of the recently described species are geographically related to existing or recent distributions of temperate, tropical and sub-tropical rainforest.

Baker (2010) described a new subgenus, Arachnocampa (Lucifera) subgen. nov., to include the Tasmanian species, *Arachnocampa tasmaniensis* and the Mount Buffalo glow-worm, *Arachnocampa buffaloensis*. The new subgenus is separated from the subgenera Arachnocampa (Arachnocampa) and Arachnocampa (Campara) by differences in wing venation (Baker 2010; Baker *et al.* 2008). The subgenus Arachnocampa now includes only the New Zealand species *A. luminosa* which differs from species of Lucifera and Campara by its method of vertical pupal suspension (Richards 1960).

Larvae, the glow-worm stage, have the longest life span of all of the life stages. Egg development in *A. flava* takes 10 days with the larvae then living for up to 1 year depending on food availability (mainly small flies and longtails) and environmental conditions. The larvae then pupate for 7–9 days and emerge as adult flies. In contrast to larvae, the adults have very short life spans with females living for 2 days and males no more than 6 (Baker 2003).

A critical factor in the distribution of *Arachnocampa* species is that the adults are poor flyers which therefore restricts their ability to colonize new areas (Baker *et al.* 2008; Baker and Merritt 2003; Richards 1960). However, just as significant is the fragmentation of rainforest habitat occupied by *Arachnocampa* as well as the isolation of wet cave systems. Further rainforest fragmentation as a result of clearance or environmental change would clearly increase pressure on glow-worm populations. The Mount Buller glow-worm is likely a relict from when the area was covered in temperate rainforest.

TOURISM

The notion of charismatic fauna and flora is often applied to species that serve as flagships for conservation efforts (Hall *et al.* 2011;

Table 14.1 *Arachnocampa* spp. *in Australia and New Zealand*

Region	Species	Comments
New Zealand	A. luminosa	Found throughout New Zealand on both the North and South islands. Its Māori name is titiwai, meaning 'projected over water'. The Waitomo Caves on the North Island is an internationally recognized attraction, although there is also significant glow-worm tourism on the South Island.
Tasmania	A. tasmaniensis	Endemic to Tasmania (as the name suggests). The glow-worms in the Marakoopa Cave, Mole Creek Karst National Park, which is part of the Tasmanian Wilderness World Heritage Area, have been a tourist attraction since the late nineteenth century.
South-East Queensland, north-east NSW	A. flava	The Natural Bridge in Springbrook National Park inland from the Gold Coast is one of the most accessible colonies open to visitors.
NSW, southern rainforest	A. richardsae	The Newnes glow-worm tunnel, a disused railway tunnel in the Wolgan Valley in the Wollemi National Park, part of the Greater Blue Mountains World Heritage Area, is one of the most well-known colonies available to visitors.
Victoria, Mount Buffalo	A. buffaloensis (Mount Buffalo glow-worm)	Consists of a colony of Arachnocampa found in a sub-alpine cave on Mount Buffalo in Victoria, part of Mount Buffalo National Park. Because of its restricted distribution the Victorian Government listed it (calling it the Mount Buffalo glow-worm) as a threatened species.
Western Victoria, Otway ranges	A. otwayensis	Described 2010.
Eastern Victoria, Gippsland	A. gippslandensis	Described 2010.
North Queensland tropical rainforest	A. tropica	Described 2010.
South-East Queensland, northern NSW	A. girraweenensis	Described 2010.

Source: derived from Baker *et al.* 2008; Baker 2002b, 2003, 2010; Meyer-Rochow 1990; Pugsley 1983; Wilson *et al.* 2004.

Leader-Williams and Dublin 2000), including insects (Lemelin 2007; New *et al.* 1995). In a broader sense, charisma is closely associated with the attractiveness of something and implies the possession of special qualities (Hall *et al.* 2011). Lorimer (2007) developed a three-part typology of non-human charisma, consisting of ecological, aesthetic and corporeal charisma, which was applied to an exploration of how flagship species in the UK were utilized to achieve biodiversity conservation goals. According to Lorimer (2007) ecological charisma refers to both the frequency at which humans encounter a species and the combination of properties that make an organism readily identifiable to humans. Ecological charisma is therefore an indication of familiarity with an organism. Aesthetic charisma is defined by an organism's appearance and behaviour, and the emotional responses that may be triggered by these in humans (Lorimer 2006). Finally, corporeal charisma refers to the prolonged and specialized interaction of particular groups of people with specific organisms. Glow-worms fulfil all of these aspects of non-human charisma, they are identifiable and have long been recognized by humans, the light from glow-worms means that they are also regarded as an attractive curiosity (Wilson *et al.* 2004). Their common name of glow-worm also makes them potentially more attractive as well as contributes to their conservation. As Baker (2003: 13) commented, 'Being fly larvae, a more literal description of this organism would be "glow-maggot", however there is little chance this name will catch on in the tourism market'.

Glow-worm viewing in its natural habitat is an activity that occurs in a cave or in a rainforest at night-time. However, because of the potential to attract greater number of visitors during the day and to make glow-worms more accessible, Wilson *et al.* (2004) note that some entrepreneurs have created artificial habitats for glow-worms to attract day-time (fee-paying) visitors (Department of Environment and Resource Management (DERM) 2010; Wilson *et al.* 2004).

The only significant Australian study of glow-worm tourists' motivations and experiences is Wilson *et al.*'s (2004) study of visitors to the Natural Bridge glow-worm colony in Springbrook National Park, Queensland, although the region was also subject to a visitor capacity assessment by the Queensland Park and Wildlife Service (DERM 2010) in May 2007. According to Wilson *et al.* (2004) visitor numbers brought to Natural Bridge by commercial tour operators averaged more than 50 000 a year during the study period 2001–03. In addition to the mostly Asian visitors brought by commercial tour operators, the number of independent visitors was around 13 000 per annum for the study period. This is a significant figure and comparable to other commercial species-focused tour operations in south-east Queensland, such as whale watching in Hervey Bay (Wilson *et al.* 2004).

As of 2007, 13 commercial operators held day-time commercial activity permits and seven commercial operators held night-time permits to the Natural Bridge and Mt Cougal sections of Springbrook National Park and Joalah (Curtis Falls) section of Tamborine National Park in the Gold Coast hinterland of south-east Queensland (DERM 2010). Table 14.2 provides an overview of the glow-worm tourism sites available in the region as of May 2007. The most visited of these by far is Natural Bridge where the cave is home to one of the largest glow-worm colonies in Australia (DERM 2010). As of 2007 Natural Bridge received approximately 280 000 visits per year. Of these about 180 000 or 65% were made during the day and 99 000 or 35% at night. Most day visitors were not part of commercial groups. In contrast, most night visitors are part of commercial tours. 'At peak times the site can have up to 260 persons at one time, and on a peak day as many as 1800 people may have visited, arriving by car, motorbike and various buses up to 50-seater coaches' (DERM 2010: 18). In 2005–06, commercial operators conducted night tours for over 80 000 visitors, a significant increase over the Wilson *et al.* (2004) figures. Changes in commercial visitation levels reflected seasonal peaks in the Asian tourism market. In contrast, commercial day use of Natural Bridge (23 000 visits) was much less but was still significant and growing despite a generally static regional tourism industry (DERM 2010). According to DERM (2010: 6): 'In addition to the fees paid for access, Tourism operators have contributed to the better management of the national parks by sponsoring research and contributing to site facilities and supervision'.

The local significance of glow-worm tourism in the Gold Coast hinterland is moderated by the finding by Wilson *et al.* (2004) that only a very small number of visitors stayed in Springbrook because of the glow-worms and no extra days were spent in the area because of their presence. The amount of money spent by surveyed tourists within a 25 km radius of Springbrook was also small and the economic impacts from this form of tourism is not significant to the immediate local economy. However, Wilson *et al.* (2004) did note that although visitor figures were not available, there were two private properties in Springbrook that offer glow-worm viewing facilities during the day and at night they charge a fee and maintain a restaurant and conduct other activities (e.g., pottery displays) to 'add value' to their business

Table 14.2 *Glow-worm sites in the Gold Coast Hinterland as of May 2007*

Location	Location	Description	Viewing Time
National Parks			
Natural Bridge	Springbrook National Park	The most popular national park site for night-time glow-worm tourism because the display of bioluminescence is reliable all year. This feature makes the site highly viable for tour operations marketed in advance and overseas. Entering a cave to see glow-worms is also a highly desirable experience that 'adds value' to the tour. In summer some colonies visible along banks to side of walking track. Access via 1 km return circuit bitumen walking track. Adjacent picnic area. Receives about 280 000 visits per year. Of these, about 180 000 or 65% were made during the day and 99 000 or 35% at night. At peak times, the site can have up to 260 persons at any one time, and on a peak day as many as 1800 people may have visited.	Night viewing. Thirteen commercial tour operators with day permits and seven with night permits.
Mt Cougal	Springbrook National Park	Glow-worms can be seen at night in the earthen banks to the side of a rainforest walking track. Limited display. There is no primary glow-worm concentration. A 1.1 km return on bitumen walking track. About 60 000 people visit Mt Cougal each year – almost all general visitors. At peak times, there can be up to 300 people in a day, and up to 80 people at one time.	Night viewing. Very few commercial tour operators visit at night and during 2004–06 no passenger returns were lodged for night tours.
Joalah	Tamborine National Park	Glow-worms occur in the banks under and to the left of Curtis Falls. They can be seen from the viewing platform which looks out to the falls. Because they are far more exposed than at Natural Bridge, their prevalence is affected by weather and season. The display can vary from a few lights to many. Graded earth walking track, non-circuit, 1.6 km return.	Night viewing. Not used for commercial tours as of 2007.

		About 173 000 people visit Joalah each year with tour operators or as general visitors. At peak times the site can have up to 250 persons at once, and the total for a peak day could be 800 people.	
Palm Grove	Tamborine National Park	Glow-worms in earth bank on rainforest walk. Limited display. More a spot-light walk with the chance of seeing some glow-worms. A 3 km graded earth walking track circuit. Small picnic area with BBQ and seating.	Night viewing.
Box Forest Track	Lamington National Park	Steep gully just before Picnic Rock, 3 km from O'Reilly's Guesthouse. Glow-worm City (marked on old maps) on the Box Forest Track 3.5–4 km from O'Reilly's.	Night viewing.
Private natural conditions			
Star Pools	Tamborine Mountain Road, North Tamborine	Natural glow-worm colony on private land.	Night viewing.
O'Reilly's	Morans Creek	Three spots on high banks of Morans Creek. One spot is 120 m above the Wishing Tree, one 180 m below it and another a further 80 m downstream. The latter now has a micro- watering/misting system to maintain glow- worm numbers.	Night viewing.
Private artificial conditions			
Springbrook	Forest of Dreams	Artificial cave for day and night glow-worm viewing.	Day and night viewing by appointment.
Tamborine	Cedar Creek Estate Vineyard and Winery	Artificial cave with reversed day–night cycle for daytime glow-worm viewing. 10 am – 4 pm	

Source: Derived from DERM 2010.

operations. Of the 18.6% of visitors who spent money within 25 km of the site, the maximum amount spent per person was AUS $40 and the minimum amount was AUS $1.70. The average amount was AUS $12.20. Nevertheless, as Wilson *et al.* (2004: 21) noted, 'this does not mean that free glow worm viewing at Springbrook National Park is not a valuable economic asset. Commercial tour operators' fees for the excursion that involves glow worm viewing are high and judging by the number of visitors...the revenue generated by this activity in the Gold Coast region/southeast Queensland region must be considerable'.

It was found that the main reason for visiting the glow-worms was to entertain visitors (25.5% of respondents). Curiosity was ranked second (25.1%) followed by the star-like features of the glow-worm display (22.3%). The vast majority (98%) of respondents were satisfied by their visit. The knowledge of glow-worms of the majority of visitors was also reasonably high (although there was uncertainty regarding some specific questions such as 'what do the adult glow worms look like'. Importantly for future conservation efforts, the majority of visitors were willing to purchase a booklet that explained the biology and ecology of glow-worms (Wilson *et al.* 2004).

Doorne (1999, 2000) undertook a study of the visitor experience at Waitomo Glowworm Cave. According to Doorne (1999) only 54% of the respondents to his survey knew anything about the Glowworm Cave prior to visitating. Doorne's (1999) results suggested a high level of satisfaction with the Glowworm Cave and glow-worms but there was significant dissatisfaction with other elements of the product including the number of groups in the cave (e.g., 45% of respondents that were surveyed in summer were not satisfied); waiting for other groups during the tour (e.g. 44% in summer were not satisfied); and facilities such as toilets and the shop. Asian groups (Korea, Japan, Taiwan) were the least satisfied with waiting either before the tour, during the tour or waiting for the boat). In general, satisfaction levels were lower during the summer period than in winter. This was especially noticeable for elements relating to the level of visitation including:

- waiting before the cave tour;
- waiting to buy cave tickets;
- the size of the cave tour group;
- the number of groups in the cave; and
- waiting for other groups in the cave.

Such results highlight the importance of developing good management practices not only for glow-worm conservation but also for the visitor experience.

Although the overall number of tourists who engage in glow-worm tourism is unknown there is a large number of tourism businesses which promote glow-worm tours. Table 14.3 provides an overview of glow-worm tourism businesses via a search of New Zealand tourism websites undertaken in June 2011. The table indicates that glow-worm tours are located in five locations in the South Island, of which Te Anau is the most significant in terms of visitor numbers and economic impact (Warren *et al.* 2000). Ten North Island locations promote glow-worm tourism, of which Waitomo is the most important. Waitomo attracts of the order of 450 000–500 000 visitors a year, although estimates vary (Doorne 1999; Pavlovich and Kearins 2004). For example, Tourism Holdings Limited (THL) which operates the Waitomo Glowworm Caves, Ruakuri Cave, Aranui Cave and The Legendary Black Water Rafting in the Waitomo region, claims on their corporate website that these businesses host 'close to a million international and domestic tourist to the area each year' (THL 2011). However, Waitomo is the only destination for which glow-worms are the primary attraction, as Table 14.3 indicates for many businesses and regional tourism organizations offer a much wider range of products in order to remain commercially viable. Nevertheless, the numbers of tourists and tourism operations that promote glow-worm tourism in New Zealand and Australia does highlight the need for the development of appropriate visitor management strategies.

VISITOR MANAGEMENT AND CONSERVATION

Glow-worms can be affected by tourism in a number of different ways:

- direct disturbance of glow-worms
- impact on glow-worm prey
- effects on glow-worm habitat and environments.

These effects can occur at both the larval and fly stages.

Baker (2003) reported that experiments indicated that larvae were negatively affected by torchlight. Larvae moved away from the torchlight while switching off their own light source. Larvae took up to 10 minutes to turn their light back on, thereby decreasing the

Table 14.3 *New Zealand tourism businesses promoting glow-worm tourism as of June 2011*

Location	Business	Other products
North Island		
Kaitaia, Northland	Eco Valley	Bushwalks, encounters with glow worms, weta, freshwater crayfish and kiwi birds
Lake Karapiro, Waipa District	The Boatshed Kayaks	Kayak tours, kayak rentals, kayak sales
Mangakino, Waikato	Tranquility Tours	Kayak hire, guided trips, instruction and courses, fitness and exercise training, overnight camping
Pauanui Beach, Coromandel	Wincorp Adventures	Scenic coastal tours, nature walks, 4WD safaris, guided hunting trips, outdoor adventure course
Tauranga, Bay of Plenty	Waimarino	Kayak shop, kayak tours, kayak school, after school programme, adventure park
Tauranga, Bay of Plenty	Canoe and Kayak Ltd	Online shop, kayak club and magazine, courses and licensing
Tauranga, Bay of Plenty	Adventure Bay of Plenty	Team building activities, school programme, rotorua tours, wildlife tours, kayak tours, blokart, mountain biking tours, cruise ship tours, boat cruises
Te Kuiti, Waikato	Green Glo Eco-adventures	Caving, abseiling, rock climbing
Thames, Coromandel	Te Mata Lodge	Swimming and kayaking, birdwatching, rockhounding and gold panning, shop
Waiomio Valley, Northland	Kawiti Caves	

Waitomo, Waikato	Discover Waitomo (owned by Tourism Holdings Limited) (Includes Waitomo Glowworm Caves, Ruakuri Cave, Aranui Cave, The Legendary Black Water Rafting Co.)	Visitor centre, restaurant, café, gift shop, theatre, exhibition centre, conference facilities
Waitomo, Waikato	Waitomo Adventures	Abseiling, caving, tubing, blackwater rafting, rock climbing
Waitomo, Waikato	Spellbound Glowworm and Cave Tours	
Waitomo, Waikato	Cave World	Abseiling, tubing
Wellington	Wellington Botanic Gardens	
South Island		
Catlins, Southland	Catlins Tours (owned by Pounawea Grove Motels)	Accommodation, sightseeing tours, kayaking, penguin excursions
Catlins, Southland	Catlins Adventures and Personalised Tours	Scenic tours, kayaking adventures, yellow eyed penguin tour, catlins river walk
Dunedin, Otago	Glowworm Tours (Otago Cycle Hire)	Cycle Hire and bike tours, backpacker camping packages
Karamea, West Coast	Oparara Guided Tours	Guided bush walks, guided kayak tours
Te Anau, Fiordland	Real Journeys Ltd	Cave tour includes cruise across Lake Te Anau and informative displays at Cavern House
Westport, West Coast	Norwest Adventures Limited	Underworld rafting, adventure caving, Charleston Nile River rainforest Train

density of the display for following tourists. Correlations between climatic data and the number of glow-worms glowing at both a heavily visited tourism site and a non-visited site revealed similar overall fluctuations, indicating weather was the major factor involved in population crashes at particular times of the year (Baker 2003). Population change appears closely related to seasonal and climatic conditions including rainfall, temperature and relative humidity (Baker 2002a; Driessen 2009). In Tasmania, Driessen (2009) monitored glow-worm populations for 24 months in both Exit and Mystery Creek Cave. He found a strong seasonal pattern, with pupae and adults most common in spring and summer. The increase in numbers of pupae and adults coincided with an increase in the number of prey caught in silk threads produced by the larvae. Larvae were present throughout the year but the number glowing varied both seasonally and spatially. In Mystery Creek Cave, the number of larvae glowing was generally highest during summer and autumn and lowest in winter and early spring. In Exit Cave, there was no consistent seasonal pattern in the number of larvae glowing among sites, and overall there was less variation between monthly counts than at Mystery Creek Cave. Driessen (2009) speculates that this difference in seasonal patterns between the two caves may be due to climatic difference, with Mystery Creek Cave possibly experiencing a greater drying out of the cave air in winter than Exit Cave. However, such research also highlights the need for careful management of visitation to glow-worm caves because of the potential impacts on temperature and humidity, as well as disturbance of prey.

In the case of Waitomo Caves Schmekal and de Freitas (2001) found that air exchange between the cave and the outside can increase or decrease condensation rates depending on warm-season and cool-season differences and by the opening and closing of the solid door at the upper entrance to the cave. As a result of their research a set of management guidelines was developed to maintain appropriate temperatures, condensation and humidity levels in different zones of the cave. In non-cave environments visitors will have very little effect on temperature and humidity levels. Instead, general climatic and weather conditions will have the most impact. In such circumstances the greatest tourism-related risk is from inappropriate tourist behaviour (Baker 2002b, 2003) as well as poor design of tourist access and facilities. For example, at Natural Bridge in Queensland, 'Damage in the past has or may have been caused by inappropriate [general public visitor] behaviours such as fires in the cave and the unwitting use of insect repellents' (DERM 2010: 26).

In the case of Queensland the overall objectives for the management of glow-worm commercial operations on National Parks are to:

- provide fair and equitable access to the cultural and natural resources of glow-worm sites to all user groups;
- ensure visitor safety and amenity;
- ensure that commercial guided tours are ecologically sustainable and promote the conservation values and objectives of the protected area;
- develop a regulated, self-managed system for commercial operations which provides maximum visitor through-put, subject to environmental and social quality and sustainability;
- provide a level of market security and return to commercial operators;
- ensure that a variety of quality goods and services are provided by commercial tour operators;
- continue communication between [the Queensland Parks and Wildlife Service] and tour operators to improve the awareness of roles, needs and problems of each party (DERM 2010: 6).

With respect to Natural Bridge in Springbrook National Park probably the most visited glow-worm colony in Australia, for example, site management recommendations to underpin its sustainable visitor capacity numbers covered seven key areas (DERM 2010):

- Glow-worm habitat protection
- Visitor safety
- Size and layout of the carpark
- Flow of visitors at the site
- Activities permitted at the site
- Managing impacts on neighbouring properties and the community
- Managing general public visitors.

In the case of Natural Bridge, although the annual use levels of the site at night was considered sustainable, DERM (2010) recommended changes in the number of people at the site at the same time (Persons at One Time or PAOT) should be reduced: 'To achieve a more predictable and even PAOT it is recommended that Commercial Tour Operator (CTO) groups and General Public Visitor (GPVs) arrivals at night be scheduled between 6 and 10 pm with a peak commercial period between 7 and 9 pm. This creates quieter shoulders from 6–7 and 9–10pm which will be promoted to GPVs as a more suitable time to visit' (DERM 2010: 2). Importantly, such measures may also have a positive benefit for tourism because of the potential to increase levels of tourist satisfaction by minimizing the likelihood of overcrowding (Doorne 1999). In the case of the Waitomo Caves Doorne (1999) recommended a number of basic approaches to try and manage congestion issues including developing a visitor management process driven by quantifiable objectives (and surveys); introducing logging and reporting systems for guides and staff; and standardizing traffic management practices.

CONCLUSION

Luminescent insects such as glow-worms are a form of charismatic micro-fauna that come out of the pages of Hans Christian Andersen or hark back to myths, legends and stories of enchanted forests and caves. The continuing attraction is witnessed in the large numbers of tourists who visit glow-worm colonies in Australian and New Zealand caves and forests. However, the development of glow-worm tourism presents the same problems to managers and conservationists as any other form of nature-based tourism in that there is a need to ensure that visitation does not damage the very species and biodiversity that attracted tourists in the first place (Hall 2010).

In Australia, in particular, substantial efforts are being put in place to manage visitor access to glow-worm colonies, many of which occur in national parks. This includes sponsorship by governments and the tourism industry of biological research to develop appropriate management strategies (Baker 2002b), that have proven significant for improving the sustainability of glow-worm populations as well as commercial operations (DERM 2010). In New Zealand research has been more focused on a single location, Waitomo Caves (Doorne 1999; Schmekal and de Freitas 2001), which although significant because of its large numbers of tourists represents only one of a number of areas in the country where glow-worm tourism is promoted. Indeed, arguably a longer-term task in glow-worm conservation is going to be developing management strategies in forest and riverbank locations in which conservation agencies have a lower presence and consequently greater difficulties in controlling access. Yet the interest of many people in glow-worms and their somewhat romantic nature may well prove the key to their survival with their long-term potential for direct and indirect tourist expenditure to provide financial support for their conservation being critical. In a world where the vast majority of species are not conserved for their intrinsic right to exist but for the values they provide to humans, the glow-worm is likely to not only remain the subject of fairy tales but also have a happy ending.

ACKNOWLEDGEMENT

With great thanks to Jody Cowper-James for assisting with the survey of New Zealand tour operators.

REFERENCES

Baker, C. H. (2002a) Dipteran glow-worms: Marvelous maggots weave magic for tourists. *Biodiversity*, **3**(4), 23–27.
Baker, C. H. (2002b) A biological basis for management of glow-worm populations of ecotourism significance. University of Queensland Honours Thesis, Gold Coast: CRC Sustainable Tourism, Common Ground Publications.
Baker, C. H. (2003) Australian glow-worms: managing an important biological resource. *Australasian Cave and Karst Management Association Inc. Journal*, **53**, 13–16.
Baker, C. H. (2010) A new subgenus and five new species of Australian glow-worms (Diptera: Keroplatidae: Arachnocampa). *Memoirs of the Queensland Museum*, **55**(1), 11–41.
Baker, C. H. and Merritt, D. J. (2003) Life cycle of an Australian glow-worm *Arachnocampa flava* Harrison (Diptera: Keroplatidae: Arachnocampinae). *Australian Entomologist*, **30**(2), 45–55.
Baker, C. H. and Merritt, D. J. (2004) Management of Australian glow-worms (Diptera: Keroplatidae: *Archnocampa* spp.): identification of threats and types of glow-worm tourism. In *XXII International Congress of Entomology. Entomology Strength in Diversity*, ed. J. LaSalle, M. Patten and M. Zalucki. Brisbane, Australia, 15–21 August. Brisbane: Carillon Conference Management.
Baker, C. H., Graham, G. C., Scott, K. D. *et al.* (2008) Distribution and phylogenetic relationships of Australian glow-worms *Arachnocampa* (Diptera, Keroplatidae). *Molecular Phylogenetics and Evolution*, **48**(2), 506–514.
Department of Environment and Resource Management (DERM) (2010) *Gold Coast Hinterland Glow-worm Sites. Sustainable Visitor Capacity, May 2007.* Brisbane: Queensland Parks and Wildlife Service Department of Environment and Resource Management
Doorne, S. (1999) *Visitor Experience at the Waitomo Glowworm. Cave.* Science for Conservation 95. Wellington: Department of Conservation.
Doorne, S. (2000) Caves, cultures and crowds: carrying capacity meets consumer sovereignty. *Journal of Sustainable Tourism*, **8**, 116–130.
Driessen, M. M. (2009) *Baseline Monitoring of the Tasmanian Glow-worm and Other Cave Fauna Exit Cave and Mystery Creek Cave, Tasmania.* Nature Conservation Report 09/02, Hobart, Australia: Biodiversity Conservation Branch, Resource Management and Conservation Division, Department of Primary Industries and Water.
Hall, C. M. (2010) Tourism and biodiversity: more significant than climate change? *Journal of Heritage Tourism*, **5**(4), 253–266.
Hall, C. M., James, M. and Baird, T. (2011) Forests and trees as charismatic mega-flora: Implications for heritage tourism and conservation. *Journal of Heritage Tourism*, **6**(4), doi:10.1080/1743873X.2011.620116.

Leader-Williams, N. and Dublin, H. (2000) Charismatic megafauna as 'flagship' species. In *Priorities for the Conservation of Mammalian Diversity: Has the Panda had its Day?* ed. A. Entwistle and N. Dunstone. Cambridge: Cambridge University Press.

Lemelin, R. H. (2007) Finding beauty in the dragon: the role of dragonflies in recreation and tourism. *Journal of Ecotourism*, **6**, 139–145.

Lorimer, J. (2006) Non-human charisma: which species trigger our emotions and why? *Ecos*, **27**(1), 20–27.

Lorimer, J. (2007) Nonhuman charisma. *Environment and Planning D: Society and Space*, **25**, 911–932.

Meyer-Rochow, V. B. (1990) *The New Zealand Glowworm*. Waitomo, New Zealand: Waitomo Caves Museum Society Inc.

New, T. R., Pyle, R. M., Thomas, J. A., Thomas, C. D. and Hammond, P. C. (1995) Butterfly conservation management. *Annual Review of Entomology*, **40**, 57–83.

Pavlovich, K. and Kearins, K. (2004) Structural embeddedness and community-building through collaborative network relationships. *Management*, **7**(3), 195–214.

Pugsley, C. W. (1983) Literature review of the New Zealand glowworm *Arachnocampa luminosa* (Diptera: Keroplatidae) and related cave-dwelling Diptera. *New Zealand Entomologist*, **7**, 419–424.

Richards, A. M. (1960) Observations on the New Zealand glow-worm *Arachnocampa luminosa* (Skuse) 1890. *Transactions of the Royal Society of New Zealand*, **88**, 433–436.

Schmekal, A. A. and de Freitas, C. R. (2001) *Condensation in Glow-worm Cave, Waitomo, New Zealand: Management Guidelines*. Department of Conservation Science Internal Series 15. Wellington, New Zealand: Department of Conservation.

Tourism Holdings Limited (THL) (2011) THL Businesses. Online. Available at: www.thlonline.com/THLBusinesses/default.aspx, accessed 18 September 2011.

Warren, J., Taylor, N. and McClintock, W. (2000) *Resource Community Formation and Change: A Case Study of Te Anau*. Working Paper 27, Christchurch, New Zealand: Taylor Baines & Associates.

Wilson, C., Tisdell, C. and Merritt, D. (2004) Glow Worms as a Tourist Attraction in Springbrook National Park: Visitor Attitudes and Economic Issues. Working Papers on Economics, Ecology and the Environment No. 105, St. Lucia: School of Economics, University of Queensland.

Part IV Conservation Frontiers

15

May you live in interesting times: technology and entomology

FORREST L. MITCHELL

INTRODUCTION

Anyone involved in technology would agree that such times are upon us. Whether it is a curse or not is, of course, a matter of opinion but the scope of change due to the application of science in our societies is unprecedented. Opportunities to incur change have never been in the reach of more participants or presented to a larger – and growing – audience. Any field of endeavour is subject to the sweep of information technology and the presentation of entomology and insects is prospering. Humanity and insects are involved on many different levels and the association has not often been a pleasant one. Insects have been getting a bad press since biblical times, where the book of Exodus records three of the ten Egyptian plagues being caused by insects (gnats, flies and locusts). It is hard to overcome the innate fear many people have of insects or just creepy-crawlies in general. In the past this suspicion bled off in some measure to the people who studied them, making entomologists almost as much of a curiosity as the insects. The popular view of entomologists as cartoon characters come alive, complete with aerial nets and pith helmets, is disappearing (mostly) to be replaced with or at least complemented by forensic investigators, environmental activists, molecular biologists, agricultural consultants, pest managers, ecologists of all stripes and a multitude of other highly trained specialists who rely on the science of entomology. The classical practitioners of systematics and taxonomy supported by large, well-curated insect collections are still very much in need, though even they spend more time in the laboratory than in the field.

Entomology as a pastime activity is often found associated with insect collecting and insect collections have been the point of engagement for many students (see Pearson, this volume). The well-appointed Victorian household could be expected to have a glass-topped case of showy butterflies or beetles mounted somewhere in the sitting room or library (see Dodd, this volume). Nets have been constructed of just about every cloth available. Many a seamstress' supply of sewing pins has been raided by budding collectors, eager to acquire a collection themselves. Unfortunately, this is often where the passion was cut short. Catching insects is the easy part though it may not seem like it at the time. Dispatching, mounting and maintaining them in a collection requires training and until the recent past, finding a mentor could be difficult. The Lepidoptera are usually the first target of nascent collectors and these are about the most difficult of the insects to properly kill, pin and spread. Reading how to do this in a book is not the same as having someone to demonstrate and to assist in locating or building the necessary pinning boards and display cases. Failure to understand just what dermestid beetles are capable of or how the smell of mothballs can permeate a sibling's room can instigate abrupt shifts to more passive interests, like stargazing. Fortunately, personal instruction is where information technology has made some of the aforementioned sweeping changes in culture and the topic of collections can serve as the illustration.

Since the early part of the twentieth century, the Boy Scouts of America (BSA) have provided structured studies in natural history and offered a series of merit badges for scouts to advance in rank up to the final goal of Eagle Scout. In 1923 a merit badge for Insect Life was offered and involved, among other tasks, the collection and presentation of 50 insects of different species. 'I can remember our frantic quest of bugs, cyanide jars in hand as we wandered through the woods,' reminisces Martin J. Hillenbrand (1998) in his autobiography. 'Our objective was the Insect Life merit badge, one of the most difficult in the panoply of available merit badges.' Ambassador Hillenbrand's fond memories, in this case from the late 1920s and early 1930s, were shared by many young men over the years. Instruction in collecting insects and creating and maintaining the personal collection came via a guide provided by the BSA and mentoring from its leaders and members.

In 1985 the merit badge changed from Insect Life to Insect Study and in 2009 the collection requirement was dropped in favour of a scrapbook that requires the scout to perform a detailed study of 20 different insect species in their habitats. This is to include photographs, illustrations, sketches and articles. This change in approach is due in part to the disfavour that collecting of any sort has aroused in society, but also reflects the increased information available for identification and study of many of the insects encountered by new students of entomology. Rather than incorporating an interesting insect in a collection, the student has the option of photographing the specimen in the field or catching, photographing and releasing it. Good images can easily identify a specimen to order and often to species via entomology-oriented websites, online keys or downloadable applications. Parents also have peace of mind knowing that their children are not doing anything with cyanide.

In the past, finding a mentor to develop a collection was difficult, but getting identification of an insect species could be even more so. A common recourse was to the county extension agent, who might be able to provide an identification or who sponsored 4H educational meetings. 4H is a nationwide organization in the US meant to develop school age students' interest in a number of science, agriculture and health programmes. The extension services of many states are involved with this (details can be found at www.4-h.org). In instances where the agent was at a loss, there was a network of extension specialists affiliated with a state university to fall back on. This could become a slow process especially with uncommon species, and was not likely to be encouraged extensively. In some instances, perhaps, there was an identification guide for support. Most books relied on skilled technical illustrators working from preserved specimens when there were species illustrations at all. Standouts such as Holland's (1908) *Guide to the Moths* are still being reprinted. For less popular orders of insects sometimes the only available text was a technical work from the local library and so even an elementary grasp of the insect group was kept tantalizingly out of reach of the amateur. Details on rearing or keeping insects would be sketchy or absent. For someone interested in entomology, an insect collection was therefore likely to be the best prospect for getting to know the insects, likely assembled over the course of work for a BSA merit badge (which left out half the population) or a county 4H group, specimens identified to order and then the collection set aside.

For many decades then entomology did not receive popular attention among the general public due to being hampered by the lack of information developed for a beginner, the imagery needed to bring it to a wider audience and the support of like-minded individuals. The arrival of the information age, fuelled by the rise of computational capability and driven by digitally linked imaginations, began a movement across multiple fronts: easily accessed information via hyperlinked transport protocols, quality imagery from inexpensive and filmless cameras, email, cellular connectivity and a myriad other means of allowing groups of enthusiasts to first find and then communicate their collective knowledge.

Digital imagery is the driving technological element behind the rising popularity of entomology in public appreciation. Fortunately, it is not so much the technology, but the insect itself that may be presented in a new light or at a new level of detail or caught in interesting behaviours that catches someone's eye or captures their imagination. Until recently and in spite of the ubiquity of insects, few people had a chance to identify even common species, let alone take a close-up view. When presented with sharp, colourful imagery of the insects around them and the detail in compound eyes, wing scales or antennae, the observer may be seeing something new that subtly alters their perception of insects (see Figure 15.1). In the aggregate and over time, these changes can transform a negative opinion or bring a positive opinion into fascinated engagement.

Of the many means of presenting insects, photography is notable for the large amount of technological change that has occurred in the last 15 to 20 years. Digital cameras, once an unaffordable and specialized item are now common and have revolutionized the spread of entomology into the popular consciousness. Rather than simply being assigned as a pest or put into general categories (bees, butterflies, dragonflies), insects are now routinely captured into image files and posted on websites around the world. Images are reviewed by everyone from experts in taxonomy to someone whose picture of the same species was identified also and are anxious to spread the knowledge. Interest groups form, especially on sites devoted to gardening, and observations proliferate. Where information on a relatively common but poorly covered insect species such as a crane fly (Diptera: Tipulidae) could once be difficult to locate, now there is an immense amount of imagery and easily accessible information as close as the nearest web-enabled appliance.

Cameras fall into two general groups: point and shoot cameras that automatically control the many needed inputs for a desirable image and 35 mm single lens reflex (SLR) cameras. A third type, compact mirrorless cameras that have interchangeable lenses, are just now

Figure 15.1   Illinois river cruiser (*Macromia illinoiensis*).

being introduced. Point and shoot cameras are capable of capturing impressive images of insects and other arthropods. Originally considered little more than expensive toys, camera control software and lens quality have improved dramatically. Current resolution reaches 16 megapixels (Mpx) and more than 10× actual zoom factors (as opposed to a digitally interpolated, often degraded zoom). Now that most cellular phones and personal music players have such cameras embedded, anyone with expertise at insect identification will find themselves popular as more and more insects are serendipitously imaged. Just as a figurine collection in a curio cabinet invites more additions from friends and relatives, entomological knowledge invites imagery. Positive feedback changes the casual observer to the curious observer and now the camera is more likely to come out when anything remotely insect-like crosses their path. So much so that insect photography can become a recreational activity or part of another recreational event or activity.

In the case of a digital SLR, manufacturers adapted 35 mm optics technology pioneered for film cameras to modern charge coupled device (CCD) based cameras. Along with the supporting software and microelectronics, these cameras are continually updated and capture images of ever increasing size and quality. Current cameras average

10–36 Mpx, up to as much as 80 Mpx in a single image. Digital SLR cameras have become more accessible to the average budget and because the cost and time constraints imposed by film are not involved, less intimidating. The technical hurdles to getting a good result remain, but the image is available for immediate review and camera settings changed on the spot for improvement on the next try, rather than waiting sometimes for days to see that everything was perfect, except for an errant finger in front of the lens.

One technological advance of interest to entomologists is the reduced signal to noise ratio in camera CCD chips. This has been vastly improved so that sensors are less noisy in low light, meaning an image taken in low light will be less grainy. Currently, ISO settings to over 100 000 are possible in high-end cameras, and near noiseless images at ISO settings in four digit ranges can be obtained. Higher ISO settings mean that faster shutter speeds with smaller f-stop apertures and deeper focus can be used with less compromise in image quality. Insects are often in motion and many are the images that recorded merely a blur or were compromised by wind blowing an otherwise still specimen on a grass stem or leaf. Even on a bright sunny day, using tight f-stops of 16 or higher to try and capture an active insect with a macro lens may require so fast a shutter speed that a low ISO of 100 or 200 may yield a dark image. Being able to raise the sensitivity past ISO 400 without additional electronic noise to weed out in post-processing is very useful.

Another means of achieving high levels of detail and deep fields of view is by image stacking. This technique is gaining popularity as a recreational outlet because of the lower cost of computing power and the development of optimized algorithms for handling the huge number of calculations needed to remove unfocused pixels from an image. A group of images, referred to as a stack, are taken at sequential depths of field, aligned and recombined by software (see the resources section) into a single, focused image. Though a rather specialized leisure activity, enthusiasts across the world are engaged in using microscopes, camera lenses on bellows, macro lenses and creative lighting to obtain images of spectacular clarity, colour and depth of field (see Photo Macrography and Amateur Micrography websites). Subjects at high magnification are normally difficult to observe due to the shallow depth of field. By taking a series, sometimes hundreds, of digital images through a focal range, a stack is obtained for processing into a single, focused image. Here the technique is as much the point of the exercise as the photographed subject and if the subject is an insect or other arthropod, its identity often is not known to the photographer. The submitted results might be for identification as well as viewing by the rest of the world. The photography equipment, arrangements and lighting are as individual as the photographers. Photos of the elaborate setups are studies in the leading edge of image acquisition. Very often they are ad hoc contrivances, and may include plastic Lego™ blocks and duct tape in the building materials, but there is no arguing with the resulting imagery. For the less adventuresome there are commercial turnkey products that mount on a standard tripod and are equipped with stepper motors that advance the camera in small increments (see the resources section). The controllers for these units can be programmed or connected to a computer running the stacking software. Images are stored in the onboard camera memory card, although some cards may also have an embedded network transmitter that can send images directly to the computer drives for subsequent use, thus reducing the steps between acquisition and processing (Eye-Fi™ SDRAM cards discussed below).

Image source is irrelevant to the software and excellent results can be returned from an inexpensive point and shoot camera on a small tabletop tripod, although a remote shutter release and a focusing rail are useful accessories. The camera is focused through the subject in small steps and an image acquired at each step to obtain a stack. Pinned insects work well as subjects. Living insects seldom hold still long enough without cooling or other pacifying means, although moths waiting out the day on a wall or porch screen are an exception.

While advancements in image stacking are driven mainly by user groups and companies, another means of making a single image via a collection of multiple images has a more centralized origin and a clearer trail in the scientific literature. GigaPan technology (Nichols *et al.* 2009) was used to create the immersive panoramic scenes widely publicized during exploration of the Mars surface by the unmanned NASA rovers Spirit and Opportunity. In this instance, individual photographs of the area around the rover and extending to the visible horizon were stitched together to make a single image. The viewer can pan the finished frame backwards to make a single, visible scene or pull in the view and continue forward until fine details of the individual frames emerge and minute exploration of rocks or soil is possible. Though developed for space exploration, this technology has many uses on the home planet (Frenkel 2010). One of the biggest hurdles to conquer is the large size of the resulting images. Nichols *et al.* (2009)

provide illustrative detail: in their study a 704 image pan took approximately 7 hours for a desktop workstation to compile and a further 3 hours of upload time via a 1.5 megabit internet connection to the online server. Serving and using such huge images are beyond the average user and present special problems that were overcome by the grant funded GigaPan Project. Established at the GigaPan website, it provides free storage and web serving for these large files. The viewing software is supported by the website, but web location coordinates for the images can be embedded in external sites and image creators may present the results on their own web page. Nearly 55 000 images are currently available for public viewing and over 1000 contain the keyword 'insect'.

While it is possible to obtain the images manually, part of the GigaPan Project was to develop a low cost motorized camera mount that removed price as an impediment to using the technique (Nourbakhsh 2010). The mounts, which are referred to as GigaPan robotic imagers (Nourbakhsh 2010), have onboard controllers that rotate the camera vertically and horizontally. Different models are available that support a range of cameras from simple point and shoot models up to heavily loaded SLR systems. The size and detail of the final images are determined by a number of user selections, both on the mount and the camera. Among the largest posted so far is a panorama composed of *c.* 12 000 images and containing over one terapixel.

In addition to developing the mounts, software (GigaPan Stitch) for compiling the images into a result compatible with the GigaPan website has also been developed. Similar tools are also available in popular image management software such as Adobe Photoshop™.

Panoramic landscape views were the original goal of this technology and that application is still a useful choice where entomology is concerned. Insects and invertebrates are part of the landscape and because component images are captured at full zoom, it is possible to see insects and other invertebrates in GigaPan images that on the surface appear to be on far too large a scale. For example, an image taken at the Cockrell Butterfly Center and Insect Zoo in Houston Texas has two obvious tropical walking stick insects (Order Phasmida) in the frame, but there are also butterflies to be seen on the foliage and flowers.

Panoramas do not have to be on grand scales. An innovative use of this technology is employed by North Carolina State University. Whole insect drawers from the university entomology museum are being captured into single GigaPan images. The museum drawers are sorted phylogenetically so users may more easily find the group of interest. Once the drawer is identified, individual specimens in the collection may be viewed in fine detail. Though there are limitations (collection labels tags cannot be seen for most specimens), the images are freely accessible and the collection can be perused without having to pull and open drawers.

Individual insect specimens may also be subjected to the GigaPan process. A cherry gall azure butterfly (*Celastrina serotina*) has been imaged to micron level detail using a new prototype system. The level of detail is extraordinary and is as sharp as images acquired through the stacking process. Using the GigaPan servers, both the top and bottom of the spread butterfly are available for viewing and may be selected from the user interface. Other entomological applications, such as detailed views of honeycomb affected by colony collapse disorder may be seen. References to other entomological images may be found on the Small World Explorations website and in Frenkel (2010).

Another, more specialized means of insect image acquisition is by use of flatbed scanners (Mitchell and Lasswell 2000, 2005). It is specialized because, unlike a camera, there is no shutter for quick image acquisition and the specimen has to hold still for the duration of the scan. The depth of field is limited to approximately 8 mm, the usual distance from the top of an insect pin to the top of the pinned specimen. Within these constraints however, images of remarkable clarity can be acquired. The technique was used to obtain images of dragonflies (Odonata: Anisoptera), a group notoriously difficult to capture digitally. In the wild, specimens are seldom perpendicular to a camera lens so that capturing a close-up image completely in focus is rare. Further, the intricate venation of the wings is often lost in the background or obliterated by glare. Dragonflies are not showy specimens in collections because the colour fades quickly, especially in the eyes. Scans of living individuals result in capture of wing detail, colour and eye detail. With care, the subject can be released if appropriate. A dorsal scan and a side scan usually are sufficient to identify the dragonfly to species. The scans are also consistent in detail from one individual to the next, unlike photography. Further, since the pixels (dots) per inch or DPI is known, physical measurements can be taken from the scans. Scanning obtains a very clear, uncluttered image. The colours are life-like, detail is minute and many of the technical requirements to obtain a similar photograph are automatically handled by the software. The aesthetic of the final result is all centred on the insect as a specimen and while it can be clinical in its starkness, the beauty of

Figure 15.2   Blue-eyed darner (*Rhionaeschna multicolor*).

the insects draws the eye of the viewer (see Figure 15.2). Photography of insects is more versatile and can capture not only the subject, but behaviour, environment, background and other ecological information that the photographer may not have even noticed at the time. The scanning protocols discussed here were developed in the 1990s, before the advent of modern cameras that handle as many of the technical elements as the user chooses. Would-be photographers whose creative streak was inhibited by the details needed to effectively operate a film camera could find at least a partial outlet with an affordable scanner. Scanning insects is not an impulsive activity, however, and one does not run for the scanner rather than the camera. A combination of scanning and photography together can provide a significant amount of information. With practice and patience, it is possible to build a digital insect collection that requires backup rather than mothballs. As with traditional collections, specimens must still be acquired, arranged and labelled.

One of the most valuable assets in a collection is the specimen label, without which much information of scientific interest is lost. Location and date of capture are very important in developing knowledge about insect species. Image cataloguing programs such as Adobe Lightroom™ and Apple Aperture™ have significant documentation capabilities that may be used when the images are downloaded into their databases. Many cameras record a host of details in EXIF files embedded with the image. These can be preprogrammed in the camera to record additional information, such as expedition location. The increased availability of global positioning (GPS) sensors has been taken advantage of by some camera makers to allow geotagging of images with exact longitude and latitude data, a development far more refined than the simple country, state and county information on a label. Though still in its infancy, the technology is poised to become a standard feature or upgrade for many cameras. One example is the Nikon GP-1 GPS unit which adds satellite based GPS coordinates to a range of cameras in the Nikon line and is accurate to within 10 m. Another protocol for obtaining location coordinates involves identifying a Wi-Fi transmitter site or hotspot whose location is recorded in a central database. The Sandisk Eye-Fi™ SDRAM card contains an embedded wireless transceiver and uses this means of geotagging images captured by the host appliance. Image files may be uploaded immediately to websites that display their locations on a map. The card will also transmit to appliances with Wi-Fi capability, such as a smartphone, tablet or notebook computer. As this technology matures, combinations of cell towers, Wi-Fi hotspot location references and dedicated GPS chips will inform a multiplicity of software, hardware and third party add-ons and plug-ins that will make geographic location information as readily available as date and time stamps on every digital image captured.

On seeing an eye-catching insect, the obvious next question is that of identity (Penev *et al.* 2009; Walter and Winterton 2007). In an age where conservation and environmental ethics are permeating more extensively through human society, people now more than ever are looking for such answers. It is a question under debate in scientific circles as well, where the perceived slow speed of traditional species descriptions and lack of biosystematic support, especially in the tropics, is referred to in frustration as the 'taxonomic impediment' (Cigliano and Eades 2010). Even when an insect has been described, attempting to ferret out obscure species names can be difficult for a trained entomologist, let alone someone who is not. For the nature lover or vacationer attempting to put names to insects seen during a trip, this can seem almost impossible.

Traditional couplet keys are a mainstay of entomology (Walter and Winterton 2007). The user is given a pair of morphological state or state combinations to choose from and advances stepwise into the identification process. Such keys may fall short when the choice of states can only be determined by a user with advanced training or when a wrong choice is made. Difficulties can also occur when an incomplete specimen is all that is present or perhaps an image that only presents a dorsal or lateral view. One new approach that can assist in narrowing the search for an insect's identity is a matrix key (Walter and Winterton 2007). Rather than a paired choice approach, matrix keys work through a number of character states simultaneously and are flexible enough to accommodate a wrong choice or partial information. A matrix key to the orders of insects is presented on the Discover Life website. It contains 21 states that have from two to eight choices for each. The choices have numbers beside them indicating how many orders contain the state and small illustrations that help in making a decision. In choosing a housefly (*Musca domestica*) as an example, it can take up to seven selections to narrow the options to a choice between Diptera and Hemiptera if the antennae are absent. If the antennae are present, and the correct choice made, then only one selection is needed to find the order.

Knowing the order of an insect has more value than having the same information for a vertebrate order, where fewer species are present to choose from. Being told that a bat is in the order Chiroptera is not really going to help in the search for a species, while knowing a fly is in the order Diptera will prevent much fruitless searching among the other orders of insects. Achieving a family level identification will help even more. Having an ordinal identification in hand, a person trying to make a determination can turn to a field guide if one is available or to one of the many websites devoted to insects and entomology. Bug Guide is one such resource and provides interaction between people at all experience levels of familiarity with insects and arthropods. Instead of keys, the Wiki approach is employed and relies on the dedication of participants to be effective. Starting with the order or family of the insect, the user screens through images until an insect similar to their own appears. The disadvantage of the picture-screening route of identification is that there may be many images to go through and many similar looking species, especially if the unknown specimen is small. On the other hand, many new and interesting insects will be seen.

The increasing demand for open access journals such as ZooKeys (Penev *et al.* 2008) means that more professional articles are available to the general public via web access. Such articles have the flexibility of linking maps, images, descriptions, DNA sequences and other relevant information to names or terms within the body of text. Employing these assets collectively in the discussion of taxa and description of species is termed cybertaxonomy (exemplified by Winterton 2011). The majority of such publications may be beyond the interpretation of the average person, but open access journal articles will still appear in search engine results where someone has succeeded in identifying their insect and is looking for more information. Distribution maps and images linked to a species name or genus can be of value even to relatively naïve users.

In addition to insect collecting or imaging, there is also the build-your-own option for the enterprising individual. Modelling in general is a human passion that extends to many areas of interest and entomology is no exception. Miniaturization of electronics and integrated circuits continues to the point that small model insects are now feasible. The forefront of robotic insect development is military in nature, with small flying models patterned on hawkmoths (Lepidoptera: Sphingidae) and dragonflies (Odonata) being envisioned as scouting tools or even small combatants. Even so, there are a number of recreational outlets involving microelectronics as insect models. One design plan that has wide acceptance is the BEAM philosophy (Hrynkiw and Tilden 2002). BEAM is an acronym: Biology, Electronics, Aesthetics, Mechanics. It was developed in an attempt to simplify robotics and bring an imposing technical field to a broader lay audience. It also appeals in the use of biology as a pattern and the arthropod body as a common blueprint.

First developed in the early 1990s, small, autonomous insect-like rovers were introduced as means of planetary survey for NASA. The bugbot Attila and its hexapod successor Genghis were microprocessor controlled robots that reacted to their environment rather than needing AI to direct their motions (Brooks and Flynn 1989). Though outsized for an insect, they were tiny by comparison to the popular idea of robots as upright canisters. Each of the six legs was controlled by a microprocessor that received feedback from the leg as it attempted to traverse its environment. By using low-level responses in hierarchical order, the robots were able to move over obstacles autonomously and explore the area around them. Public interest was high and the small rovers captured imaginations. Today, in addition to bugbots, there are bristlebots, junkbots, vibrabots, solarbots, beambots, photovores, solar rollers, symmets, headbots and bots on wheels. Numerous clubs and groups have formed and are devoted to building and developing small-scale robots, many with an arthropod-like appearance.

One of the simplest examples is a bristlebot that uses a detached toothbrush head for a body, a tiny pager motor for movement and a small coin battery for power. Also known as a vibrabot, these small assemblages skitter across smooth surfaces, bouncing off obstacles and heading in unpredictable directions. Observing one in action brings to mind a small sea creature or perhaps a silverfish scurrying about its business. Various toothbrush bristle conformations envisioned for entirely different reasons by manufacturers will change the movement characteristics of the robot.

A major goal in robot design is self-sufficiency and batteries are a weak, if not the weakest, link. One approach to circumventing the problem is by building the robot around solar engines. These are modules made from solar cells measuring as little as 9 cm^2, although some cells may be as small as 25 mm^2. Solar cells of this size do not normally supply a steady power source for even the smallest of motors, though some motors are so small that they can use cylindrical fuse holders as motor mounts.

The quality and strength of incidental light affect the output of the solar engine. When exposed to full sunlight they produce relatively large amounts of power, while those under incandescent lights produce power more slowly. Capacitors are used to store the trickle of energy coming from the engine and then discharge it at the capacitance rating, providing a burst of energy that is predictable in amount, if not time. The energy is used to directly power the motors, normally one on each side of the chassis. This affects the motions of the robot, which moves in quick, jerky bursts, very like those of insects and spiders. By changing the surface area of the solar cell and the size of the capacitors, designers can build solar engines to change the movement characteristics or size of the robot.

A common function built into robots with solar engines is light seeking. Using photodiodes as optical sensors on each side of the chassis, power to the capacitors may be modified such that those nearest the light source fill more slowly, causing motor(s) on the brighter side to lag behind those on the dimmer side. The robot then begins to edge toward the light source. Since the motors on each side are not synchronized, the robot approaches the light in a zigzag motion as the sensors on each side are swung back and forth into the light path.

Obstacle avoidance is another function important to robotic movement. A simple form is to have stiff wires protruding from the robot that are connected to a switch. When an obstacle pushes against the wire, it engages a contact switch and power to the motor(s) opposite the switch are cut or reduced. The motor on the same side of the chassis as the switch will remain active and swing the robot away from the obstacle. The wires may be short and stubby or long and sweeping, but either way they resemble antennae and give the robot a much more insect like resemblance.

The pleasure taken from robots like these may be more in the making than in the observing, as the simplicity or slow movement may not inspire more than passing interest in some people. For those not interested in soldering, wing-flapping, radio controlled dragonfly and butterfly models are currently on the market, but flying or agile scale-model insects are still in the conceptual or prototype stage, even for military-sized budgets. Tantalizing hints appear in articles and webpages from time to time, making such robots seem more likely to be available for popular entertainment than the robotic butlers and imaginative servants populating the pages of magazines from the last century. With the pool of resources and imaginations growing, the only real question left is when the next one will crawl into public perception like its forebears, Attila and Genghis.

Examples in this chapter present interesting case studies in the speed of technological change. Even a short time ago, techniques such as image stacking or GigaPans would not have been available except to a specialized few. Moore's Law (Moore 1965) has long predicted the exponential increase in processing power, but the benefits realized from these increases also spill into other areas of electronics, such as the development of the CCD chip. The relatively low cost of a robotic imager or a camera and lens or perhaps a microscope (found as surplus on eBay) allows high quality image acquisition systems to be assembled. The personal computer, memory, video card, video display and disc storage needed for the images and to process images is also within reach of individuals. Software is reasonably priced or even free. The combination of these three sets of technologies in the hands of the general public would have seemed impossible less than a decade ago. The many ways these multitudes of technologies and their interactions have played out could scarcely have been imagined and are even now proceeding at a prodigious pace. Application of new technologies, faster hardware and increased bandwidth will take unique and unexpected forms that will be developed, enhanced and then popularized through the social aspects of the internet.

A recent article in the popular press concerning a prehistoric huntsman spider (Araneae: Sparassidae) hidden for millennia in a dark piece of amber was impressive not only for the technology, in this instance X-ray tomography, but the fact that it is reported in layman's terms and the images accompanying the article presented in high resolution three-dimensional detail. The author of the piece is accessible by email and may be followed on Twitter. All the information can be accessed by a smartphone in the middle of a national park over a cellular network. It speaks volumes that this capability is already taken for granted and that tomorrow it will be passé. What will come next?

REFERENCES

Brooks, R. A. and Flynn, A. M. (1989) Rover on a chip. *Aerospace America*, **27**(10), 22–26.

Cigliano, M. M. and Eades, D. (2010) New technologies challenge the future of taxonomy in Orthoptera. *Journal of Orthoptera Research*, **19**(1), 15–18

Frenkel, K. A. (2010) Panning for science. *Science*, **330**, 748–749.

Hillenbrand, M. J. (1998) *Fragments of Our Time: Memoirs of a Diplomat*. Athens GA: University of Georgia Press.

Holland, W. J. (1908) *The Moth Book: A Popular Guide to a Knowledge of the Moths of North America*. New York: Doubleday.

Hrynkiw, D. and Tilden , M. (2002) *JunkBots, Bugbots, and Bots on Wheels: Building Simple Robots with BEAM Technology*. Maidenhead, UK: McGraw-Hill Osborne Media.

Mitchell, F. L. and Lasswell, J. L. (2000) Digital dragonflies. *American Entomologist*, **46**(2), 110–115.

Mitchell, F. L. and Lasswell, J. L. (2005) *A Dazzle of Dragonflies*. College Station, Texas: Texas A&M University Press.

Moore, G. E. (1965) Cramming more components onto integrated circuits. *Electronics* **38**(8), 114–117.

Nichols, M. H., Ruyle, G. B. and Nourbakhsh, I. R. (2009) Very-high-resolution panoramic photography to improve conventional rangeland monitoring. *Rangeland Ecology and Management*, **62**(6), 579–582.

Nourbakhsh, I. R. (2010) Gigapixels for science. *IEEE Robotics & Automation Magazine*, doi: 10.1109/MRA.2010.936948.

Penev, L., Erwin, T., Thompson, C. *et al.* (2008) ZooKeys, unlocking Earth's incredible biodiversity and building a sustainable bridge into the public domain: From 'print-based' to 'web-based' taxonomy, systematics, and natural history. *ZooKeys*, **1**, 1–7.

Penev, L, Sharkey, M. and Erwin, T. *et al.* (2009) Data publication and dissemination of interactive keys under the open access model. *ZooKeys*, **21**, 1–17.

Walter, D. E. and Winterton, S. (2007) Keys and the crisis in taxonomy: extinction or reinvention? *Annual Review Entomology*, **52**, 193–208.

Winterton, S. (2011) Revision of the stiletto fly genera *Acupalpa* Kröber and *Pipinnipons* Winterton (Diptera, Therevidae, Agapophytinae) using cybertaxonomic methods, with a key to Australasian genera. *ZooKeys*, **95**, 29–78.

RESOURCES

Image stacking

http://www.photomacrography.net/ (accessed 21 September 2011).
http://www.amateurmicrography.net/ (accessed 21 September 2011).
Cognisys Inc. Automated Macro Photography (Stackshot Rail) http://www.cognisys-inc.com/stackshot/stackshot.php (accessed 21 September 2011).

Zerene image stacking software: http://zerenesystems.com/stacker/ (accessed 21 September 2011).
CombineZP image stacking software http://www.hadleyweb.pwp.blueyonder.co.uk/ (accessed 21 September 2011).
Helicon Focus image stacking software: http://www.heliconsoft.com/ (accessed 21 September 2011).

GigaPan

GigaPan website: www.gigapan.com (accessed 23 July 2012).
GigaPan Robotic Imagers: http://www.gigapansystems.com/cms/shop/software/ (accessed 23 July 2012).
GigaPan Stitch software: http://gigapansystems.com/cms/shop/software/ (accessed 23 July 2012).
Cockrell Butterfly Center Image: http://www.gigapan.org/gigapans/31802/ (accessed 23 July 2012).
North Carolina State University Entomology Museum: http://insectmuseum.org/ (accessed 23 July 2012).
Cherry gall azure butterfly: http://gigapan.org/gigapans/64267 (accessed 23 July 2012).
Honeycomb: http://www.gigapan.org/gigapans/fullscreen/58526/ (accessed 23 September 2012).
Small World Explorations: http://smallworldexplorations.com (accessed 23 July 2012).

Geotagging

Nikon GP-1 review: http://www.digitalreview.ca/content/Nikon-GP1-D90-GPS-Accessory.shtml (accessed 21 September 2011).
Eye-Fi SDRAM cards: http://www.eye.fi/ (accessed 21 September 2011).
Insect identificationDiscover Life website http://www.discoverlife.org (accessed 21 September 2011).
ZooKeys: http://www.pensoft.net/journals/zookeys/ (accessed 21 September 2011).
Bug Guide http://www.bugguide.net (accessed 21 September 2011).

Robotics

Bristlebot video: http://www.youtube.com/watch?v=rUSTXUis_ys&feature=related (accessed 21 September 2011).
Solar Robots: http://www.robotroom.com/Solar-Robots.html (accessed 21 September 2011).
http://www.youtube.com/watch?v=YzFCA-xUc8w&feature=relmfu (accessed 21 September 2011).
Solar cells and other parts: http://www.solarbotics.com/ (accessed 21 September 2011).

Huntsman spider article

http://www.livescience.com/14261–49-million-year-spider-face.html (accessed 21 September 2011).

16

Citizen science and insect conservation

KELSEY JOHANSEN AND ALAINE AUGER

INTRODUCTION

The roots of citizen science extend to the beginnings of modern science and the amateur scientists of the 1700s and 1800s in England and Europe (Silvertown 2009). However, the Christmas Bird Count, organized by the National Audubon Society in 1900, and the British Trust for Ornithology, founded in 1932, are generally considered the earliest documented citizens science projects (Hand 2010; Silvertown 2009). The popularity of citizen science projects has been increasing steadily throughout the late twentieth and early twenty-first centuries, and Cohn (2008) now estimates there are between 500 to 1000 such projects under way throughout North America alone (Scientific American 2012). Today citizen science partnerships use volunteers to collect data for projects on studies ranging from climate change to invasive species, conservation biology and ecological restoration, to air and water quality monitoring, to population ecology (Cohn 2008; Silvertown 2009). With the advent of new technologies, individuals can use laptops, tablets and iPhones to participate in these studies (Kim *et al.* 2011). iPhone applications like *Creek Watch*, suggest Kim *et al.* (2011), have removed some of the past physical barriers associated with citizen science, and in some cases, promoted the participation in these studies, by new volunteers.

In terms of insect conservation, thousands of people across North America gather each year to monitor monarch butterfly migration (Cohn 2008; Oberhauser and Prysby 2008; Wells 2010). In Europe,

The Management of Insects in Recreation and Tourism, ed. Raynald Harvey Lemelin. Published by Cambridge University Press. © Cambridge University Press 2013.

volunteers monitor ants (Marrou 2008), while in Asia citizens monitor cicadas and ladybugs (Hatsuyado 2002). Elsewhere, the Lost Ladybug Project (The Lost Ladybug Project 2011), the Urban Ant Collector (AntWeb 2012) and Odonata Central (Odonata Central 2012), all use technology to connect researchers and citizen scientists across the world. Location is less relevant for those projects that use technology to enlist and assist volunteers with collecting geographically disperse data. The goal of all these projects is to establish inventories and monitor insect activities, foster a dialogue between researchers and non-researchers, as well as engaging and educating participants through experiential activities.

In this chapter we will explore the history, and development, of modern-day citizen science. We then discuss the benefits of citizen science and examine some of the challenges. Next, we explore the benefits of citizen science for enhancing research related to insects, including an examination of a sample of contemporary insect-related citizen science projects. We conclude by discussing the implications of citizen science in insect conservation.

DEFINING CITIZEN SCIENCE

According to Silvertown (2009: 1) 'A citizen scientist is a volunteer who collects and/or processes data', in this way they act as a research or field assistant to professional scientists engaged in a variety of scientific studies and enquiries (Cohn 2008). In other words, citizen science is the process of using volunteers to conduct scientific projects (Dickinson *et al.* 2010). These volunteers in turn collect data from across diverse locations and habitats over prolonged periods of time (Bonney *et al.* 2009) and even sometimes process data (Silvertown 2009). Modern-day citizen scientists are mostly amateur volunteers, with at least a basic understanding of the scientific process (Cohn 2008). They can also include members of conservation groups, hobbyist, amateurs, outdoor recreationists and even in some cases volunteer tourists who pay to participate in such activities, all of whom have a vested interest in conservation (Cohn 2008).

Silvertown (2009) notes three main factors responsible for the recent growth in citizen science. The first is the ease of use of the internet and other means for informing and recruiting volunteers, and the availability of technical tools for gathering data (Silvertown 2009). Secondly, professional scientists are realizing that volunteers, once provided with proper instruction, represent a free source of labour, including data collection, computational power and even in-kind financial resources (Silvertown 2009) that can be leveraged to secure other funding. Third, Silvertown (2009) notes that the growth of citizen science is likely to continue, particularly as funding bodies like the Natural Science and Engineering Council (Canada), National Science Foundation (United States) and the Natural Environment Research Council (UK) begin to require recipients to engage in scientific outreach as part of the dissemination process. Even more interesting is the fact that organizations like the National Science Foundation, who provides funding for citizen science projects under their mandate 'to increase public awareness of and participation in science … are more interested in the educational values than the research results' (Ucko cited in Cohn, 2008: 193).

To accomplish project goals, most of today's citizen scientists work with professional scientists and research teams; these 'projects … have been specifically designed or adapted to give amateurs a role, either for the educational benefit of the volunteers themselves or for the benefit of the project. The best examples benefit both' (Silvertown 2009: 467). Irrespective of the programme's goals, most projects have professional scientists that in addition to providing training, develop standard procedures, analyse data, and publish associated project findings (Oberhauser and Prysby 2008).

WHY USE VOLUNTEERS? BENEFITS TO SCIENCE, RESEARCHERS AND CITIZENS

Cohn (2008) notes several benefits of citizen science for both the research and volunteer community, including broadening the scope of a study or extent of data collection (researchers), and the development of ecological literacy (citizens). The benefits to science and the research community are extensive. 'Citizen science projects have been remarkably successful in advancing scientific knowledge' (Bonney *et al.* 2009: 977), sometimes providing in excess of tens of millions of observations for single projects on an annual basis (Bonney *et al.* 2009). Geographically disperse volunteers that participate in citizen science projects also provide the opportunity to collect data over larger areas and to monitor species over longer periods of time (Cooper *et al.* 2007; Hand 2010) thus broadening the scope of research (Howard and Davies 2009; Snäll *et al.* 2011). The increased amount of information in turn helps researchers to spot discrepancies in the data when they compare results across geographic areas; it also helps them to distinguish trends, while providing insight into a geographically distinct subpopulation's differences (Bonney *et al.* 2009; Cohn 2008). Additionally, financial constraints can be negotiated, especially by departments or projects that cannot afford expensive graduate students or field technicians (Cohn 2008). Efficiency is also improved through the reduced costs associated with using volunteers to collect data (Crall *et al.* 2010; Ingwell and Preisser 2010) and by receiving data from multiple sources (Dickinson *et al.* 2010; Snäll *et al.* 2011). There is also the potential to receive information from various locations faster due to the use of more data collectors (Rotman *et al.* 2012; Wiggins and Crowston 2012a).

According to Brossarda *et al.* (2005), many citizen science projects are designed with experiential educational goals. As such, benefits for citizens include the provision of numerous educational opportunities and increasing public awareness (Jordan *et al.* 2011; Losey *et al.* 2007), learning how to use scientific equipment and technology (Cohn, 2008), learning more about the scientific process (Cooper *et al.* 2007; Oberhauser and Prysby 2008) and learning data collection protocols (Braschler *et al.* 2010; Oberhauser and Prysby 2008). Despite evidence indicating that public awareness can be increased through these types of projects, there is a lack of information identifying the extent to which public behaviour actually changes with increased education and awareness (Jordan *et al.* 2011).

BUT IS IT REAL RESEARCH? QUESTIONS OF INCLUSION
AND VALIDITY

Despite its history, the expansion of citizen science beyond a few key areas is relatively recent and is accompanied by an associated deficit of representation within the academic literature. Two potential causes for this deficit include the relatively recent advent of the terms 'citizen science' and 'citizen scientists' and that while a number of scientific papers have included data collected by citizen scientists, for example the data generated from Christmas Bird Count, they have not been recognized as such (Silvertown 2009). In fact, researchers at Cornell University, home of the Laboratory of Ornithology and 'one of the leading users and promoters of citizen science' (Cohn 2008: 193) did not begin using the term until the 1990s (Cohn 2008). Secondly, Silvertown (2009) notes that researchers associated with projects that fail to conform to the standard model of hypothesis-testing often experience difficulty in publishing, and report the findings in project reports or government publications. Silvertown (2009 :470) concludes that 'good science can, [and] indeed often does, come from surveys that would probably never be approved by peer-review committees that demand rigorous tests of explicit hypotheses'.

In addition to questions of inclusion, those scientists who conduct or lead citizen science projects are often faced with questions of validity and data quality (Cohn 2008; Crall *et al.* 2010; Dickinson *et al.* 2010). In response to these questions, most projects have improved the training provided to volunteers as a means to reduce common errors associated with observation and identification (Conrad and Hilchey 2011; Dickinson *et al.* 2010). Cohn (2008: 194) notes that the easiest way to ensure that citizen science is accepted by the larger scientific community is to 'write study protocols that take citizen scientists into account' and then test their results for validity. Silvertown (2009) suggests that scholars go a step further and implement several stages to increase the validity of data collected by citizens. These include: well designed and standardized data collection techniques; creating explicit assumptions and articulating a basic hypothesis; conducting external, and expert, validation of the data; and providing feedback to volunteers about their data collection and the project goals in recognition of their participation and contributions (Bonney *et al.* 2009; Devictor *et al.* 2010). 'Pair training' citizen scientists with staff and comparing the data sets collected by each party is one way to test, and improve, the validity of the data collected by citizen scientist (Cohn 2008). Bonney *et al.* (2009) also include the importance of having a strong team of scientists, educators, technologists and evaluators to facilitate the project, as well as the inclusion of both specific and measurable public education goals and educational support materials, in their criteria for sound project planning.

Oberhauser and Prysby (2008) note that even non-citizen science projects, such as those involving data collection by students with varying degrees of investment in the project and its accuracy, can have questionable validity. Improving data validity through standardized protocols and filters is recommended as a means to address data quality concerns (Bonney *et al.* 2009; Dickinson *et al.* 2010; Mayer 2010). Schwartz (cited in Mayer, 2010) states that most of the individuals that participate in citizen science projects do a great job collecting data when given clear instructions because the individuals that volunteer are highly motivated to do the job well so that their contributions are used. Some individuals are so dedicated to the research, and there is such high demand to participate, that they may even pay the researchers to be a part of the process (Mayer 2010); this is especially true if voluntary data collection is combined with recreational or tourism opportunities. Mayer (2010) states that even without standard protocols, the data collected by amateurs was historically considered fairly reliable.

Current research on the actual level of error and bias involved in using citizen scientists is not well understood or addressed (Dickinson *et al.* 2010). A greater understanding of the differences, if any, between the validity of the data collected by professional scientists, students and volunteers, and how to proactively address these concerns, is needed (Bonney *et al.* 2009; Cohn 2008) as are increased means for disseminating the results of research conducted using citizen scientists (Kim, *et al.* 2011). Lepczyk (2009: 316–317) noted that in the research presented at the 2008 Ecological Society of America Annual Meeting 'several speakers presented data from their research illustrating that the quality of the data collected by citizen scientists is of the same quality or better quality than that collected by professional ecologists'. Conrad and Hilchey (2011) reviewed a decade of environmental citizen science and concluded that the benefits of citizen science programmes far outweigh the concerns over data quality.

BENEFITS AND APPLICATIONS FOR INSECT CONSERVATION

There are numerous goals that can be, and have been, achieved through projects that use citizen science. As the list of citizen science projects grows, projects are beginning to focus on a broader array of topics, as well as species (Table 16.1). Despite the fact that insects provide valuable services to the ecological community (Gullan and Cranston 2010; Hoffman Black 2011) substantial gaps exist within the academic literature related to their conservation and management (Hughes *et al.* 2000, Johansen and Lemelin 2010), population sizes and locations (Wild 2011). As such, citizen science projects provide opportunities to advance insect research in ways that were previously inaccessible due to constraints imposed by large geographical research areas and a lack of funding (Araya *et al.* 2009; Newberry 2011).

Table 16.1 identifies in alphabetical order, a sample of the insect-related citizen science projects presently under way. Recognizing that this internet search was by no means exhaustive, 55 citizen science projects, most of these being undertaken in North America, are nevertheless highlighted. The purpose of Table 16.1 is to provide a brief overview of the citizen science projects that are directly related to insects.

Table 16.1 *Citizens science insect projects*

Insect order	Insect(s) involved	Project name (approximate start date)	Country	Project type	Project goals
Coleoptera	Asian longhorned beetle	Beetle Busters	United States	monitoring collect photographs	distribution public awareness
	Fireflies	Firefly Watch (2007)	North America	observing sighting counts	distribution activity conservation
	Ladybugs	The Lost Ladybug Project	North America	sightings photograph discovery	map status conservation
	Ladybugs	Osaka Museum of Natural History	Japan	monitoring observation	education public awareness
	Viburnum leaf beetle	Viburnum Leaf Beetle Project	North America	sighting monitoring	conservation management learning about species
Diptera	Fruit fly	Spotted Wing Drosophila Volunteer Monitoring Network	United States	monitoring	track movement and seasonal biology
Hemiptera	Cicadas	Osaka Museum of Natural History	Japan	monitoring observation	education public awareness

Hymenoptera	Ants	Antareas (2008)	France	monitoring geolocation counts	identify distribution population
	Ants	Antweb- Urban Ant Collector (2010)	United States	monitoring geolocation counts	identify distribution population
	Ants	Bay Area Ant Survey	United States	locating identifying collecting	baseline data
	Ants	Iimbovane Outreach Project	South Africa	monitoring	population diversity and dispersal impact of built environment
	Ants	School of Ants	Global	collect	map native and urban ants science
	Bees	Bee Hunt	North America	photograph rearing	better understanding inventory species distribution
	Bees	Bee Spotter	United States	monitoring photograph	understand bee demographics baseline info on population educate public
	Bees	Black Hills Bee Project	United States	monitoring collecting observation photograph	contribute to understanding discover species

Table 16.1 (*cont.*)

Insect order	Insect(s) involved	Project name (approximate start date)	Country	Project type	Project goals
	Bees	Bumblebee Conservation Trust	UK	sightings surveys photographs rearing	conservation education
	Bees	Bumblebee Nest Survey	North America	surveys sightings	science conservation
	Bees	The Great Sunflower Project	North America	identifying environment development for breeding or feeding	map status decline influence food supply
	Bees	Honey Bee Net	North America	scale hives monitoring	monitoring changes nectar flow plant pollinator interaction relationship to climate change
	Bees	Texas Bee Watchers	United States	environment development for breeding or feeding	awareness knowledge
	Wasp	Wasp Watcher	North America	monitoring	engage and support public with discovery and monitor

Lepidoptera	Butterflies	Big Butterfly Count (2010)	UK	identifying counts	conservation management education
	Monarch butterflies	Bring Back the Monarchs	United States	environment development for breeding or feeding	environmental restoration
	Butterflies/moths	Butterflies and Moths of North America	North America	sighting photograph	inventory species Identify species, location and number promote knowledge
	Butterflies	Butterflies I've Seen	North America	sighting locating identifying	conservation distribution abundance
	Butterflies	Cascades Butterfly Project	United States	monitoring identifying	monitor response to climate change
	Butterflies	Chicago Park District Butterfly Monitoring Program	United States	monitoring identifying	monitor health population help for management
	Monarch butterflies	Florida Butterfly Monarch Network	United States	observation monitoring location description	population status, trends, distribution conserve and mgmt strategies phenology

Table 16.1 (cont.)

Insect order	Insect(s) involved	Project name (approximate start date)	Country	Project type	Project goals
	Butterflies	Illinois Butterfly Monitoring Network	United States	monitoring identifying counts	population trends help for management increase knowledge status distribution geographic priorities
	Butterflies	Maine Butterfly Survey	United States		
	Monarch butterflies	Monarch Larvae Monitoring Project	North America	observation monitoring	status monarch population scientific research appreciation, understanding of insects and scientific process
	Monarch butterflies	Project Monarch Health	North America	monitoring sample collection rearing	help track spread of protozoan parasite
	Butterflies	Project Butterfly WINGS	United States	monitoring counts	compare and trend inquiry, analysis conservation
	Monarch butterflies	Monarch Watch (1992)	United States	tagging environment development for breeding or feeding	track pop trends prediction publish/present

	Butterflies	North American Butterfly Association Butterfly Count	North America	monitoring counts	distribution public awareness
	Butterflies (monarch)	Western Monarch Thanksgiving Count	United States	counts	size of colonies, census monarchs
	Moths	National Moth Recording Scheme (2007)	UK	sightings	baseline
	Moths	Moth Night (2000)	UK	counts	raise awareness
	Butterflies	United Kingdom Butterfly Monitoring Scheme (1976)	UK	collect data support studies monitoring	assess butterfly trends assess impact of climate change snapshot of status awareness identify species, location and number research, QOL
	Butterflies (Vanessa)	Vanessa Migration Project	North America	observation sightings	distribution migration trends
Odonata	Dragonflies/ Damselflies	Arizona Odonates	United States	photograph	inventory species range flight seasons grow body of knowledge

Table 16.1 (*cont.*)

Insect order	Insect(s) involved	Project name (approximate start date)	Country	Project type	Project goals
	Dragonflies	Dragonfly Monitoring Network	United States	monitoring identifying	monitor population health response to land management techniques greater knowledge
	Dragonflies	Dragonfly Swarm Project	Global	monitoring observation sighting	swarm behaviour publish data disseminate info about behaviour
	Dragonflies	Migratory Dragonfly Partnership (2012)	North America	monitoring geolocation observation	science conservation education
	Dragonflies/ Damselflies	Northwest Dragonflier	North America	counts observations identification specimens	distribution migration awareness science
	Dragonflies/ Damselflies	Odonata Central	Global	photograph	distribution biogeography, biodiversity Identify species, location and number map generation
		Ohio Odonata Society Dragonfly Monitoring	United States	monitoring photograph specimens	inventory species Identify species, location and number promote knowledge

Other – Invasive	Invasive insects species	Catskill Regional Invasive Species Partnership (2012)	United States	observation monitoring	distribution protection public awareness
	Multiple Insect Species	First Sightings National Biodiversity Data Centre	UK (Ireland)	sighting counts monitoring	distribution population size phenophases to indicate trends and changes in population size and dispersal
	Invasive insects species	My Invasive	United States	photograph geolocation	map distribution of invasive species
	Invasive insects species	Track Invasive Species	United States	monitoring	phenology predictive modelling
Other	Benthic macroinvertebrates	Jug Bay Macroinvertebrate Sampling	United States	monitoring	water quality
	Mutiple insects species	Nature's Notebook Wildlife Phenology Program (2009)	United States	counts observing	phenology
	Multiple insects species	OPAL Bugs Count	UK	counts geolocation	science distribution how built environment may affect insects
	Benthic invertebrates and larvae	Waterworx Bug Hunts (2000)	United States	identifying counts collecting	snapshot of status environmental health

Table 16.2 *Summary of citizen science insect projects*

Country	#	Insect involved	#	Project type	#	Project goals	#
United States	26	Lepidoptera	20	monitor, observe	40	data collection	71
North America	16	Hymenoptera	14	count, identify	25	education, science	29
UK	7	Odonata	7	photograph	11	conservation	25
France	1	Coleoptera	6	rearing	8		
South Africa	1	diverse insect species	6	geolocation	7		
Japan	1	invasive species	3	specimen, tag	6		
Global	3	other orders	2				

These projects are further broken down in Table 16.2, revealing that the indicator species (Coleoptera, Hymenoptera, Lepidoptera, Odonata) were the most common species examined.

Popular insect-related citizen science projects include monarch butterfly migration monitoring from the Upper Midwest to Texas and the Atlantic Coast of the United States (Cohn 2008; Wells 2010). In 2005, *Journey North*, a citizen science programme asked volunteers 'to record sightings of overnight roosts of monarchs during their fall migration' (Howard and Davies 2009: 279). Using this data, researchers were able to map these migratory routes for the first time and on a continent-wide scale, thus providing information integral to the conservation of monarch migration habitat, and identifying potential areas to protect (Howard and Davies 2009). Insect monitoring projects have also been a component of larger studies, such as the MEGA-Transect project; citizen scientists record when butterflies and other pollinators appear, along with other factors, to monitor and create baseline indicators of the effects of global climate change and other ecological changes along the Appalachian Trail (Cohn 2008). Alternatively, they can also be used exclusively to assist with conservation goals by discovering indicators for climate or ecological change (Dickinson *et al.* 2010; Mayer 2010). Indicator insects (Coleoptera, Hymenoptera, Lepidoptera and Odonata), due to their sensitivity to environmental changes, can act as early warning of ecological changes within an ecosystem (Big Butterfly Count 2012; The Great Sunflower Project 2012).

Citizen science projects have resulted in positive conservation outcomes through increases in volunteer knowledge about the value of insects and insects' interactions with the world and with humans (Braschler *et al.* 2010; Devictor *et al.* 2010). Citizen science projects can accomplish this by increasing citizens' understanding of insect species, providing educational opportunities aimed at increasing public awareness to enhance insect-specific conservation strategies, and by increasing recreational and tourism opportunities associated with insects (Araya *et al.* 2009; Lemelin 2009). Furthermore, the use of volunteer citizen scientists helps to reduce the cost of underfunded research projects (Boettner *et al.* 2000; Hughes *et al.* 2000) that focus on monitoring large insect populations.

Citizen scientists have also been used to conduct insect population counts (Hand 2010; Lemelin 2009) and to 'catch, bag, and tag native bees in a pilot study designed to begin gathering data on bee numbers and status' (Cohn 2008: 196). Population and distribution of insect species, according to the literature, is relatively unknown (Crall *et al.* 2010) and with the potential of citizen science projects to reach further geographic areas without the cost, the gap of knowledge about insect populations, species, and distribution can be addressed (Dickinson *et al.* 2010; Heimbuch 2010). Some citizen science projects have even re-discovered rare insect species, such as the nine-spotted lady beetle, *Coccinella novemnotata*, found 'in early October 2006 in Arlington, Virginia. The specimen ...[was] the first individual collected in eastern North America in over fourteen years and ... [was] only the sixth of its species known to be collected anywhere in North America in the last ten years' (Losey *et al.* 2007: 415).

The United States Department of Agriculture (2012) has used technology to engage youth and encourage participation in citizen science projects through educational online tools. Mueller (2011) states that today's generation spends more time with technology than outdoors in nature, and suggests that the solution is to get youth involved in citizen science projects which may instil an understanding of ecological processes and encourage civic engagement. With the current advances in technology (see Mitchell, this volume), items like iPods, tablets, digital cameras and smart phones, most of citizen science data can be collected remotely and reported virtually to scientists; this, therefore, reduces gaps in access to geographically expansive locations inhabited by insects (Jentsch 2011; Wiggins and Crowston 2012b). Scientists are taking advantage of this technology to gather information about insects; for example, *Antweb* has designed an *Urban Ant Collector* application for Android phones to send pictures and relevant data (Wild 2011). Cornell University recently started collecting data captured on volunteers' iPhones because images transmitted for identification are embedded with geolocation information allowing scientists to map insect species distribution (Jentsch 2011). The Natural History Museum in Ottawa Canada, also has a Bugs Count application for mobile devices which features an identification guide, a species quest (a search for six identified species and submitting photographs using the application) and a gallery of insect pictures (Natural History Museum 2012). The United States Department of Agriculture (2012) is targeting youth involvement in the reporting of infestations of the Asian long-horned beetle through their website (www.BeetleBusters. info). Beetle Busters.info uses engaging tools, such as games (Freeze & Collect), videos, Facebook and Twitter to educate and encourage participation in reducing the presence of invasive species in the United States (United States Department of Agriculture, 2012). Using this technology is just the beginning of the possibilities for scientists, allowing them to tap into the public to collect data, create awareness and perhaps increase ecological stewardship.

Engaging children through public presentations about insect research also leads to improved motivation for continuing scientific education and insect conservation (Bearden 2011; The Lost Ladybug Project 2011). According to the Children's Nature Institute (2010), children's academic scores, physical and psychological health, behaviour and attitudes, and connection with nature all improved with participation in citizen science projects. Citizen science is a fantastic way to engage children in nature which encourages a healthy lifestyle and environmental stewardship, as well as inquiry-based learning (Lowman 2006; Schibsted 2007). Instilling conservation values early through hands-on learning and inquiry can create a more informed generation with greater commitment to civic engagement (Caplan 2010; Mueller 2011).

CONCLUSION

From its early roots in biology, natural history, archaeology and astronomy, citizen science has grown to encompass a variety of disciplines and diverse methodologies and to contribute to the wealth of scientific knowledge pertaining to a number of species and phenomena. The benefits to the research community include access to geographically disperse and longitudinal data in volumes impossibly expensive for projects that rely on specialized field technicians, increased means of disseminating research results, decreased research project costs, and improved opportunities for community-based science education. Volunteers who participate in citizen science benefit from access to specialized training, a sense of contributing to the larger body of knowledge and pride in contributing to personally relevant causes, as well as access to new, and alternative, recreation and tourism opportunities.

Scientists and volunteers are not the only ones who benefit from citizen science. Underfunded and under-researched species, such as insects, also benefit from the attention of citizen scientists. Gaining recognition as indicators of ecosystem health and increasing people's understanding of the value of their contributions to said ecosystems' stability, are some of the many benefits that insects and arthropods accrue as the subject of citizen science projects. Invertebrate, arthropod and insect conservation and management also face substantial barriers. These barriers, many of which stem from a lack of knowledge about insects or funding for arthropod and invertebrate research, can often be broken down through citizen science. Citizen science provides a means to overcome negative attitudes towards insects through the provision of experiential education opportunities and exposure to insects, while fostering community support for their conservation. Low labour costs associated with these endeavours also provide an opportunity for researchers to offset the costs associated with larger scale or underfunded insect projects.

All of these benefits are negligible, however, if the scientific community fails to recognize the validity of citizen science – either as a branch of science, or due to questions about the accuracy of the data collected by amateurs. Therefore, researchers must be sure that the training and education provided to potential volunteers is sufficient to provide a basic level of competency, that volunteers are asked to provide data that will answer specific research hypotheses, and that all data is verified by experts before it is interpreted. As gap analyses shed light on deficits within the literature on insect conservation, citizen science and projects that use this method as a primary data collection tool, it becomes increasingly important for researchers who engage in citizen science to disseminate their studies' findings beyond their participants and stakeholders and into the larger research community. In order for dissemination to happen, the processes used to conduct citizen science must become better known and standardized, leading to increases in the perception of the validity of their findings. Increasing the validity of their findings, and the scientific communities' perception of it, will allow researchers who, use citizen science as a data collection method, to overcome barriers to publishing the findings of underfunded research topics that rely on citizen science to offset their costs, as is often the case with research projects dedicated to insect conservation. The improved education and acceptance in turn will help in addressing issues of Arthropod Discourse Disorder, and ultimately lead to a greater body of knowledge about, and understanding of the importance of, conserving and protecting these important species.

REFERENCES

AntWeb (2012) About antweb. Available at: http://www.antweb.org/about.do, accessed 22 January 2012.

Araya, Y. N., Schmiedel, U. and von Witt, C. (2009) Linking 'citizen scientists' to professional in ecological research, examples from Namibia and South Africa. *Conservation Evidence*, **6**, pp. 11–17.

Bearden, T. (2011) Colorado kids act as citizen scientists in national lady bug hunt. *PBS Newshour Report,* 12 July. Available at: http://www.pbs.org/newshour/bb/science/july-dec11/ladybugs_07–12.html, accessed 20 January 2012.

Big Butterfly Count. (2012) About the big butterfly count. Available at: http://www.bigbutterflycount.org/about, accessed 20 January 2012.

Boettner, G. H., Elkinton, J. S. and Boettner, C. J. (2000) Effects of a biological control introduction on three nontarget native species of saturniid moths. *Conservation Biology*, **14**(6), 1798–1806.

Bonney, R., Cooper, C. B., Dickinson, J. *et al.* (2009) Citizen science: a developing tool for expanding science knowledge and scientific literacy. *BioScience*, **59**(11), 977–984.

Braschler, B., Mahood, K., Karenyi, N., Gaston, K.J. and Chown, S.L. (2010) Realizing a synergy between research and education: how participation in ant monitoring helps raise biodiversity awareness in a resource-poor country. *Journal of Insect Conservation*, **14**(1), 19–30.

Brossarda, D., Lewenstein, B. and Bonney, R. (2005) Scientific knowledge and attitude change: The impact of a citizen science project. *International Journal of Science Education*, **27**(9), 1099–1121.

Bumblebee Conservation Trust. (n.d.). Our aims. *Bumblebee Conservation Trust.* Available at: http://www.bumblebeeconservation.org.uk/who_are_we.htm, accessed 22 January 2012.

Caplan, S. (2010) Citizen science with children and families. *Kids Activities @ Suite 101* (blog) 11 April Available at: http://susan-caplan.suite101.com/citizen-science-with-children-and-families-a224252, accessed 20 January 2012.

Catskill Regional Invasive Species Partnership. (2012) New York invasive species. *Catskill Regional Invasive Species Partnership.* Available at: http://www.nyis.info/?action=prism&prism_id=4, accessed 22 January 2012.

The Children's Nature Institute (2010) Why it's important: connecting children (and adults) to nature. Available at: http://www.childrensnatureinstitute.

org/newsite/index.php?option=com_content&view=category&id=10&Itemi
d=53, accessed 20 January 2012.

Cohn, J. P. (2008) Citizen science: can volunteers do real research? *BioScience*, **58**(3), 192–197.

Conrad, C. C. and Hilchey, K. G. (2011) A review of citizen science and community-based environmental monitoring: issues and opportunities. *Environmental Monitoring Assessment*, **176**, 273–291.

Cooper, C. B., Dickinson, J., Phillips, T. and Bonney, R. (2007) Citizen science as a tool for conservation in residential ecosystems. *Ecology and Society*, **12**(2): 11.

Crall, A. W., Newman, G. J., Jarnevich, C. S. *et al.* (2010) Improving and integrating data on invasive species collected by citizen scientists. *Biological Invasions*, **12**(10), pp. 3419–3428. doi 10.1007/s10530-010-9740-9

Devictor, V., Whittaker, R. J. and Beltrame, C. (2010) Beyond scarcity: citizen science programmes as useful tools for conservation biogeography. *Diversity and Distributions*, **16** (3), 354–362.

Dickinson, J. L., Zuckerberg, B. and Bonter, D. N. (2010) Citizen science as an ecological research tool: challenges and benefits. *Annual Review of Ecology, Evolution, and Systematics*, **41**(1), 149–172.

Florida Butterfly Monitoring Network. (2006) About the FBMN. Available at: http://www.flbutterflies.net/about.jsp, accessed 20 January 2012.

The Great Sunflower Project. (2012) The backyard bee count. Available at: http://www.greatsunflower.org/, accessed 20 January 2012.

Gullan, P. J. and Cranston, P. S. (2010) *The Insects: An Outline of Entomology*, 4th edn. Oxford: Wiley-Blackwell Publishing.

Hand, E. (2010) People power. *Nature*, **466**, 685–687.

Hatsuyado, S. (2002) Friendly association of lifelong learning on nature study organized by museums. Part 1: large-scale survey with citizen participation and lifelong learning. *Chiiki no Shizen no Joho Kyoten Shizenshi Hakubutsukan Kagakukei Hakubutsukan Katsuyo Nettowaku Suishin Jigyo Hokokushu Heisei* 14nen, 34–43.

Heimbuch, J. (2010) Are fireflies disappearing? Citizen science project plans to find out. *Treehugger* (blog) 30 September. Available at: http://susan-caplan.suite101.com/citizen-science-with-children-and-families-a224252, accessed 20 January 2012.

Hoffman Black, S. (2011) Migratory dragonfly partnership: Using research, citizen science, education and outreach to understand North American dragonfly migration and promote conservation. Available at: http://www.migratorydragonflypartnership.org/MDP_Fact.pdf, accessed 19 January 2012.

Honey Bee Net. (2010) Honey bee net objectives. *NASA Goddard Space Flight Center*. Available at: http://honeybeenet.gsfc.nasa.gov/, accessed 22 January 2012.

Howard, E. and Davies, A. K. (2009) The fall migration flyways of monarch butterflies in eastern North America revealed by citizen scientists. *Journal of Insect Conservation*, **13** (3), 279–286.

Hughes, J. B., Daily, G. C. and Ehrlich, P. R. (2000) Conservation of insect diversity: a habitat approach. *Conservation Biology*, **14**(6), 1788–1797.

Ingwell, L. L. and Preisser, E. L. (2010) Using citizen science programs to identify host resistance in pest-invaded forests. *Conservation Biology*, **25**(1), 182–188.

Jentsch, P. J. (2011) Tracking invasives using your iPhone: the BMSB citizen science project. Hudson Valley Regional Fruit Program. Highland, NY. Available at: http://hudsonvf.cce.cornell.edu/bmsb1.html, accessed 19 January 2012.

Johansen, K. and Lemelin, R. H. (2010). Recognizing arthropod discourse disorder in multidisciplinary journals: a content analysis. Conservation for a Changing Planet: International Congress for Conservation Biology (ICCB), University of Alberta, Edmonton, AB, July 3–7, 2010.

Jordan, R. C., Gray, S. A., Howe, D. V., Brooks, W. R., and Ehrenfeld, J. G. (2011) Knowledge gain in behavioural change in citizen-science programs. *Conservation Biology*, **25**(6), 1148–1154.

Journey North (2012) Track spring's journey north. Journey North: a global study of wildlife migration and seasonal change. Available at: http://www.learner.org/jnorth/, accessed 20 January 2012.

Kim, S., Robson, C., Zimmerman, T., Pierce, J. and Haber, E. M. (2011). Creek watch: pairing usefulness and usability for successful citizen science. *Proceedings of the 2011 Annual Conference on Human Factors in Computing Systems*, Vancouver, BC, 7–12 May 2011. New York, NY: ACM, pp. 2125–2134.

Lemelin, R. H. (2009) Goodwill hunting? dragon hunters, dragonflies and leisure. *Current Issues in Tourism*. **12**(3), 235–253.

Lepczyk, C. A. (2009) Symposium 18: Citizen Science in ecology: the intersection of research and education. *Ecological Society of America*, **90**, 308–317.

Losey, J. E., Perlman, J. E. and Hoebeke, E. R. (2007) Citizen scientist rediscovers rare nine-spotted lady beetle, *Coccinella novemnotata*, in eastern North America. *Journal of Insect Conservation*. **11**(4), 415–417.

The Lost Ladybug Project. (2011) Why we need you. Available at: http://www.lostladybug.org/participate.php, accessed 22 January 2012.

Lowman, M. (2006) 'No child left indoors' is a worthy goal for parents and educators. *Herald-Tribune*, 22 October 2006. Available at: http://www.heraldtribune.com/article/20061022/COLUMNIST18/610220517?p=2&tc=pg, accessed 22 January 2012.

Marrou, J. (2008) The antares project brief history. Antares: study, identification, distribution, localization of ants French metropolitan forum (blog, translated from French) 25 May. Available at: http://antarea.fr/antarea/viewtopic.php?f=18&t=53, accessed 22 January 2012.

Mayer, A. (2010) Phenolgy and citizen science. *BioScience*, **60**(3), 172–175.

Monarch Watch (2010) The future of monarch watch. *Monarch Watch*, (blog) 30 January. Available at: http://monarchwatch.org/bring-back-the-monarchs/, accessed 22 January 2012.

Mueller, M. (2011) Citizen science can renew a child's love of nature. *Education.com*. Available at: http://www.education.com/reference/article/citizen-science-childrens-love-nature/, accessed 20 January 2012.

Natural History Museum. (2012) OPAL Bugs Count. Available at: http://www.opalexplorenature.org/bugscount, accessed 20 January 2012.

Nature Watch (2009) Irish Phenology Network. *National Biodiversity Data Centre*. Available at: http://phenology.biodiversityireland.ie/introduction/why-monitor-phenology/, accessed 22 January 2012.

Newberry, D. (2011) On the trail of the black petaltail. *The Xerces Society News* (blog), 10 October. Available at: http://www.xerces.org/2011/10/10/on-the-trail-of-the-black-petaltail/, accessed 19 January 2012.

North American Butterfly Association (2011) New butterfly counts. Available at: http://www.naba.org/counts/start.html, accessed 22 January 2012.

Oberhauser, K. S. and Prysby, M. D. (2008) Citizen science: creating a research army for conservation. *American Entomologist*, **54**(2), 103–105.

Odonata Central. (2012) About. Available at: http://www.odonatacentral.org/index.php/PageAction.get/name/About, accessed 22 January 2012.

Project Budburst. (n.d.). Phenolgy defined. *Project Budburst*. Available at: http://
neoninc.org/budburst/phenology_defined.php, accessed 22 January 2012.

Project Butterfly WINGS. (n.d.). What is Project Butterfly WINGS 'citizen science'? Florida Museum of Natural History. Available at: http://www.flmnh.
ufl.edu/wings/aboutWINGS.asp, accessed 22 January 2012.

Rotman, D., Preece, J., Hammock, J. *et al.* (2012) Dynamics changes in motivation in collaborative citizen-science projects. In *Proceedings of Computer Supported Cooperation Work (CSCW) 2012*, Seattle, WA, Feb 11–15, 2012. Available at: http://hcil.cs.umd.edu/trs/2011-28/2011-28.pdf, accessed 22 January 2012.

Schibsted, E. (2007) Kids count: young citizen-scientists learn environmental activism. *Edutopia*. Available at: http://www.edutopia.org/service-learning-c itizen-science, accessed 20 January 2012.

Scientific American (2012) What is citizen science? Available at: http://www.
scientificamerican.com/citizen-science/, accessed 29th December 2011.

Science for Citizens LLC (2011) Jug Bay Macroinvertebrate Sampling. Available at: http://scistarter.com/project/298, accessed 23 January 2012.

Silvertown, J. (2009) A new dawn for citizen science. *Trends in Ecology and Evolution*, **24**(9), 467–471.

Snäll, T., Kindvall, O., Nilsson, J. and Pärt, T. (2011) Evaluating citizen-based presence data for bird monitoring. *Biological Conservation*, **144**(2), 804–810.

UKBMS (United Kingdom Butterfly Monitoring Scheme) (2012) Butterflies and indicators. United Kingdom Butterfly Monitoring Scheme. Available at: http://www.ukbms.org/butterflies_as_indicators.htm, accessed 20 January 2012.

US Department of Agriculture (2012) Be a beetle buster. *Beetle busters.info*. Available at: http://www.beetlebusters.info/beABeetleBuster.php, accessed 22 January 2012.

Wells, C. N. (2010) An ecological field lab for tracking monarch butterflies and their parasites. *The American Biology Teacher*, **72**(6), 339–344.

Wiggins, A. and Crowston, K. (2012a) Goals and tasks: two typologies of citizen science projects. *45th Hawai'i International Conference on System Science (HICSS-45)*, Maui, HI, Jan 4–7, 2012. Available from: http://crowston.syr.edu/system/files/hicss-45-final.pdf, accessed 19 January 2012.

Wiggins, A. and Crowston, K. (2012b). Describing public participation in scientific research. *iConference 2012*, Toronto, Ontario, Canada, 7–10 February 2012. Available from: http://crowston.syr.edu/system/files/iConference2012.
pdf, accessed 19 January 2012.

Wild, A. (2011) Antweb expands its citizen-science program. *Myrmecos:Alex Wild on Insects, Science, and Photography* (blog), 25 March. Available at: http://
myrmecos.net/2011/03/25/antweb-expands-its-citizen-science-program/,
accessed 20 January 2012.

17

The institutionalization of insect welfare: the cultural aspects of establishing a new organization dedicated to conserving invertebrates

MATT SHARDLOW

Organizations dedicated to saving invertebrates are still very rare so it has been an honour to have a key role in establishing one – Buglife – The Invertebrate Conservation Trust (www.buglife.org.uk). The charity has grown from the employment of the first staff member in February 2002 to an organization with 20 staff and a conservation programme of £1 million per year in February 2012. During those first 10 years we have worked towards our objective – to stop invertebrate extinctions and maintain sustainable populations of invertebrates – by surfing in on waves driven by the people's appreciation of invertebrates, but we have also encountered and tackled walls to progress built of negative, or disinterested, attitudes.

Buglife covers all invertebrates, and it is worth noting that there is an immense range of cultural contexts and associations linked to the segments of the wide taxonomic spray of organisms that this definition covers. While butterflies and lice stimulate very different emotional and intellectual responses, other non-insect groups such as worms and spiders have their own, even more diverse, cultural associations (see Lemelin, this volume). Despite international commonalities in how cultures view and treat types of invertebrates, there are also subtle and occasionally stark differences between cultures. Buglife functions at a European policy level and has run projects overseas in Sri Lanka, St Helena and South Georgia; while our work in these spheres gives hints of the cultural spaces that invertebrates inhabit in other cultures, primarily Buglife's perspective is a British and European one.

Buglife is not a static platform from which to view culture; its purpose means that it is constantly trying to change culture, to make invertebrate conservation a mainstream activity by fostering the reasons for invertebrate conservation so that they move towards being accepted societal norms. It is this focus on bending society towards invertebrates that defines the difference between Buglife and all other wildlife conservation organizations operating in Europe.

THE BRITISH ORGANIZATIONAL SCENE

Britain has a good and rich history of societies, institutes, agencies and charities dedicated to aspects of nature conservation or invertebrate interests (see Dodd, this volume). However, in the late 1990s it was not clear if there was sufficient overlap between conservation and invertebrate interests to ensure that invertebrates were being adequately and appropriately conserved. There was a good range of societies dedicated to studying different groups of invertebrates, producing the species distribution and status information that forms the bedrock of nature conservation. But were the nature conservation organizations responding and progressing invertebrate conservation at a reasonable pace?

There was one point in the spectrum where it was clear that there was a good overlap between conservation and invertebrate interests. Butterfly Conservation was established in 1968; local committees of volunteers had taken valuable action to conserve butterflies at many localities across the UK. The mid-1990s was a period of change for Butterfly Conservation, a professional core of conservation staff was developed under the leadership of Martin Warren and this included the first national moth conservation officers, brought in to take action to conserve the species listed on the UK Biodiversity Action Plan (Anonymous 1995a, b).

The UK Biodiversity Action Plan was a response to the Biodiversity Convention signed in 1992 (UNEP 1992). The existence of the convention appeared to change subtly our society's priorities, opening doors that may previously have been shut. Not only did the convention enhance the significance of conservation issues generally, the term biodiversity was a clear departure from the terms wildlife and nature. The word wildlife was linked to what we perceive, relate to and appreciate; it had become predominately associated with birds, mammals, flowers and trees. Nature is a highly complex term (Shardlow 2006) and often incorporates a wilderness versus artificial axis that is only partly correlated with the axis of conserving the full range of species and failing to do so. The word biodiversity escapes the baggage of history; it is by definition a level playing field – at each taxonomic level one unit of biodiversity is equivalent to another in the cause of sustaining the variety of life. What then matters in determining the priority for conservation action is the severity of threat faced by the units of biodiversity – pragmatically measured at the level of species.

In light of the shift towards biodiversity conservation it was legitimate to explore how egalitarian British conservation was, and how level was the taxonomic playing field.

The WWF was set up as the World Wildlife Fund in Britain in 1961. This was a period where the focus of conservation was overseas and pandas and rhinoceroses were writ large in the early campaigns. By the 1990s the name had changed to World Wide Fund for Nature and it was a large, global organization with hundreds of staff members and a focus on large mammals, non-temperate zones, and increasingly on issues related to the effects of consumption and unsustainable development on the environment. This meant that WWF was neither on the invertebrate conservation playing field in the UK, nor a significant force in shifting the balance of resource expenditure or public perception towards the less historically favoured taxonomic groups. This transfixion with big charismatic animals is an example of a phenomenon that has become known as 'institutional vertebratisim' (Leather 2009).

Other nature conservation charities such as the Wildlife Trusts, RSPB and National Trust, major land owners in the UK, took their invertebrate conservation responsibilities seriously. Regularly, although not always routinely, doing surveys and managing habitats focused at least in a significant part on the needs of invertebrate species of conservation significance on their reserves. In some cases these charities were involved in experimental and trailblazing action to help conserve particular invertebrate species. However, outside the nature reserves the benefits to invertebrates provided by the scientific research, policies and campaigns of the big conservation charities were incidental. Issues having a specific impact on invertebrate conservation were not of directional significance for the large non-governmental organizations (NGOs). Insect conservation in the UK was recognized as a 'neglected green issue' (Fry and Lonsdale 1991).

The invertebrate specialist societies were all focused primarily on supporting the study of their particular groups. Many had conservation committees or conservation officers, volunteers putting their spare time into efforts to promote and develop invertebrate conservation. Only the British Dragonfly Society appeared to want to move into the realm of professional invertebrate conservation activities, an ambition achieved in 2001 with the employment of a conservation officer.

Government nature conservation agencies in England, Wales and Scotland had dedicated invertebrate conservation officers and species conservation programmes. Despite the successes of these programmes the overall trend in the agencies was away from species focused work and towards statutory duties, broad-brush habitat approaches and agri-environment schemes. Invertebrates featured little in the statutory duties, were viewed as detail by those promoting broad-brush habitat approaches, and were not well served by agri-environment schemes that were dominated by measures aimed at reversing the declines in farmland birds – a metric that had become a government target. Some of the shift in the focus of statutory agencies is attributable to declines in the proportion of staff with detailed natural history knowledge. All organizations have to have consistent standards, protocols and policies, the coarseness of the grain of these depends on the ability of the bulk of relevant officers to understand and translate them into appropriate action. When invertebrate conservation knowledge is uncommon the fine grain advice and policy implementation necessary to retain and foster the delicate ecological niches of many invertebrates cannot be achieved.

Government agencies are sometimes able to campaign for change, but are always to some degree restrained in their positions because they must have regard for how their particular agency will fare in the next budget round, or worse in the next 'bonfire of quangos'. Hence while doing good work on the ground, the motivation for Government agencies to extend the scope of conservation activity further into the realm of invertebrate conservation was limited.

The development of the UK Biodiversity Action Plan in the mid 1990s enabled a coming together of the wildlife NGOs and invertebrate specialists as they collaborated to define lists of priority species. This continued after the plan was published with a series of conferences that addressed the apparent invertebrate conservation shaped gap in the activities of existing organizations. As a result of this process Buglife – The Invertebrate Conservation Trust was set up in 2002 to fill the gap (see Stubbs and Shardlow 2012 for more detail on the establishment of Buglife).

Invertebrate conservation activities took another step forward in 2006 with the establishment of the Bumblebee Conservation Trust. Based in Stirling, Scotland, the new trust quickly gathered public support and employed researchers and conservation staff.

PUBLIC AND PRIVATE VALUES

Public attitudes to invertebrates may be widely perceived, but they are rarely quantified. This is even more the case in relation to the public's attitudes towards invertebrate conservation.

An individual's motivation to conserve natural resources arises from one or more personal values. These values can be subdivided into two categories, instrumental or utility values and intrinsic value (Soulé 1985). In its simplest form anything that is of value to the valuer is a utility value. This value may derive from a direct financial benefit, for instance a commercial fishery; an externalized use, such as an ecosystem service like pollination; or from a sense of pleasure or wellbeing, as might be experienced in a meadow full of butterflies and grasshoppers. The latter value is often split from other utility values and referred to as the aesthetic value. The second main category, intrinsic value is harder to define. It is the imbuing of the object with an existence value, in this context usually a species (although nature, natural processes and individual organisms can also be attributed with an intrinsic value). It maintains that there is value to a species beyond the benefits it provides to us, or our descendants. The difficulty of attaching a measurement to intrinsic value has led to questions about the usefulness of intrinsic value as a separate category or even as a definable concept at all (Justus *et al.* 2009). However, intrinsic value is clearly evident as a belief that the future of life on Earth should not be moulded purely on the basis of the needs and whims of *Homo sapiens*, and that it is wrong for one species to casually cause the extinction of another. Appreciating an intrinsic value in species is closely linked to a point of view that recognizes rights of existence for a species.

Legal systems are slowly moving in the direction or providing recognition of rights for non-humans, with the Convention of Biological Diversity (UNEP 1992) setting a strategic context and implementation coming through legislation such as the EU Habitats Directive (European Union 1992) which legally embeds the right for selected threatened species to have their wild populations maintained in a favourable conservation status. The selection of the species for the Habitats Directive lists was doubtlessly affected by utility and aesthetic values and of course for many species that might otherwise have been listed there was simply not enough data to enable an assessment of conservation need. However, the list is taxonomically broad and includes a good many species with minimal utility and aesthetic values. For instance, the charisma of the little whirlpool ram's-horn snail (*Anisus vorticulus*) is only appreciated when you have a knowledge of its sensitivity and picky habitat requirements, otherwise, it is tiny, brown, exceedingly rare and looks more or less exactly like a closely related snail. The inclusion of the little whirlpool ram's-horn snail and the lynx together on this most significant list, according them both the highest level of conservation priority, can only be explained by placing a significant value on the intrinsic right of the little snail to exist.

For a new charity with an uncharted cause to develop it was important to understand better the scale and shape of the task of persuading the public to provide more support for invertebrate conservation. What proportion of the public thought that conserving invertebrates was a worthy cause? Where might there be easy wins for Buglife, subsets of society that may be preconditioned to understand the significance of invertebrates to ecosystems, their importance to humans or their intrinsic value? Indeed which of these premises had the greatest resonance with the public?

In March 2002 Buglife worked with the countryside Agency to explore public attitudes to invertebrate conservation. An omnibus telephone survey of 1682 people was undertaken (BMRB Social Research 2002). Respondents were asked if they thought that it was important to protect or keep specified features of the English Countryside.

A total of 84% of respondents thought it was important to protect or keep insects and creatures such as beetles, snails, bees and worms. However, other features were considered important by more people – wild birds 97%, fields and hedges 97%, wild flowers and grasses 96% and wild animals such as deer, hedgehogs and field mice 96%. Interestingly 89% of respondents from rural areas thought it was important to protect or keep invertebrates, compared to 83% of respondents from urban areas.

The 84% of respondents who thought protecting invertebrates was important were then asked to give reasons why it was important to protect insects and creatures such as beetles, snails, bees and worms. The most frequently cited reasons are shown in Table 17.1.

The reasons given defy a simple split into utility, aesthetic or intrinsic values. Many of the reasons were less outcome focused,

Table 17.1 *Most frequently cited reasons to explain why it is important to protect insects and creatures such as beetles, snails, bees and worms in 2002 omnibus telephone survey (BMRB Social Research 2002)*

Reason	% of people giving this response	Value category
Have an important role/ purpose	26%	E/N
Help natural ecosystem	17%	E/N
Part of the food chain	16%	E/N
Keep balance/mustn't upset balance of nature	15%	E/N
Food for birds/other animals	14%	E/N
Life cycle	12%	E/N
Good for soil	9%	U
They have a right to live	8%	I
They will become extinct/ die out	7%	I
Natural	5%	E/N
Pollination	4%	U
Essential for Earth	3%	E/N
Maintain diversity	2%	I
Beautiful to look at	2%	A
Education for children	1%	U
Honey from bees	1%	U
Of great interest	1%	A

Value categories: E/N = ecosystem or nature, U = utility value, A = aesthetic value and I = intrinsic value.

expressing the perceived importance of ecosystems or nature without defining a specific attribute of actual value. This category of reason, 'ecosystem or nature', could be derived from the respondents intrinsic values, utility values, a mixture of both, or a desire to give what they hope will be perceived as the right answer. Someone actually reasoning that invertebrates are important as food for birds may be doing so because they like to see birds or because they think those birds have a right to existence.

The top six reasons given were 'ecosystem or nature' responses, indeed 76% of all reasons given were vague recognitions of the importance of invertebrates in ecosystem function and in nature. Only 12% of responses were clearly related to intrinsic value, a surprisingly low

11% of answers recognized their usefulness, and their aesthetic value registered at a measly 2% of reasons provided.

The implication was that a message that insects were beautiful was only likely to resonate strongly with a very small number of people, whereas most people would relate to a message conveying the importance of invertebrates to healthy ecosystems. It was clear that the point of greatest consensus between Buglife and the public was that invertebrates were important in ecosystems. Educating and persuading people about the intrinsic value and many utility values of invertebrates would be important themes and would fall on many sympathetic ears, but the strap line the charity selected was 'Conserving the small things that run the world' (a homage to Wilson 1987).

Buglife staff are typically very very enthusiastic about invertebrates, it is a discipline to remember that while we may think (and say) that a particular rare beetle is beautiful and amazing, by also putting it in a context of its ecosystem role and significance it is likely that the beetle will get more sympathy and attention from the public.

To help establish a better connection between the public and invertebrates the decision was taken to use English names for animals whenever possible, fostering new ones where appropriate. Scientific names are very off-putting to most people and have minimal instant impact. Many entomologists resented Buglife's use of common names, seeing it as imposing an unnecessary dual naming system; however, most education and publicity officers understood that reducing the height of the language barrier between invertebrate experts and the public was important for the promotion of invertebrate conservation. This stance was quickly vindicated when local people came to the aid of the sorrel pygmy moth, which was threatened with Scottish extinction by habitat destruction, only after Buglife had revealed this to be the common name of *Enteucha acetosae* (Stubbs and Shardlow 2012).

CONSERVATION SECTOR ATTITUDES

A shift in public attitude does not automatically transcribe into changes of attitude within the conservation sector, but effort can be productively put into improving the ability and willingness of the conservation sector to take the appropriate invertebrate benefitting actions.

Successful conservation delivery depends on the acquisition and availability of data and the production and dissemination of advice. Developing and providing these elements is the daily work of an invertebrate conservation charity: supplying the information that should enable conservation practitioners and policy makers to take decisions that preserve the planet's biodiversity.

At a deeper level progress towards conserving invertebrates meets obstacles associated with certain preconceptions and prejudices. Many a budding invertebrate conservation initiative has not reached fruition due to an encounter with an unsound belief that 'there are rare invertebrates everywhere' or that 'that particular species is probably common on other sites in the vicinity, or in other regions'. Invertebrates have to compete in budgeting rounds with species that have a far greater level of public awareness and aesthetic appeal. Hence a county will write an action plan and employ an officer to conserve the widespread and charismatic otter, but a beetle that has disappeared from the general countryside and that now survives only in that county and on just a handful of small sites does not merit a mention.

The conservation movement in the UK has established a principle that if more than 1% of the British breeding population of a bird species occurs in one locality then that place is eligible for designation as a protected area – a Site of Special Scientific Interest (SSSI) (JNCC 1989). Yet the same movement tolerates a situation where there are many invertebrate species with 25–60% of their UK populations on individual sites that are not SSSIs. Indeed in the case of the Croston worm (*Prostoma jenningsi*) 100% of the known world population occurs in a single non-SSSI pond. This example illustrates the huge discrepancy between the conservation support extended to different taxa, but it is hardly unexpected, prior to Buglife there was no professional advocate to articulate the case for fair treatment for invertebrate species in the UK. Part of the reason for existence of an invertebrate conservation charity to be the ever present, affirmative, answer to the question 'Who cares about the invertebrates?'

For the sector to fully adopt an invertebrate conservation agenda requires the adjustment of several accepted norms. Some significant causes of extinction of invertebrate species are not issues for other groups of organisms. In particular, many vulnerable invertebrate species are narrow habitat specialists, even to the point of being associated with one foodplant or host. Destruction or mismanagement of the niche or the decline or extinction of the host results in the disproportional extinction of such species (Dunn 2005). Invertebrate conservation reveals new extinction threats that require new solutions.

Some habitats are important sanctuaries for many rare and endangered invertebrate species but do not rate on botanical or vertebrate conservation priorities. Coastal soft rock cliffs form a jumbled mix of flowers, bare ground and seepages. Often with a sun-facing southern aspect they are havens for invertebrates with very narrow ecological niches, but are not regularly home to rare plants or nesting seabirds (or other conservation priorities). Saline lagoons with their harsh fluctuating salinity are very species poor but support the entire British populations of over 20 specialist worms, crustaceans, insects and other invertebrates as well as a handful of algae species (Bamber *et al.* 2001). Some insect specialist habitats such as deadwood have become more highly valued in recent years, not simply due to their recognition of their importance to invertebrate biodiversity, but also because of the usefulness of standing timber to various nesting birds and the realization that deadwood in water courses helps to create a healthy river structure and ecosystem, fostering commercially valuable fish stocks as well as the occasional endangered cranefly or hoverfly species.

While the conservation movement has now broadly accepted the importance of deadwood habitats, in recent years one invertebrate habitat has had to challenge a more fundamental conservation value. Brownfield land is land that has been previously developed. This development can include mining activities, dumping of dredging, factories or housing or even development as a theme park or military-training area. Hence alongside bare concrete, the definition includes quarries, spoil-heaps and a range of extensively developed areas, all of which can incorporate large areas where, due largely to being ignored or abandoned by humans, wildlife thrives.

The varied habitats on many brownfield sites – often incorporating bare substrates, abundant pollen and nectar rich flowers, scrub and ponds or other wet features – mimic habitats associated with dynamic processes of erosion and low nutrient soils. However the coasts and rivers have been tamed and the land is being fertilized by agricultural and aerial depositions. The non-human factors causing fresh disturbed habitats on a range of substrates are in decline. In addition to specialist species, brownfields have become refuges for other species such as the Shrill-carder bee (*Bombus sylvarum*) extirpated from the countryside by intensive farming practices (Shardlow 2008a). Brownfields also play an ecological and conservation function in enabling the spread of species. Brownfield sites appear to provide a readily colonized first British foothold for some species in the process of moving north, probably in response to climate change.

On any assessment based on conserving biodiversity the importance of brownfields is highly significant, across the board they are home to more red-listed invertebrate species than ancient woodland (Gibson 1998) and individual brownfield sites support so many rare species that several are in the top ten of sites with the greatest number of rare UK species; and a preponderance of these in the Thames Gateway to the East of London (Harvey 2000; Jones 2008).

However, in an assessment based on conserving naturalness brownfields are at the bottom of the list, as they are highly artificial. Naturalness and wildlife value are conflated in the public perception and in policy. For instance the level of naturalness is one of the tests in the SSSI designation criteria, which also discuss when it might be acceptable not to exclude artificial habitats from an SSSI designation footprint (JNCC 1989). None of the most important brownfield sites in the Thames Gateway were protected as SSSIs in 2002. 'Brownfield' is also juxtaposed with 'greenfield' in spatial planning with greenfield being conceived to be the good land to be retained and brownfield being the neglected land to be developed.

Highlighting the importance of brownfield sites and getting them managed for their contribution to sustaining biodiversity has been challenging, for instance the proposals to build a warehouse and lorry park on an expanse of pulverized fuel ash in the Thames Gateway that supported 36 red-listed species went through a protracted and high profile court case before permission was finally given to allow the development (Shardlow 2008b).

While there have been several important brownfield sites destroyed in the last decade, there has also been some real progress. In 2005 Canvey Wick, also in the Thames Gateway, was designated as an SSSI, providing some protection to its 32 red-listed species, and in 2007 Open Mosaic Habitat on Previously Developed Land was adopted as a priority Biodiversity Action Plan habitat and given protection on the NERC Act S.41 list of habitats. This is the lowest level of statutory protection; it means that the conservation of the habitat must be 'regarded' by public bodies. A management handbook for brownfield habitats, primarily focused on creating the right habitats and niches to support their important invertebrate assemblages has been produced (Buglife 2009).

However, the shift from just conserving land we think is deserving, to also conserving land that the endangered species need has not fully occurred. While the issue of the biodiversity significance of brownfield sites is now discussed with government ministers, the realization that providing the future habitats that will enable biodiversity to thrive is likely to depend in part on how humans manage our ground disturbing activities so as to maintain a refugee fauna, has not yet been strategically addressed.

CONCLUSION

Coverage of Buglife related stories has been consistently high, generating about 20 media occurrences a month, almost half of which have national impact. News items about invertebrates and their conservation have frequently been given high profile coverage. It seems unlikely that all the newspaper column inches generated by the work of Buglife, Butterfly Conservation and the Bumblebee Conservation Trust are simply replacing other invertebrate conservation stories. So it seems probable that there has been a significant increase in awareness of invertebrate conservation issues.

Also of note was the invertebrate-dedicated David Attenborough 2005 television series 'Life in the Undergrowth' which was regularly watched by 4 million people and got the highest audience appreciation figures in the history of British television. The series ended with David Attenborough, a figure with great public trust, pronouncing that:

> If we and the rest of the backboned animals were to disappear overnight, the rest of the world would get on pretty well. But if they were to disappear, the land's ecosystems would collapse. The soil would lose its fertility. Many of the plants would no longer be pollinated. Lots of animals, amphibians, reptiles, birds, mammals would have nothing to eat. And our fields and pastures would be covered with dung and carrion. These small creatures are within a few inches of our feet, wherever we go on land – but often, they're disregarded. We would do very well to remember them. (Attenborough 2005)

Hopefully this and other awareness raising activity has continued to reinforce the belief that invertebrates should be protected because of their important role in ecosystems. The Countryside Agency ceased to exist in 2006 and the countryside omnibus survey is no longer active, hence the 2002 survey questions have not been repeated. Given all of the newspaper articles, radio interviews, television coverage, the production of this edited volume and increasingly visited web pages there is good reason to hope that invertebrate conservation compatible intrinsic, utility and aesthetic values are being nurtured in the population.

It is difficult to relate the spreading of invertebrate conservation messages directly to measured changes in the public's willingness to support invertebrate conservation. However, one measure is the number of people joining up and paying to become a member of Buglife in support of the charity's work. While membership levels have grown over the last decade to a little over 1000 individuals, the cause has a way to go before it can compare with the 1 000 000 members of the comparable bird conservation organization, the RSPB. If enough people visibly support a cause then another value comes into play, people start to take appropriate action not because of the value they place on wildlife but because they value the opinions of other people and do not want to be perceived as unethical. It would even be conceivable to make a conservation action fashionable and attract to the cause categories of people who tend not to be motivated by environmental concerns.

It has been very encouraging that new funding streams for conservation such as the BBC Wildlife Fund have taken deliberate positive action to provide funds for the less charismatic species and thereby helped start to level out the emphasis on different taxonomic groups.

In convincing the conservation sector to place a greater priority on the conservation of invertebrates, the levelling arguments of intrinsic value and fairness have been just as, and probably more, effective in stimulating change than expounding utility arguments. This is counter to what might be expected as over the last decade there has been an attempt by the UK Government to refocus conservation efforts away from species and towards valuable ecosystem services. The negative attitude towards conserving species has had a detrimental effect on aspects of invertebrate conservation, reducing the funding going into work focused on recovering populations of Biodiversity Action Plan listed species for instance. Unfortunately, on the other hand the lack of any plan or clear programme of work to halt pollinator declines, and the absence even of a metric to assess pollination service levels are testament to the fact that the new emphasis on ecosystem services is yet to produce significant tangible benefits to relevant invertebrate populations.

In times of economic hardship established models of human values would favour more selfish behaviour as people struggle to meet more fundamental needs and think less about the common good of looking after their environment (Maslow 1943; Schwartz 1992). Perhaps the Government's slant towards utility arguments is an expression of this effect. Against this backdrop it is encouraging that British invertebrate conservation NGOs have continued to thrive in recent years.

The 'it's good for the ecosystem' proposition is certainly useful, particularly in diffusing public concern about conflict species such as social wasps and ants, and it can create useful common ground. However, a question mark remains as to whether just believing that a species is an important part of nature is capable of motivating people to take affirmative action on its behalf. Perhaps for an individual to be willing to take action a more profound level of appreciation is required, be that based on the beauty, intrinsic value or usefulness of the species.

REFERENCES

Anonymous (1995a) *Biodiversity: The UK Steering Group Report. Volume1: Meeting the Rio Challenge*. London: HMSO.
Anonymous (1995b) *Biodiversity: The UK Steering Group Report. Volume 2: Action Plans*. London: HMSO.
Attenborough, D. (2005) *Life in the Undergrowth*. London: BBC/Open University, episode Supersocieties.
Bamber, R. N., Gilliland, P. M. and Shardlow, M. E. A. (2001) *Saline Lagoons: A Guide to their Management and Creation (Interim Version)*. Peterborough, UK: English Nature.
BMRB Social Research (2002) *Adult Survey Report*. Cheltenham, UK: The Countryside Agency.
Buglife (2009) *Planning for Brownfield Biodiversity: A Best Practice Guide*. Peterborough, UK: Buglife – The Invertebrate Conservation Trust.
Dunn, R. R. (2005) Modern insect extinctions, the neglected majority. *Conservation Biology* **19**(4), 1030–1036.
European Union (1992) Council Directive 92/43/EEC of 21 May 1992 on the conservation of natural habitats and of wild fauna and flora, Brussels, Belgium: European Union.
Fry, R. and Lonsdale, D. (eds.) (1991). *Habitat Conservation for Insects: A Neglected Green Issue*. Middlesex, UK: Amateur Entomologists' Society.
Gibson, C. W. D. (1998) *Brownfield: Red Data: The Values Artificial Habitats have for Uncommon Invertebrates*. Peterborough, UK: English Nature.
Harvey, P. R. (2000) The east Thames corridor: a nationally important invertebrate fauna under threat. *British Wildlife*, **12**(2), 91–98.
JNCC (1989) *Guidelines for the Selection of Biological SSSIs* (1998 revision). Peterborough, UK: NCC.
Jones, R. A. (2008) *Caught in the Greenwash: What Future for Invertebrate Conservation on the Brownfield Sites of the Thames Gateway*. Peterborough, UK: Buglife – The Invertebrate Conservation Trust.
Justus, J., Colyvan, M., Regan, H. and Maguire, L. (2009) Buying into conservation: intrinsic versus instrumental value. *Trends in Ecology and Evolution*, **24**(4), 187–191.
Leather, S. R. (2009) Institutional vertebratism threatens UK food security. *Trends in Ecology and Evolution*, **24**(8), 413–414.
Maslow, A. H. (1943) A theory of human motivation. *Psychological Review*, **50**(4), 370–396.
Schwartz, S. H. (1992) Universals in the content and structure of values: theory and empirical tests in 20 countries. *Advances in Experimental Social Psychology*, **25**, 1–65.
Shardlow, M. (2006) Is nature conservation natural? *ECOS*, **27**(1), 2–8.

Shardlow, M. (2008a) Brownfields as refuges. *Wings: Essays on Invertebrate Conservation*, Spring 2008, 19–23.

Shardlow, M. (2008b) Defending the bugs of West Thurrock Marshes. *Atropos*, **35**, 45–51.

Soulé, M. E. (1985) What is conservation biology? *Bioscience*, **35**, 727–734.

Stubbs, A. and Shardlow, M. (2012) The development of Buglife: the Invertebrate Conservation Trust. In *Insect Conservation: Past, Present and Prospects*, ed. T. R. New. Berlin: Springer, pp. 75–105.

UNEP (1992) *Convention on Biological Diversity. United Nations Environmental Program. Environmental Law and Institutions Program Activity Center.* Nairobi: UNEP.

Wilson, E. O. (1987) The little things that run the world (the importance and conservation of invertebrates). *Conservation Biology*, **1**, 344–346.

18

Insects in education: creating tolerance for some of the world's smallest citizens

CRYSTAL M. ERNST, KRISTEN M. VINKE, DONNA J. GIBERSON
AND CHRISTOPHER M. BUDDLE

WHY INSECTS?

In December 2010 a research paper entitled *Blackawton Bees* was published in the well-respected science journal, *Biology Letters*. The article, which documented a well-crafted and thoughtful examination of visual perception, memory and foraging strategies in bumblebees (Blackawton *et al.* 2011), was an example of good science. The paper has been considered remarkable because the research was designed, conducted and written by a group of 8- to 10-year-olds, under the light supervision of a research scientist and a primary school teacher. The idea that children could do real, relevant, novel entomological work was perhaps rattling for some, but it was certainly affirming for others.

Given their diversity of form, function, habits and habitats, arthropods (insects, spiders and their relatives) have long been recognized as useful models for teaching general scientific and biological principles (Bergman 1947: 24; Fischang 1976: 204). As the Blackawton study demonstrates, insects also represent an excellent opportunity for young people to become engaged in 'real science' and to practise the process of scientific discovery. For many who choose natural sciences as a profession, childhood encounters with insects can often be traced to the origins of a life-long curiosity about nature and a passion for learning and science (C. M. Buddle *et al.*, unpublished data).

The Management of Insects in Recreation and Tourism, ed. Raynald Harvey Lemelin. Published by Cambridge University Press. © Cambridge University Press 2013.

For young people, working with arthropods can be a novel, memorable and exciting experience: these animals can simultaneously thrill, chill and enthral. 'Children seem to be particularly open to the wonders of six- or eight-legged creatures' (Gray 1976: 211). For example, a Madagascar hissing cockroach or a stick insect brought into a classroom elicits squeals of delight, awe and excitement or exaggerated expressions of fear. When invited to approach and touch the insects, students' reactions range from tentative to enthusiastic, but there is always tremendous curiosity and a steady stream of questions for the insect's handler: 'Where do they come from? What do they eat? How do they move? Why do its feet tickle my skin? Will it bite me?' Hands-on classroom encounters with insects do more than engage and entertain. These first experiences of observation and questioning stimulate critical thinking and hypothesis formation. Insects are particularly useful for this because, from a practical standpoint, they are one of the few types of animals that can be brought into a classroom. They are also readily encountered and captured outdoors with minimal equipment, in all types of urban, naturalized and undisturbed environments.

Entomology, the study of insects, has a long history of participation by amateurs, particularly studies of moths, butterflies and beetles. In recent decades, 'entomology became more of an academic pursuit, with most major contributions coming from university researchers' (Mathews 1988: 157). Entomological research, however, is once again being embraced by a broader audience in the form of citizen scientists and amateur naturalists as well as teachers (see Johansen and Auger, this volume). Entomological training opportunities are routinely offered to elementary and high school teachers by university faculty and graduate students, forming important bridges between university researchers and the interested public. The courses or workshops expose teachers to knowledge and techniques that can facilitate the use of insects in elementary or high school classrooms (e.g., Boardman *et al.* 1999; Haefner *et al.* 2006). A tremendous number of insect-related teaching tools are available both in print (see, for example, the review in Matthews *et al.* 1997) and on-line (e.g., a simple Google search for 'insects lesson plan' yields well over a million web pages).

This burgeoning interest in insects has potential benefits for students, their schools and their communities, but also for scientists and their research programmes. Unfortunately, opportunities to work with teachers and youth are rarely considered by academic researchers, who 'tend to write and speak to other entomologists, rather than to general audiences' (Fischang 1976: 204). We argue that partnerships between academic entomologists and school programmes can stimulate excitement in science (and therefore work as a recruiting tool), but also yield valuable data that might not otherwise be available. Such opportunities can and should be pursued in any geographical area, but we have found these partnerships to be particularly rewarding in First Nations or Aboriginal communities, which have special ties to the land and its wildlife. Here, we discuss some of the challenges and opportunities for using both terrestrial and aquatic insects in educational research activities in such communities, and highlight two case studies of new programmes being developed and implemented with students in remote regions of northern Canada.

TEACHING AND LEARNING 'ON THE LAND' IN
NORTHERN COMMUNITIES

There is a growing trend in 'traditional' (First Nations/Aboriginal, or 'non-Western') societies 'to restore and foster connections with the land by using the outdoors as a teaching and learning environment' (Takano 2005: 464). Teachers in northern communities often express the value of learning experiences that take place 'on the land' (i.e., in nature, in the field), where the teaching methods can be more in line with historical teaching practices (C. Ernst, pers. obs.). In an outdoor environment, students learn through observation of Elders, teachers or experts in a community-based setting. Here, 'students can help each other and work cooperatively, which is a more traditional approach to teaching and learning' (Takano 2005: 469, 471; Roberts and Clifton 1988: 122). In at least one instance, this approach has been incorporated directly into curriculum. In the Northwest Territories (NWT; in northern Canada), high school students spend considerable time outdoors during the school year learning field techniques and collecting and interpreting data in ways that combine 'southern science' and traditional knowledge. This project has resulted in a series of specially commissioned textbooks incorporating experiential learning in terrestrial, marine and freshwater systems (Campbell *et al.* 2008, 2009), which have been produced with input from local Elders and scientific consultants, and tested extensively by teachers in the Northwest Territories.

In other localities, the incorporation of experiential learning is not as institutionalized, but still forms a critical part of the learning process. For example, teachers in Kugluktuk, Nunavut Territory (a small, mainly Inuit community on the coast of the Arctic Ocean in northern Canada) were quick to point out that students are more cooperative, task-oriented and focused when they are learning outdoors (C. Ernst, pers. obs.). Entomological work is a natural fit for outdoor education and experiential learning opportunities: what better place to find and work with insects than out on the land?

Before discussing some specific examples, it is important to consider some of the challenges of carrying out collaborative insect-based projects with schools in the remote north. The first challenge is to engage and train science educators so that they recognize the value of these projects, in regions where large mammals, birds and fish are highly valued and are studied more frequently than insects. Northern schools may not have access to field guides, microscopes and other tools needed for natural history study; however, meaningful, exciting and instructive work can still be undertaken on the land and in the classroom using little more than forceps, plastic containers with lids, inexpensive hand lenses and entry-level insect nets. Additionally, recent initiatives to bring satellite-based high-speed internet to classrooms, even in the far north, have opened up access to the resources on the worldwide web. This provides remarkable access to a wealth of information, activities and expertise on insects (e.g., databases, research papers, images, online field guides), regardless of geographic location (Zenger and Walker 2000).

The unpredictable nature of outdoor education in the north can be a challenge: inclement weather, unreliable group transportation or the presence of unfriendly bears can derail planned activities. Flexible lesson plans and teaching approaches and a willingness to work 'on the fly' at times help ensure a successful programme. Another challenge comes from the brief Arctic summer: much insect activity occurs when children are not in school. In many northern communities, the local Hunter–Trapper Organizations (HTOs) are instrumental in setting up 'traditional knowledge camps' and other opportunities for students to be involved in experiential learning throughout the summer months as well as during the school term (D. Giberson, pers. obs.). Collectively, these have resulted in numerous opportunities for southern-based university scientists to get involved in community (school) partnerships that engage students, build local capacity in science, and provide local data on the insects of remote northern locations.

The subject matter itself (i.e., insects) can present challenges in northern communities. Popular portrayals of wildlife in Arctic regions tend to glorify large, charismatic and iconic animals such as polar bears, Arctic fox, caribou and muskoxen. The smaller animals of the north have received considerably less attention than they deserve. Roughly 2000 species of insects and other arthropods have been identified in Arctic ecosystems (a sharp contrast to the mere 20 or so species of mammals), where they provide many important ecological services such as pollination, decomposition and food provision for other animals (Danks 1981). Public perception of insects in the north, however, often focuses on the relatively few species of biting flies, which can be present in high numbers, and overshadow the diversity and importance of other groups.

Inuit folklore is permeated with rich *qupirruit* ('insect') symbolism (an excellent review is provided in Laugrand and Oosten 2010). Some of these small animals are shamanistic helper spirits, and may be sewn into amulets to lend strength or invincibility to their wearers. Much of the insect imagery in Inuit culture centres on the ability of these animals to survive and emerge from the long, cold winter months. Not surprisingly, then, some insects are seen as symbols of survival, rejuvenation and reanimation. However, since insects are considered 'masters of life and death' that are able to 'control the transformation from life to death and vice-versa', they also engender feelings of mistrust and suspicion (Laugrand and Oosten 2010: 3).

Indeed, there seems to be an overarching fear (or at least dislike) of insects in many northern communities. This should be a familiar concept, since it is a pervading sentiment in many human societies; insects bite, sting, creep, crawl, invade living spaces, cause damage, spread disease and have unusual appearances. However, the fear or dislike of insects seems to be more extreme than expected for people who have such close ties to nature. In one field study in Rankin Inlet, Nunavut, local Inuit were horrified to find out that aquatic insects lived in streams that they used for drinking water (D. Giberson, pers. obs.). In another example, while conducting fieldwork in Kugluktuk, Nunavut, a locally hired field assistant pointed at the contents of an insect trap and remarked, 'The [local] people, they're afraid of those beetles.' The beetles in question were tiny, innocuous, dark-bodied ground beetles (Coleoptera: Carabidae), not usually considered to be frightening. The field assistant then told the tale of an elderly hunter who, having accidentally laid his bedroll atop one such beetle, awoke to find that he had been deafened in one ear as his eardrum was being loudly consumed ('Crunch, crunch, crunch!') by the insect. It was eventually removed but the damage was done (K. Kuodluak, pers. comm.). This story is similar to the familiar horror stories about earwigs one hears in southern Canada, demonstrating that the fear of insects is not limited to

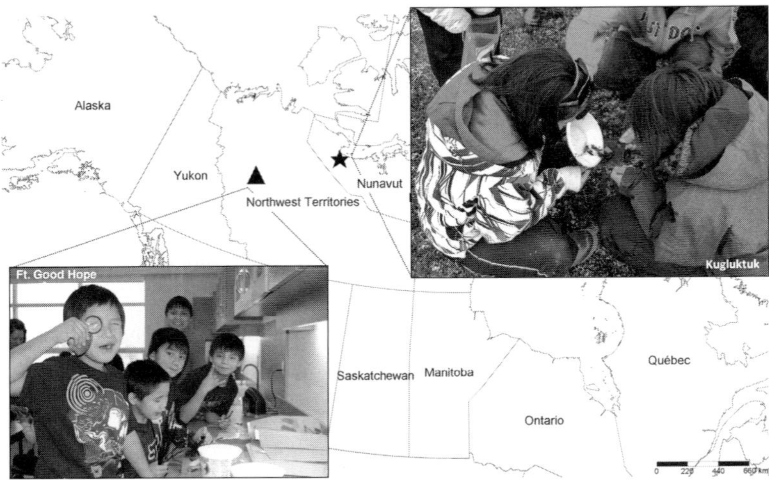

Figure 18.1   Insect education in northern Canada.

northern communities. Most entomologists working in the north are familiar with the, at times, extreme responses of community members to insects. Field assistants and other locals repeatedly ask whether 'new' insects they encounter are dangerous, or whether they bite, or are poisonous. The situation is exacerbated by the presence of insects in their communities that have never been seen before, having come in accidentally in cargo or on planes, or perhaps through natural range extensions through climate change. These suspicions and general feelings of unease about insects are common in adults and children alike; they must be acknowledged, addressed and overcome in order to work with insects successfully in a classroom setting. The simple act of learning about insects through classroom and field exercises is helpful in this regard. In the next sections, we present two case studies where insect studies are being incorporated into school programmes in Northern Communities such as Fort Good Hope (see Figure 18.1).

CASE STUDY 1: INSECTS AS FORECASTERS OF CLIMATE CHANGE IN THE NORTH: A CAREER AND TECHNOLOGY COURSE AT KUGLUKTUK HIGH SCHOOL

In Kugluktuk, Nunavut, a partnership between the local HTO and an environmental consulting firm (Golder, Inc.) led to a multi-year programme incorporating summer 'camps' where southern-based scientists were brought to the community to work with local students and

Elders. This project led to collaboration between author D. Giberson and the HTO to provide student training in aquatic insect identification and water quality monitoring, leading ultimately to a much-expanded partnership. The partnership project with Kugluktuk High School described below was a component of a large collaborative project on northern arthropods (The Northern Biodiversity Program (NBP), www.northernbiodiversity.com), and followed a summer collecting insects and other arthropods in the Kugluktuk area.

CTS at Kugluktuk High School

Teachers in Kugluktuk develop and facilitate week-long career and technology studies (CTS) courses in collaboration with outside experts, Elders and other community leaders. These courses are delivered during the first 2 weeks of school to stimulate scholastic interest and achievement. CTS courses expose students to different career possibilities in fields that are broadly applicable to their communities, and cover diverse subjects including media and film studies, global issues, cooking and sewing. Students begin their school year in early August, so teachers can take advantage of outdoor opportunities in environmental studies subjects such as GPS and mapping, outdoor survival skills ('land skills') and plant identification.

'Insects and climate change': a CTS course

In 2010, a CTS course with the theme of insects and climate change was developed in collaboration with author C. Ernst and two local teachers (D. Frenette and S. Durdle), and was delivered to approximately 30 students in grades 9 to 12 at Kugluktuk High School. Climate change is a growing global concern that has particularly significant implications for communities and ecosystems in the extreme north. Although global surface temperatures are expected to rise several degrees during this century, this warming trend will be amplified in the Arctic (IPCC 2007: 30). Insect life cycles, activities and relationships with other organisms are expected to be broadly affected by climate change (Leborgne *et al.* 2011).

Inuit people have already noted the effects of climate change on local fauna, including changes in insect populations. In particular, many 'different' or 'new' insects have been seen where they were not previously known (e.g., certain dragonflies, bees, beetles and wasps on Arctic islands and northern coastlines), more biting flies have been observed, and 'mosquitoes have been said to emerge earlier in the spring in some regions' (Ashford and Castleden 2001: 5). The possible effects of these changes 'are seen as potentially positive (e.g., the larger numbers of insects are associated with increased bird populations as well as the appearance of new bird species, which could translate to potential tourism benefits) as well as potentially negative (e.g., more biting flies could increase the incidence of disease in both animals and humans)' (Downing and Cuerrier 2011: 65). Insects are clearly a useful and readily observed indicator of environmental change in the north; monitoring insects and their kin will be an important tool for both identifying and predicting effects of environmental change.

For five days, students participated in a series of in-class and outdoor activities. The learning objectives for students in this course were to:

1.   explore the ecological and cultural roles of insects in their community;
2.   discover what insects can reveal about current and future climate change in the north;
3.   develop skills and techniques for collection, identification and analysis of insects;
4.   use knowledge acquired about insects to study general biological principles, such as classification, nomenclature, biodiversity, anatomy and adaptation.

Introducing the subject

Through in-class discussions, the teachers introduced the subject of climate change. Students were asked to think about things they had seen, or might expect to see, in their communities, as a result of climate change. 'Changes in the insects' was a subject that came up quickly, and students were able to identify several life-history traits that might be affected. 'Insects' were defined biologically/taxonomically, and were compared to other types of arthropods found in the Arctic, such as spiders. With this definition established, insects could be used to illustrate and explore science concepts such as classification, nomenclature, identification, and biodiversity.

Accessing local traditional knowledge

Students were instructed to conduct interviews with their parents, grandparents (or other family members) and Elders, in which they were to gather traditional stories or information about insects. Their findings were shared with their classmates at the end of the week.

Skills development

Students spent a morning in the classroom with the visiting entomologist (C. Ernst), who described her research activities in Kugluktuk and demonstrated different techniques used to collect insects in the field. Actual field equipment was used so that students could learn and practise collection methods before going into the field. Time was later spent outdoors with the equipment and students were encouraged to experiment with different collection techniques and tools. Back in the classroom, students discussed the efficacy and 'user-friendliness' of the different methods and had a chance to examine the specimens they collected. Students identified their specimens using field guides and computers with internet access. To further engage their interest, students each selected one insect of special interest to them. Over the week, students used the insect in a series of activities involving art (students created realistic pencil drawings of their specimens), science (students researched and labelled anatomy) and literacy skills (students researched and wrote brief reports about the biology and natural history of their insect).

Participation in a 'real' science project

This first year of this CTS course coincided with a research project in the Kugluktuk area aimed at collecting a wide variety of terrestrial and aquatic insects (i.e., the NBP). This presented a unique opportunity to expose students to an established research project and to work with a professional entomologist. The next phase of the course included a day at one of the NBP research sites. In preparation for this exercise, several sets of 'practice traps' were established at the study site to provide the students with hands-on training on the sampling protocol being used by the larger project. Students worked in supervised groups until they mastered the skills required to empty traps, label their specimens and reset the traps for future sampling. The students were given the same equipment used by researchers; this helped reinforce the notion that they were, indeed, doing actual science. Once their collection skills were sufficiently developed, the student groups serviced 'real' traps being used in the study, thereby collecting data that would contribute to the research programme.

Although the specimens collected from the 'real' traps were kept for formal identification for the research programme, those collected from the 'practice' traps were brought back to the classroom so students could put into practice their new identification skills as well as some basic data analysis.

Outcomes of the project

Training was provided to teachers that would permit the programme to continue without the need for an entomologist to be on site each year. This is necessary, since funding is rarely available to bring an entomologist to the community. Ongoing long-distance communication and partnership with the school will help preserve this programme for the students, and also may result in continued scientific collaborations (i.e., collection of specimens). With the help of these students and other members of the community, a collection of charismatic and culturally significant insects was initiated. The insects are mounted and labelled and are being stored at the high school for future use at the school and in the community at large. Research efforts in the region, and continued partnership with the school, will add to this collection in the future.

CASE STUDY 2: A SUSTAINABLE STUDENT STREAM INSECT MONITORING PROGRAMME FOR SAHTU COMMUNITIES IN THE NORTHWEST TERRITORIES

In the NWT of northern Canada, students are being trained in environmental, wildlife and geological techniques, as part of an experiential learning programme designed to engage youth in the sciences. The environmental programme includes terrestrial, marine and freshwater modules, with the freshwater modules currently under construction (S. Daniels, Northwest Territories Department of Education, pers. com.). As part of this project, sampling protocols for insects living in northern streams and rivers are being developed and integrated into stream bioassessment programmes within high schools across the NWT. In this section, we report on a partnership with schools and the local Renewable Resources Board from one district of the NWT to develop and test protocols that could be incorporated into the experiential learning programme while providing a long-term water quality monitoring programme for the region.

Context

Community-based environmental monitoring programmes, particularly those including local youth, have been gaining popularity in small communities of northern Canada. These answer a number of local needs: (1) to build capacity in small northern communities which lack access to environmental monitoring infrastructure; (2) to provide rapid access to monitoring information on a local scale; (3) to instil a sense of data ownership in the local community; and (4) to engage local youth in positive and productive learning activities. The Sahtu region of the NWT is located along the middle section of the Mackenzie River and includes the communities of Norman Wells, Deline, Tulita, Colville Lake and Fort Good Hope. The Sahtu Renewable Resources Board (SRRB) has been working with students and researchers for many years to investigate and monitor wildlife and water quality in the region.

The Sahtu district encompasses a large area rich in natural resources. Effective monitoring is needed in this area to determine the effects of resource development activities (logging, mining, etc.) on streams, rivers and lakes. One of the on-going projects in the Norman Wells area has focused on Bosworth Creek. The SSRB has been monitoring water chemistry, fish and aquatic invertebrates as indicators of water quality, with the help of local students and researchers from various government agencies and universities (SRRB 2007). Most recently, the focus has been on developing a sustainable water quality programme using aquatic insects that could be incorporated into the school curriculum. An inventory of aquatic insects can be useful for monitoring large scale patterns of change in northern insect communities (similar to the terrestrial monitoring programme described earlier).

In 2008, a partnership between the SRRB and author D. Giberson led to local workshops and student training in aquatic insect identification. The success of these ventures resulted in discussions about how to incorporate insect monitoring into the school curriculum as a long-term learning and monitoring programme. A graduate student project (K. Vinke) was subsequently established to carry out baseline monitoring and to design and test protocols for use in the schools.

Long-term monitoring programmes in the north are hampered by access to remote sites, the high cost of shipping supplies, and the unavailability of trained local scientists (Environment Canada 2008;

Giberson and Guthrie 2010). At the same time, teachers in this region have expressed a strong desire for more hands-on and relevant science activities for their students. Recent restoration activities in Bosworth Creek provide a unique opportunity to incorporate a local water quality monitoring programme in the curriculum. By entrusting the research programme to students and their teachers, and by linking the material to the science curriculum, two key outcomes are achieved: (1) insect samples that can be used in biomonitoring research are collected in a consistent manner each year, and; (2) students are provided with hands-on scientific training and are provided with a unique field-based learning experience. Specific learning objectives for this programme include:

1. understand why healthy streams and their insect fauna are important to the community and local culture, and their links to traditional knowledge;
2. be able to describe some of the human activities that can affect streams, and their implications for insects and other wildlife;
3. explore stream protection, mitigation or restoration possibilities;
4. discover how local species inventories can be used to track larger-scale and long-term trends;
5. engage directly in the methods of ecological monitoring (using benthic invertebrate tolerance scores and community composition as an indicator of stream health);
6. apply new skills and knowledge by conducting a basic stream assessment in the field; and
7. communicate study results to classmates and other people in the community.

Establishing baselines for the field and the classroom

The research project started in the summer of 2010, with baseline collections of insect samples from creeks in Norman Wells and the surrounding area using standard protocols (i.e., Environment Canada 2008). Different methods and tools were tested in the field (e.g., mesh sizes, sampling periods and types of habitats) to determine their practicality and usefulness for allowing students to obtain accurate samples. A local Dene youth was employed to assist with this preliminary sampling and lab work and to provide feedback.

Testing field protocols with school students

Introducing the subject

In the autumn of 2010, 30 students (grades 5–12) from Mackenzie Mountain School in Norman Wells embarked on a field excursion to Bosworth Creek. Beforehand, students received in-class instructions and a briefing on the project and protocols so that they knew what to expect in the field. Equipment was demonstrated, and students had an opportunity to try out equipment in a controlled classroom environment.

Testing field protocols

In the field, students were divided into three groups: (1) water chemistry, (2) physical habitat characterization and (3) stream insect sampling. In addition to measuring water turbidity, dissolved oxygen and stream flow, students practised collecting a standardized sample of the insects and identifying some of the animals in the field. This excursion allowed programme leaders to watch the students working through the various activities, allowing them to (1) assess classroom dynamics, (2) determine whether the instructions were sufficient to keep the students on track and (3) observe how well students adapted to the protocols (i.e., their user-friendliness). Students were asked to provide feedback by filling out forms upon completion of the field activities. These observations will be used to improve the programme.

These students demonstrated familiarity with terrestrial 'pest' insects, such as Caribou parasites and mosquitoes, but the concept of creek-dwelling insects was new to most of them. There were notable exceptions, however, as some students in Norman Wells had previously received training in aquatic insect sampling as part of other research projects conducted in the region. Those students helped their inexperienced classmates, thereby demonstrating the value of student mentors during this kind of project work. Regardless of their level of previous experience, students were largely interested and cooperative during the field trip.

Involving other schools in the Sahtu district

The ongoing Bosworth Creek monitoring programme has been important in facilitating partnerships between insect researchers and the local schools. Other goals of the project include providing curriculum materials and generating interest in the monitoring programme throughout the Sahtu district. Another annual science-based project in the region, the Sahtu School Tour, was instrumental in addressing these other goals. Organized by the University of Calgary Veterinary College and the Norman Wells staff of Environment and Natural Resources, this annual trip brings wildlife researchers to all five communities in the NWT (Norman Wells, Tulita, Colville Lake, Fort Good Hope and Deline). Students are engaged in entertaining classroom-based wildlife activities in order to enhance their interest in science. In February 2011, an insect component was included in the tour, organized by K. Vinke. In addition to delivering planned school activities, meetings were held with teachers in each community to discuss the monitoring project. The activities included activities and crafts designed to help students learn about stream insects and their physical and behavioural adaptations. The trip also allowed the programme leaders to assess the interest and resources available in each school for the proposed future stream biomonitoring programme.

The future of student-run stream biomonitoring in NWT

Based on analyses of the insect samples that were collected during 2010 and considering the practicality of different methods, a demonstration video and a draft set of protocols for insect sampling, water chemistry testing and physical habitat descriptions are being developed for use in the student-run biomonitoring programme. These protocols will be tested in the summer of 2012 in field trials with students in the Sahtu. Based on the feedback obtained from these trials, and from consultations with other members of the community, a final set of guidelines will be produced. These guidelines will provide links to the science curriculum, will allow teachers and students to sample their local streams and will provide a framework for data management and application of these data (e.g., adding to local species inventory, contribution to northern biodiversity monitoring and assessment of stream conditions).

CONCLUSION

We have discussed how educational opportunities involving insects engage youth and provide a tangible link to more formal science training and inquiry, and provide benefits for students and researchers. In additional to longer-term programmes, informal or impromptu learning/teaching opportunities are abundant and require little effort from scientists to find and exploit them. Such opportunities could be as simple and brief as a chat with a local who happens to stop and make an inquiry about the researcher's work, or a quick display of sweep-netting to curious children. These impromptu teaching/learning moments take little time or effort, yet can make a profound impression on the participants, and help foster strong and positive relationships within the community. Activities that require only a slightly greater time commitment might include agreeing to an interview with local media, spending a few hours collecting 'creepy-crawlies' with children in a summer camp, helping with entomology-based science fair projects, reading an insect-themed book to a school group, hiring a high school student to work as a field assistant or taking older youth on field excursions. Many opportunities exist to partner with other education-based programmes. For example, the non-profit group *Actua* offers free summer science camps to youth from across Canada, (including programmes in northern communities), and they welcome enthusiastic volunteer visiting scientists who are willing to work with their campers. The 'Let's Talk Science' programme involves science students from university campuses across Canada, and links them with school groups and youth in their communities.

Scientists working in the north gain tremendous benefits from partnerships in local communities. From a scientific standpoint, local residents can advise on site selection, weather conditions, safety, transportation routes and, perhaps most importantly, traditional and local knowledge of patterns of insect diversity, emergence and distribution. Also, trained students can help collect specimens or data as they learn and receive training, and are able to continue collecting long after the researcher has departed. However, scientists who visit isolated communities, often for only brief periods, must be mindful of the social context in which they are conducting their research. They must follow ethical guidelines and be respectful of the people and the land.

The time commitment and equipment to pursue local partnerships is minimal, but the impact can be profound. We have experienced directly the benefits of using arthropods in an educational context in northern Canada, and our experiences suggest the opportunities are untapped. Given their abundance, diversity, importance in northern Canada, and ease and efficiency of sampling, arthropods are certainly one of the best 'models' for pursuing further partnerships between schools, communities and researchers.

REFERENCES

Ashford, G. and Castleden, J. (2001) Inuit Observations of Climate Change: Final Report. New York: International Institute for Sustainable Development.

Bergman, G. J. (1947) A determination of the principles of entomology of significance in general education. *Science Education*, **31**, 1.

Blackawton, P. S., Airzee, S., Allen, A. *et al.* (2011) Blackawton bees. *Biology Letters*, **7**, 168–172.

Boardman, L. A., Zembal-Saul, C., Frazier, M., Appel, H. and Weiss, R. (1999) Enhancing the 'science' in elementary science methods: a collaborative effort between science education and entomology. Unpublished report, Pensylvania State University.

Campbell, K., MacLuich, C., Wheelock, A., Williams, K. and Wunderlich, R. (2008) Experiential Science 10. Report. Government of the Northwest Territories, Department of Education, Culture, and Employment, Pacific Educational Press, UBC Faculty of Education.

Campbell, K., Freeman, P., Laite, A. W., Williams, K. and Wunderlich, R. (2009) Experiential Science 20. Report. Government of the Northwest Territories, Department of Education, Culture, and Employment, Pacific Educational Press, UBC Faculty of Education.

Danks, H. V. (1981) *Arctic Arthropods: A Review of Systematics and Ecology with Particular Reference to the North American Fauna*. Ottawa, Canada: Entomological Society of Canada.

Downing, A. and Cuerrier, A. (2011) A synthesis of the impacts of climate change on the First Nations and Inuit of Canada. *Indian Journal of Traditional Knowledge*, **10**, 57–70.

Environment Canada (2008) Overview of Arctic freshwater systems. Available at: http://www.ec.gc.ca/api-ipy/default.asp?lang=En&n=95333198-1, accessed 22 September 2010.

Fischang, W. J. (1976) Another wasted resource. *The American Biology Teacher*, **38**, 294.

Giberson, D. and Guthrie, D. (2010) Adaptation of CABIN Sampling Protocols for Use in Canada's North. Unpublished document, University of Prince Edward Island.

Gray, A. (1976) Terrestrial arthropods in the elementary classroom. *The American Biology Teacher*, **38**, 211–115.

Haefner, L. A., Friedrichsen, P. M. and Zembal-Saul, C. (2006) Teaching with insects: an applied life science course for supporting prospective elementary teachers' scientific inquiry. *The American Biology Teacher*, **68**, 206–212.

International Panel on Climate Change (IPPC) (2007) Climate Change 2007: Synthesis Report. Contribution of Working Groups I, II and III to the Fourth Assessment Report of the Intergovernmental Panel on Climate Change. Geneva, International Panel on Climate Change.

Laugrand, F. and Oosten, J. (2010) Qupirruit: insects and worms in Inuit traditions', *Arctic Anthropology*, **47**, 1–21.

Leborgne, L., Ernst, C. and Buddle, C. (2011) Shaping tomorrow's northern ecosystem: arctic insects, spiders, and their relatives in a changing climate. *Meridian*, Spring/Summer,13–17.

Mathews, J. R. (1988) Adult amateur experiences in entomology: breaking the stereotypes. *Bulletin of the Ecological Society of America*, **34**, 157–162.

Matthews, R. W., Flage, L. R. and Matthews, J. R. (1997) Insects as teaching tools in primary and secondary education. *Annual Review of Entomology*, **42**, 269–289.

Roberts, L. W. and Clifton, R. A. (1988) Inuit attitudes and cooperative learning. *McGill Journal of Education*, **23**, 213–230.

Sahtu Renewable Resources Board (SRRB) (2007) Bosworth Creek Monitoring Project 2006/2007 Annual Summary Report. Unpublished document, SRRB.

Takano, T. (2005) Connections with the land: land-skills courses in Igloolik, Nunavut. *Ethnography*, **6**, 463–486.

Zenger, J. T. and Walker, T. J. (2000) Impact of the internet on entomology teaching and research. *Annual Review of Entomology*, **45**, 747–767.

19

Discovering the microwilderness in parks and protected areas

JESSICA J. RYKKEN AND BRIAN D. FARRELL

INTRODUCTION

Millions of visitors travel each year across town, the country or even the globe to experience the natural and cultural heritage of parks, monuments, and other protected areas. Depending on the destination, the draw may be some combination of historic sites and artefacts, wild and scenic natural landscapes and iconic wildlife species. In the United States, for example, visitors travel great distances to focus their cameras on the grizzly bears and bison of Yellowstone, alligators of the Everglades and California condors soaring above the Grand Canyon. In other parts of the world, safaris lead visitors to big game animals such as rhinos and tigers, and tropical forest guides point out colourful quetzals and toucans. 'Wildlife' in this context, generally implies mammals, reptiles, birds and other vertebrate animals. Invertebrate animals, on the other hand, remain invisible to most park visitors, at best, or, if visible, are often perceived as repulsive, terrifying or harmful (Kellert 1993). For their part, park managers and staff typically know very little about their invertebrate fauna (Ginsberg 1994; Kerley *et al.*, 2003), and may even reinforce the fear that insects and their relatives are a threat to the health of humans and landscapes. It is no wonder then, especially in parks with large, charismatic animals readily viewable to the public, that neither park staff nor visitors have traditionally shown much interest in exploring what renowned entomologist and author, E. O. Wilson, has termed the 'microwilderness'.

The Management of Insects in Recreation and Tourism, ed. Raynald Harvey Lemelin. Published by Cambridge University Press. © Cambridge University Press 2013.

Certainly, there are exceptions to these generalizations of apathy and ill will towards invertebrates in parks. Where parks do promote insects to the public, these efforts have usually focused on charismatic species, or rare species associated with unique habitats. For instance, large congregations of monarch butterflies provide a spectacle for visitors to parks along their migration routes (e.g., Point Pelee National Park on Lake Erie in Ontario) or to parks providing overwintering sites in California. As Lemelin (2007) explains, equally large, colourful and charismatic are dragonflies, and this group has become a focus for leisure activities such as dragonfly counts in protected areas, as well as for conservation efforts in dragonfly sanctuaries (see also Samways this volume). Fireflies draw tens of thousands of visitors to Japan (e.g., Fussa and Shionoe firefly festivals) and to the Great Smoky Mountain National Park each year in early June to view the remarkable phenomenon of synchronous flashing. Some insects get attention for the novel places they live. The unique troglobitic arthropods (including springtails, harvestmen, cave crickets and blind beetles) found in Mammoth Cave National Park are featured on the park's website, and even Yosemite National Park, home to abundant vertebrate wildlife, highlights a newly described pseudoscorpion on its visitor website, found, as yet, only in talus caves within the park. These exceptional cases share a common theme. They are focused on one or a few insect species for which there is access to information, especially images, that bring insects up to a scale where they become interesting creatures. Information is what these exceptional insects share with vertebrates, for which there have long been field guides, videos and other images in popular culture.

PARKS PARTNER WITH SCIENTISTS FOR BIODIVERSITY INVENTORY

It is generally the case that parks must balance dual missions, and these are clearly defined in the US National Park Service Organic Act of 1916, which mandates the conservation of natural systems, wildlife and historical objects within parks to allow for visitor enjoyment now and in the future. Inherent in this mandate is the idea that parks have a solid understanding of the biodiversity that is to be conserved, and that they provide for visitor enjoyment of this natural heritage (Ginsberg 1994). To this end, it has long been the case that national parks and other federal land management agencies have hired biologists and botanists to study, manage and promote the health and diversity of birds, fish, plants and mammals within protected areas. On the other hand, they have hired entomologists primarily to control insect pests such as gypsy moths or bark beetles. With such disproportionate knowledge of, or interest in, the largest component of biodiversity under their jurisdiction, it seems improbable that parks could successfully address the essential mission of conserving and protecting natural systems and wildlife for future generations. This situation has not gone unnoticed.

Scientists have become increasingly aware, over the last several decades, of species and habitat loss worldwide (Fahrig 2003; Hanski 2011). Networks of parks and other protected areas are presumed to be key elements in biodiversity conservation strategies, even though we lack information about the full spectrum of biodiversity contained within even the smallest conservation area (Janzen and Hallwachs 1994). In an effort to fill this knowledge gap, Janzen and Hallwachs (1994) developed the idea for an All Taxa Biodiversity Inventory (ATBI), originally to be conducted in the Guanacaste National Park in Costa Rica, with the goal of documenting *all* species within the boundaries of a protected area within a relatively short period of time. This intense effort would rely on the expertise of many specialists, simultaneously focused on the biota of one place. Naturally, insects and other invertebrates would dominate the cataloguing efforts. At about the same time as the development of the ATBI concept, entomologists within the US National Park Service were making strong recommendations that an inventory and monitoring programme for invertebrates be started in the national parks (Ginsberg 1994). They argued that because invertebrates are so vastly diverse, play an integral role in ecosystem processes and also have significant impacts on human health and economics, they should receive increased attention in the parks. The time was right for taking on an inventory effort that gave proportional focus to invertebrate biodiversity. When Janzen's plans for an ATBI in Costa Rica fell through, the model was brought to Great Smoky Mountains National Park for implementation.

During the first 12 years of the inventory in Great Smoky Mountains National Park, scientists have catalogued approximately 7000 species new to the park (including more than 5000 invertebrates and only 21 vertebrates), and more than 900 species new to science. But the ATBI was envisioned as more than a numbers game. From the beginning, the overall goal has been to integrate three components: science, outreach and education and management. Along with cataloguing new species, the aim is to educate the public and generate excitement about these new discoveries, and to use biodiversity information to inform natural resource management. Great Smoky Mountains National Park has led the way on all of these fronts. Not only has this biodiversity hotspot documented an impressive number of species, but the park has also implemented educational activities and curricula at its residential learning institute and in the surrounding public school systems. The ATBI has served as a model and inspiration for other parks across the country to begin biodiversity discovery activities of their own. While most of these efforts are not on the scale of the ATBI in the Great Smoky Mountains, all of them have one thing in common: invertebrates take centre stage in the discovery process. As a result, biodiversity projects are becoming significant drivers for invertebrate-focused research, outreach, and education in the national parks.

Not all parks interested in starting a biodiversity inventory have the resources to implement a multi-year ATBI, and an increasingly popular activity that can generate a large amount of data in a very short time is the 'bioblitz'. Originating in the US, but now held throughout Australia, Canada and Europe, and in Hispaniola, bioblitzes often have a taxonomic or habitat-specific focus (e.g., beetles or the intertidal zone), and invite specialists to catalogue as many species as possible in a specified period of time, generally 24 hours to a week. A secondary objective is to involve citizen scientists in the collection and processing of data, and to educate the general public about local biodiversity (see Johansen and Auger, this volume). As an example, Acadia National Park in Maine conducts an annual bioblitz, typically focusing on one arthropod order, in an effort to complement and make comparisons with a historical survey from Mount Desert Island (Procter 1946). Trained citizen scientists work side by side with professionals to assist in field collecting and specimen processing. Whether focused on spiders of the bottomland forests of Congaree National Park in South Carolina, or on beetles in the high deserts of Great Basin National Park, bioblitzes have been an effective way to bring scientists and the public together to document insect biodiversity. Locating bioblitzes in national parks near urban areas to attract as diverse an audience as possible has been the strategy of National Geographic-sponsored events leading up to the National Park Service Centennial in 2016, including bioblitzes in Rock Creek Park (Washington DC), Santa Monica Mountains National Recreation Area (Los Angeles) and Indiana Dunes National Lakeshore (Chicago).

AN URBAN ATBI: DISCOVERING THE MICROWILDERNESS
OF THE BOSTON HARBOR ISLANDS

Inspired by successful models in the Great Smoky Mountains and on the island of Hispaniola (Farrell 2005), Boston Harbor Islands national park area launched an ATBI in 2005, with its primary mission, as envisioned by supporter E. O. Wilson and park staff, to document the microwilderness of the islands. The project is a collaborative effort between the national park area and the Museum of Comparative Zoology (MCZ) at Harvard University, where scientists have taken on the tasks of conducting and coordinating the inventory itself, maintaining a database and website, producing educational materials, and providing direct input to park-sponsored outreach and education activities (see Figures 19.1–19.3). The MCZ benefits directly from this collaboration by providing students and scientists with freshly collected specimens for morphological and molecular studies from local species otherwise little documented since the early part of the last century.

The park is comprised of 34 islands and peninsulas which lie within a few miles of downtown Boston. They have served as sites for various human activities over the centuries, including fishing settlements, farms, forts, schools, hospitals, a brothel and a landfill. Established in 1996, the park is unusual in comprising a partnership of many landowners (Boston Harbor Islands Partnership), and the islands now include conservation lands, an environmental education centre, historic sites and recreation and camping areas. Visitors to the park are thus diverse in their backgrounds and interests, and most come for its various cultural and recreational attractions: ruins of forts built during several wars, swimming beaches and visitor centres, a few campgrounds that allow visitors to 'get away from it all' without travelling too far from the city. Nature enthusiasts explore too, but while some of the islands provide important breeding sites for sea and shore birds, there is little expectation that a trip to the islands will afford great opportunities to view wildlife in the traditional sense.

One might wonder about the motivation driving an ATBI in a small, urban island park. These islands are obviously not a hotspot of biodiversity as are the Great Smoky Mountains or parks in the tropics. In fact, biodiversity is expected to be comparatively low due to the temperate bioregion, high levels of human disturbance and the limited area and isolation of the islands. From a practical standpoint, however, this can be seen as a benefit, making the daunting task of documenting *all* species in the park somewhat more manageable and feasible.

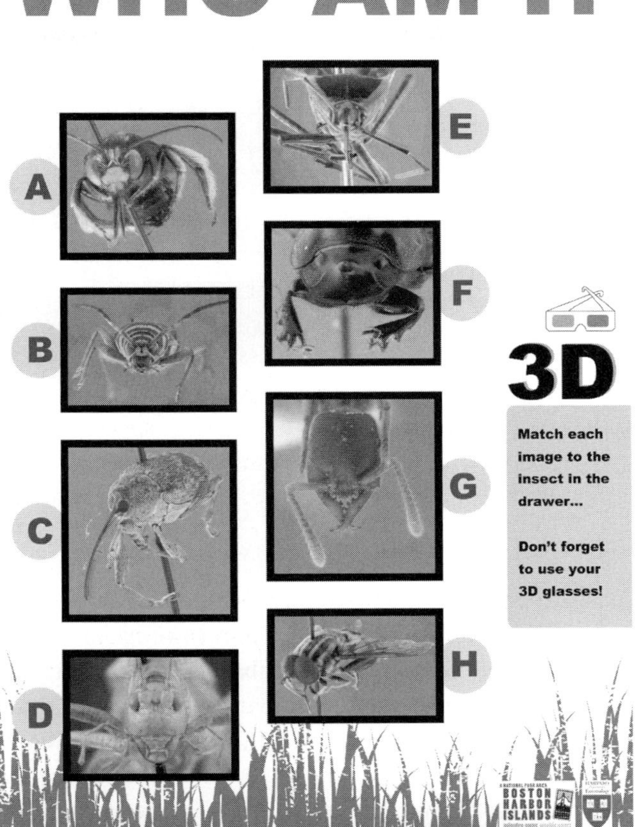

Figure 19.1   Boston Harbor Islands poster challenging participants to match 3D images with actual pinned insect specimens.

An ATBI in this landscape is also expected to provide novel information about patterns of island colonization on a small spatial scale, and species resilience to human disturbance. But those are concerns for scientists. From an interpretive perspective, the park's location, in the heart of New England's most densely populated metropolitan region, couldn't be better for accessing a large and diverse audience. And by taking advantage of the opportunities for outreach and education that this urban population affords, the ATBI can lead people on a voyage of discovery into the vast and unexplored realm of island microwilderness, while simultaneously teaching them what lives in their own back yard.

Figure 19.2   Boston Harbor Islands 'Creatures of the microwilderness' fold-out field guide (one side).

GENERATING ENTHUSIASM AND APPRECIATION FOR
THE LITTLE THINGS THAT RUN THE WORLD

Having a good project and a large audience sets the stage, but insects and their relatives can be a hard sell. From sowbugs to scorpionflies, getting park staff or the public excited about 'the little things that run the world' (Wilson 1987) is often more challenging than the process of documenting the species themselves. The Boston Harbor Islands ATBI has used two main approaches to generate enthusiasm and interest about invertebrate diversity in the park. One is to educate people about the project and its discoveries through various activities and media. The other is to involve students, volunteers, and citizen scientists in the process of discovery itself.

Bug kits and bug cubes

Essential to the success of all ATBI-related education and outreach programmes has been a close collaboration between scientists, park rangers, classroom teachers and other environmental educators. This has been especially true for the development of an insect programme that has reached several thousand students in the Boston Public Schools and surrounding school districts. Because park education rangers have long been working directly with teachers to develop and implement island-themed programmes for a diversity of students, the infrastructure was

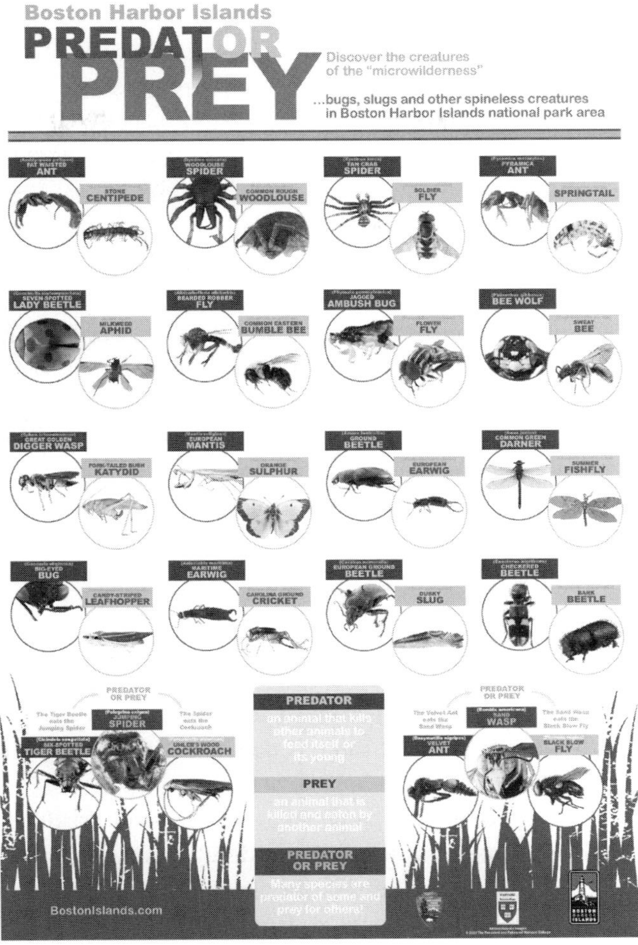

Figure 19.3   Boston Harbor Islands 'Predator prey' poster showing matched pairs from the predator–prey card game.

in place to develop a programme focused on insects and the ATBI. The first iteration trained rangers and educators to lead student groups to the islands with a bucket full of gear (nets, beating sheets, aspirators, vials, etc.) to collect insects in different habitats. The collecting is followed up with order-level field identifications using hand lenses, sorting trays and simple guides, comparing habitats and entering the information on a data sheet. Rangers emphasize to students that the specimens they contribute to the ATBI will ultimately be identified to species and integrated into the project database. Later on, students can search the online database to find their specimens, and to see high resolution images of those species. Nowadays, the 'Bug Kits' are packaged in heavy-duty rolling suitcases that can be easily hauled down city streets, on ferries and through meadows, forests and salt marshes.

The island insect collecting programme was so well-received by rangers, students and teachers alike that the next effort was to develop a classroom programme. For many years, in the winter and early spring months, park rangers have travelled to inner city public schools to present a variety of island-themed programmes in the classroom; now they were eager to include a bug unit. As students weren't out collecting specimens in the field, scientists provided the rangers with a large diversity of arthropods in glass ethanol-filled vials – replicating storage methods in the lab. This proved less than satisfactory for several reasons, among them that glass vials of ethanol have a short life expectancy among middle school students, and also the frustration felt by students when trying to view an animal spinning in ethanol through the curved sides of a vial. The next attempt proved much more successful. Using clear casting resin, an ATBI intern manufactured 160 'bug cubes' that resemble fly-in-an-ice-cube practical jokes, but feature a diversity of insects and other invertebrates from the islands. These have been a big hit with students because they are virtually indestructible, the specimen can be examined easily from all angles with a hand lens, and there is the sensation of holding a real bug in your hand without the 'ick' factor.

With bugs literally in hand then, rangers took it upon themselves, with guidance from scientists, to develop creative and focused learning activities. This included the design of colourful, simple guides to allow identification of all 'bugs' to family, order or class. The guides also provide information on feeding habits, so students can sort their invertebrates taxonomically or by functional group. Sorting and classifying generates data sheets with numbers, and these, in turn, can be used for calculating proportions and fractions, and so on. Because public school teachers are typically reluctant to devote time or effort to classroom activities that do not work towards meeting state curriculum standards, the success of this park-sponsored insect programme has relied on the close relationship rangers develop with teachers in developing activities that are fun, interesting, and incorporate explicit skills relevant to the curriculum. Ideally, most of these students will go out on an island insect-collecting expedition later in the spring, able to observe firsthand some of what they have already learned about insect diversity and ecology.

Spreading the word and the importance of stories

Getting the word out about the project and what scientists are finding in the park has been an integral component of the outreach effort. To this end, talks about the ATBI and its discoveries have been targeted to many different audiences, including local conservation groups, garden clubs, educators, national park service biologists, managers, and interpretive rangers, other scientists, youth groups and students. The talks emphasize the process of discovery (including the human effort required in collecting, sorting, pinning, labelling and identifying the tens of thousands of specimens processed in the ATBI) as well as the rich diversity of species that can be found in an urban island park. Perhaps the biggest draw of every talk, however, is the three-dimensional show of insect heads. Horse fly eyes, beetle mandibles and spider fangs the size of a projector screen zooming out at the audience in the dark elicits gasps from scientists and second graders alike. This gimmick may not go far to dispel people's fears of insects, but it does make them appreciate that insects *are* small. Another venue for showing off the project within the park is an informational display in the visitor centre, which includes a glass-topped cart with 100 common insects from the islands, and posters explaining how and why they're collected.

Although the discoveries of an inventory are easily summarized by numbers: more than 1700 new species for the park in the first phase of the ATBI; dozens of new state, regional, and country records; more than 65 000 specimens in the database – such numbers can fail to impress audiences if there is no frame of reference. A better way to captivate an audience with vast discoveries is to start at the other end of the count and tell fascinating stories about individual species. For instance, on the Boston Harbor Islands, the rediscovery of a tiny ground beetle, *Bembidion nigropiceum*, not seen in North America since 1897, has generated a disproportionate volume of excitement. The beetle is certainly not much to look at, a couple of millimetres in length, brown, just another bug scurrying in the beach gravel. Long thought to be a species new to science, then recognized as an early introduction from the beaches of northern Europe, scientists were doubtful that populations persisted to this day. When the elusive beetle showed up in pitfall traps on one of the islands, scientists were elated at the rediscovery. But the excitement reverberated much farther than the scientific community. The story of the '100-year beetle,' as it has been dubbed by park rangers, is now told regularly at donor fund-raising events, and to visitors and volunteers out on the islands. A skilled interpretive ranger sharing the excitement of scientific discovery can make even the tiniest insect loom large in the public's imagination.

Taking a closer look

By definition, the microwilderness is populated by small inhabitants. Good stories aside, it can be difficult to appreciate or generate enthusiasm about something you can hardly see, and it's no coincidence that butterflies and odonates are the darlings of what is generally perceived as a largely uncharismatic microfauna (see New and Samways, this volume). 'Insect diversity' typically doesn't instil much awe or interest when it consists of countless tiny drab creatures that look almost identical to the naked eye. To begin to appreciate the diversity among most insect species, they must be brought up to a scale where the differences in colour, form and texture of their bizarre bodies can be seen clearly (see Mitchell, this volume, for a discussion on the role of technology and the creation of awareness). To achieve this, the Boston Harbor Islands ATBI has borrowed heavily from ideas and products generated in a concurrent ATBI on the island of Hispaniola (Farrell 2005). The main ingredients are high resolution images showing off the remarkable array of tubercles, setal fringes, tentorial pits, iridescent scales and undeniably terrifying mandibles that insect biodiversity has to offer. The first place to look for these images is in the ATBI database, which is accessible to the public on the project website (see biocaribe.org). Every insect species documented on the islands, from the Colorado potato beetle to the pavement ant, has its own photo gallery. Researchers, students and interested amateurs alike can access these images, but this is just the beginning.

'Predators of the Boston Harbor Islands' is the title running across one of the dozen or so biodiversity posters featuring invertebrates of the park. But instead of wolves or alligators, lady beetles and pseudoscorpions are among the 24 invertebrate predators featured in this rogues' gallery. People can be caught off guard, incredulous, for instance, that those sweet little ladybugs could be lumped with the hungriest lions, until they think about their voracious appetite for aphids. And seen at this scale, the fearsome mandibles, armoured forelegs and vice-like pincers that adorn even the smallest creatures come into plain view. The biodiversity posters feature taxa as diverse as bees, leafhoppers, ants, and weevils, and they have been a big hit with a wide range of audiences, including park visitors, students and scientists. Teachers bring posters into their classrooms, taxonomists hang them in their labs, they're displayed in the park visitor centres and on island ferries and they are raffled off at fundraisers. Their quirky aesthetic appeal probably brings the posters into living spaces too. And even as these posters are admired indoors, the motive for their creation is clear. To know what a carpenter ant really looks like up close might also make it easier to appreciate in the field

A more practical resource to bring to the islands, of course, is a field guide. 'Creatures of the Microwilderness' is a fold-out guide featuring images of 60 commonly encountered island invertebrates. From beach hoppers to weevils to ambush bugs, the goal was to select a manageable number of diverse taxa that are large and distinctive enough to recognize with the naked eye. As with any guide to birds or plants, field guides to insects and other invertebrates in the park encourage people to look more closely at the world around them. And whether or not someone can correctly identify the creature sitting on the window of the visitor centre to family, genus, or species, matters less than the act of observing, comparing, and discovering. The guides are given to interpretive rangers during their training programme, then distributed by them to park visitors, volunteers, students and others. They are currently the *only* field guide to wildlife on the islands, other than a checklist for birds.

Games can also be an effective way to teach. Inspired by an urban youth group that works to promote the park in local schools and community centres, a collaborative team including scientists and park staff developed a card game that showcases insect diversity. The youth group initially received a small community grant with the idea to create cards featuring 'cool bugs' from the islands that could be given away to audiences after presentations. The resulting PredatOR Prey matching card game, is comprised of 20 pairs of island invertebrate predators and their prey. Along the lines of Go Fish, the fun comes in asking another player if, for instance, they have a great golden digger wasp to match the fork-tailed bush katydid in your hand (No? Then 'bug off!'). The fun, informative cards and the accompanying PredatOR Prey poster and t-shirts won a graphic design award. With several thousand decks now printed, they have also won rave reviews from educators, students and others.

Getting the public involved in the process of discovery

In addition to educating people *about* insect diversity through curricula, stories, posters, field guides and so forth, the ATBI has been an effective mechanism for getting people involved in the process of biodiversity discovery. In fact, the project relies on non-scientists as much as it does on scientists to get the work done. What's more, participants of all backgrounds have equal opportunities for making new discoveries. While some of the project's many interesting finds have come from the nets of fifth graders, valuable data for charismatic groups such as butterflies have been contributed by skilled amateurs.

Working side by side with scientists, a small army of high school and undergraduate students, as well as park-sponsored interns and volunteers, have carried out the day-to-day work of running an intensive invertebrate sampling regimen on ten islands, including various kinds of traps (e.g., pitfalls, malaise, pans, lights). For many students, collecting insects and other field data for the ATBI has been their first real-world field biology experience, and the combination of the park being local (boat docks can be accessed by public transportation) yet remote, because of the water crossing involved, has had a strong appeal. Three ATBI-related research projects culminating in senior theses have been completed by Harvard undergraduates. Often working in full view of park visitors who are curious what those nets, cups or other odd-looking traps are for, students have an opportunity to share what they're learning with the public. Rangers stationed on some of the smaller camping islands for the summer have also been recruited to tend traps, especially the light traps which are set out at night. This has been beneficial both for the efficiency of collecting samples, and because the rangers can educate island visitors about the ATBI, showing them traps, explaining how they work, and even showing them what's being collected. From a ranger's perspective, the sheer abundance of insects and the diversity of habitats in which they're found – even on an urban island – can make them ideal subjects for interpretive activities.

Most kids love bugs, and know how to hunt for them, so the ATBI has harnessed their enthusiasm and talents too. Periodically, the park hosts family-oriented ATBI collecting events, which draw groups of 5 to 50-year olds to join a scientist for several hours of observing and collecting insects. Typically, families are motivated to come by their bug-crazy kids, and the parents assume they are there in the role of disinterested or, more commonly, squeamish chaperones. It is generally the case, however, that if parents are given collecting tools and vials along with their kids, they, too, quickly become absorbed in the hunt. Given the time and the focus, apathy and general distrust of insects can be replaced by curiosity and fascination as parents tap into the enthusiasm of their children. Time is spent at the end of the event examining the catch, sharing stories about where and how insects were caught, and learning some things about their natural history. Participants learn about the goals of the ATBI and how their contributions of specimens will ultimately become part of the project database.

The ATBI has also provided more structured opportunities for citizen scientists to get involved. For instance, an urban youth group already engaged in various field projects on the islands was trained to look for additional beach sites for the recently rediscovered '100-year beetle', *B. nigropiceum*. Butterflies and dragonflies have been surveyed by skilled amateurs recruited from local clubs. With a free ferry pass and a survey protocol to ensure consistent data collection, these volunteers have contributed valuable observational data and vouchers too. Citizen scientists have also been integral to the success of a pilot long-term native bee monitoring project in the park. In addition to assisting with field work that entails setting out transects of small pan traps ('bee bowls') on warm, sunny days, volunteers also have the opportunity to spend time in the lab, sorting and processing the catch. Just a few hours at the microscope opens up a whole new world of colour and texture and form, and brings the concept of native bee diversity – more than 170 species on just a few urban islands – far beyond the familiar sights of honeybees and bumblebees. As the general population, especially gardeners and naturalists, become more interested in native pollinator diversity (see Spevak, and Daniels, this volume), the bee monitoring project is also producing a fold-out field guide to bee genera of the Boston Harbor Islands.

As any entomologist knows, collecting insects is the easy part. A single pitfall trap left open for a week can bring in scores of specimens, a malaise trap can bring in thousands of insects, a fifth-grade field trip can bring in a good haul too. The effort involved in swinging a net and tending traps in the field pales in comparison to the subsequent hours in the lab spent sorting, pinning, gluing, labelling, identifying and databasing all those collected specimens. It requires a virtual bug factory to run such a process, and with an estimated 150 000 specimens collected from the islands, the ATBI has brought in close to 40 students, interns, retirees, park staff and volunteers over the course of 5 years to accomplish this Herculean task. High school students may be the only true 'bug nuts' who have entered the lab with some previous entomological interest or experience, while interns have had professional experience as diverse as museum design, zoo keeping and remote sensing. Students come from home-school programmes, inner city public schools, prestigious academies and colleges and universities. The park superintendent has spent a day washing, blow-drying and pinning bees, and two seasonal inter-pretive rangers have worked in the lab during their off-season. While few, if any, of these people will go on to be entomologists, spending hours at a microscope examining leafhoppers, springtails, bees and weevils at 50× magnification is always a transformative experience. Literally, they are seeing a world they didn't know existed. Best of all, their new appreciation and enthusiasm for insects makes students, citizens and park staff alike ideal ambassadors and advocates for the microwilderness.

THE RELEVANCE OF BIODIVERSITY DISCOVERY WITHIN
AND BEYOND PARK BOUNDARIES

In addition to the tangible benefits that an ATBI can provide by revealing rare, new, or potentially invasive species, biodiversity discovery efforts also provide parks with superb opportunities to achieve their dual priorities of conserving natural systems and *all* wildlife for future generations to enjoy and educating visitors about this biodiversity. Yet, despite the tremendous efforts any one project might put into cataloguing invertebrate diversity and involving the public in this process of discovery, the challenge of bringing awareness, appreciation, and, ultimately, concern for invertebrates and their role in the ecosystem up to a fraction of the level garnered by bison or wolves remains dauntingly large among park resource managers and educators. Ginsberg (1994) suggests that "when the major roles played by invertebrates in natural systems are as apparent to the general public as they are to entomologists, then invertebrates will get the attention they deserve in park management." At the very least, biodiversity discovery activities can begin to make the critical connections between scientists, the public, and park staff.

Ideally, fostering public support and appreciation for invertebrates and their roles in natural systems through efforts such as ATBIs and bioblitzes will contribute to an increased public awareness that transcends park boundaries. Especially in urban parks, where millions of people may live within just a few miles of park boundaries, invertebrate inventories can help get the point across that 'biodiversity' is not a phenomenon restricted to exotic tropical jungles, or defined by colourful birds and large mammals in wild, remote parks (Kerley *et al.* 2003). In fact, thousands of fascinating creatures live in urban neighbourhoods, and a field guide to the common beetles, snails, crickets and other invertebrates of the Boston Harbor Islands will likely be just as useful in a backyard, overgrown lot or conservation area anywhere in the city and its suburbs. Ultimately, forging more knowledgeable and intimate connections between the public and their local biodiversity may result in greater public support for the conservation and protection of invertebrates in both urban and natural landscapes.

REFERENCES

Fahrig, L. (2003) Effects of habitat fragmentation on biodiversity. *Annual Review of Ecology, Evolution, and Systematics*, **34**, 487–515.

Farrell, B. D. (2005) From agronomics to international relations: building an online encyclopedia of life in the Dominican Republic. *Revista: Harvard Review of Latin America*, Fall 2004/Winter 2005, 6–9.

Ginsberg, H. (1994) Conservation of invertebrates in US National Parks. *American Entomologist*, Summer, 76–78.

Hanski, I. (2011) Habitat loss, the dynamics of biodiversity, and a perspective on conservation. *Ambio*, **40**, 248–255.

Janzen, D. H. and Hallwachs, W. (1994) All Taxa Biodiversity Inventory (ATBI) of terrestrial systems: a generic protocol for preparing wildland biodiversity for non-damaging use. Report of a National Science Foundation Workshop, 16–18 April 1993, Philadelphia, PA.

Kellert, S. R. (1993) Values and perceptions of invertebrates. *Conservation Biology*, **7**, 845–855.

Kerley, G. I. H., Geach, B. G. S. and Vial, C. (2003) Jumbos or bust: do tourists' perceptions lead to an under-appreciation of biodiversity? *South African Journal of Wildlife Research*, **33**, 13–21.

Lemelin, R. H. (2007) Dragonology 101: understanding dragon-hunters and Odonata interactions in protected areas. In *Proceedings of the 2007 George Wright Society Conference on Parks, Protected Areas, and Cultural Sites. Rethinking Protected Areas In A Changing World*. Hancock, MI: The George Wright Society.

Procter, W. (1946) *Biological Survey of the Mount Desert Region*, Part VII. Philadelphia, PA: The Wistar Institute of Anatomy and Biology.

Wilson, E. O. (1987) The little things that run the world (the importance and conservation of invertebrates). *Conservation Biology*, **1**, 344–346.

20

Conclusion

RAYNALD HARVEY LEMELIN

Waring (2004: 123–124) once noted that 'perhaps the great majority, if not all of us, are born potential entomologists. Clearly, other interests may compete for our attention, but whether we develop our entomological interest may well depend on the level of encouragement or discouragement we receive, at home and elsewhere.' This statement supports many of the conclusions reached in this book and illustrates the overwhelming evidence that our bonds with nature are fostered through learning and experiential dimensions (Louv 2011; Simaika and Samways 2010).

Far from being a fringe activity enjoyed by a few eccentrics, these case studies suggest that millions of individuals participate in a wide range of activities involving insects: creating suitable habitats for insects through gardens, ponds and protected areas; attending festivals; visiting butterfly pavilions; and partaking in citizen science projects and outreach programmes. In addition to generating millions of dollars of revenues, these activities also generate awareness, and in some cases provide valuable contributions to scientific research. The fact that so many individuals actually seek out encounters with insects suggest that many of these animals do indeed have a certain allure or a 'je ne sais quoi' (see Dodd and Hall, this volume). In some cases, a lack of charisma or the negative sublime (the attraction of 'creepy-crawlers', the 'yuck' factor) has also been successfully implemented in various experiential strategies for children in parks and protected areas (see Lockwood and Rykken and Farrell, this volume). The fact that most of these encounters occur in cities in controlled settings (pavilions, parks and zoos), suggests that the urban milieus can also become locales of positive human–insect encounters (see Daniels, this volume).

This edited volume provides one of the first interdisciplinary and international examinations of insect–human interactions in leisure and tourism settings. The findings from this multi-authored work suggest that there are various social, cultural, economic and geographical forces encouraging and discouraging human–insect interactions. In some cases, large aposematic (brightly coloured) insects like butterflies, bees and beetles attract millions of visitors to insect exhibits, gardens, pollinator parks and to entomotourism (see Spivek and Veltman, this volume). In some cases, endemic species like the sorrel pygmy moth (*Enteucha acetosae*) of Scotland can become a flagship species for the preservation of sensitive habitats (see New and Shardlow, this volume). In other situations, invertebrates such as worms and dung beetles, which are perceived to have high utilitarian values but little aesthetic appeal, can become 'likeable' suggest Brown *et al.* (2003) and Seidl and Klepeis (2011). Programmes like those provided through the Roger Williams Park Zoo in Providence, Rhode Island, and the Woodland Park Zoo in Seattle, Washington, and books like *Extreme Insects* by Richard Jones (2010) have helped to increase our understanding of lesser-known insects through entertainment or edutainment.

Research has also demonstrated intra-order preferences where organisms from the same order – Lepidoptera (butterflies and moths), Hymenoptera (bees and wasp) and even the same family Coccinellidae (the Seven Spot Ladybug *Coccinella septempunctata* and the Asian Lady Beetle or Harlequin Ladybug *Harmonia axyridis*) – can be both awesome and awe-full (Lorimer 2007; Royal Entomological Society 2008). A tourist fascinated by a dung beetle during a Kenyan safari may not necessarily be fascinated by a beetle in his/her house. A butterfly collector may not think twice about exterminating an ant colony on his/her lawn. Thus, Norton (2000) was correct when he suggested that we should be cautious of dichotomies like entomophobia and entomophilia and seek new approaches that acknowledge the nuances and inconsistencies that constitute human values of nature in general and insects in particular.

As stated in the introduction, this book was written in order to illustrate human–insect interactions from historical to contemporary perspectives, and highlight the opportunities and contributions from the realms of leisure and tourism. It is an attempt to focus on the innovators, educators, proponents and activists, who through citizen

Figure 20.1   Stick insect and young insect enthusiast.

science, experiential education, collaborative research and dedication have brought insects from the recreational fringes to the forefront of conservation and leisure and tourism initiatives. Two research questions guided the analysis of this book: (1) what are the biggest challenges to human–insect interactions? (2) what techniques, strategies, and/or activities have insect proponents (researchers, amateur entomologists, managers) used to encourage human–insects encounters in leisure and tourism settings? Both of these are discussed next.

QUESTION I: WHAT ARE THE BIGGEST CHALLENGES REGARDING HUMAN-INSECT INTERACTIONS?

Apart from habitat loss, pollutants and a lack of conservation focus, the biggest challenges to human–insect interactions include the social constructions of anthropomorphism and entomophobia, the extinction of experience, a philosophical collecting schism and the belief that science is utilitarian and positivistic. Although some contributors and insect proponents may disagree, the following discussion is rooted in my own research and years of observation and experience as a participant in this field.

The social construction of anthropomorphism and entomophobia

The increasing disconnect between humans and nature has in some instances, been brought about through urbanization and technology, which have sometimes resulted in our increasing fear of the unknown (i.e., nature) and dislikes of certain animals (Louv 2011; Van Den Born *et al.* 2001). Anthropomorphism suggests that preferred animals are those perceived to be most human-like. While anthropomorphism may provide some insights into human affinities towards certain mammals, it does fail to address the popularity and appeal of many non-human animals like birds and fish. Furthermore, it overlooks the aesthetic appeal of insects such as butterflies, dragonflies and beetles, such as ladybugs and tiger beetles (Cormier *et al.* 2000; Knegtering *et al.* 2002; Pearson *et al.*, this volume).

Kellert (1993) and Hardy (1988) argue that these particular interests in insects are rather unusual, and that in most instances our relationship to insects can be best defined as one of apathy and anxiety. Other possible factors contributing to the entomophobic tendencies include social mores and a general tendency to lump all insects in the same pest category (Samways 2005). As stated in the introduction, 97–99% of insects are disproportionately represented by a very powerful pesticide industry aimed at controlling or eliminating insects, even those that are not considered pests (Baldwin *et al.* 2008; Russell 2001). The impacts from such social constructions have been pervasive, from the passive acceptance of these concepts by researchers, to management biases, where the potential appeal of insects is discounted or refuted by managers and officials tasked with the protection and conservation of these animals (Kerley *et al.* 2003). From this perspective, entomophobia and anthropomorphism actually deter our understanding of insects and result in further taxonomic bias and Arthropod Discourse Disorder (Clark and May 2002; Leather 2009; Lemelin 2012). Considering the size, appeal, diversity and economic worth of human–insect encounters, researchers would be best served to approach these 'grand narratives' of anthropomorphism and entomophobia with some scepticism (Lemelin 2007, 2012), and conduct further collaborative research examining these concepts from inter-disciplinary and multi-cultural perspectives. It is through such approaches that we will be able to acquire a greater understanding of the myriad of emotions that human encounters with insects generate.

The extinction of experience

Current global trends (particularly related to urbanization), argue New, Samways, Kawahara and Pyle, have resulted in an estrangement from nature, undermining the 'appreciation for biodiversity and removing sources of inspiration for potential naturalists. This effect was termed the 'extinction of experience' (Cheesman and Key 2007: 340) by Robert Pyle in 1993. The term was subsequently expanded and re-defined as nature deficit disorder by Louv in 2008. In essence, the argument proposes that that as urbanization increases, people increasingly become disconnected from nature.

> Predicted consequences include declining knowledge and appreciation of biodiversity, and declining support for its conservation amongst those who increasingly accept a biologically impoverished environment as the norm. This 'ratcheting down of expectations' is likely to be most pronounced between generations, as 'the environment encountered during childhood becomes the baseline against which environmental degradation is measured later in life (Cheesman and Key 2007: 322–323).

Dutcher *et al.* (2007) and Slimak and Dietz (2006) suggest that people less experienced with nature, as compared to individuals who are active in outdoor recreation, are less likely to support conservation efforts and engage in pro-environmental behaviour. However, historical studies (see Dodd, this volume) and more contemporary studies (see Lemelin 2012) have challenged the notion that declining interest in nature and insects is solely a result of urbanization, but instead suggest that urbanization has in some situations, increased or facilitated the interest in insects.

The Pesticide-Free movement in Canada in the early twenty-first century was essentially a national grassroots movement calling for the ban of cosmetic pesticide use in urban environments (Pralle 2006a, b; Sandberg and Foster 2005). The use of cosmetic pesticides in many Canadian cities and provinces since its beginning in the late 20th century, is now largely prohibited (Pralle 2006a, b). Apart from the potential health benefits, proponents believe that pesticide-free lawns and gardens could also provide the opportunity for birds and insects to return, and perhaps as Uhl (1998) stated, revitalize the interest of urbanites in natural environments.

The creation of pollinator parks and pesticide-free parks in Seattle and Portland suggest that comparable movements have also occurred in certain cities of the American Northwest (Shepherd *et al.* 2008). Butterfly gardens, dragonfly ponds, apiaries, insect gardens and garden biocontrol are some examples of landscape modification strategies aimed at increasing preferable insect habitats near one's domicile (Evans 2008; Tallamy 2009). Similarly, many infrastructures like insectariums and butterfly pavilions (see Veltman, this volume), butterfly gardens (see Daniels, this volume) and special events like festivals (see Hvenegaard *et al.*, this volume), many of which are held in urban or semi-urban environments thereby further challenging the myth that insect interests decline with urbanization. Much like any social phenomena, the answers to understanding the so-called extinction of experience and nature-deficit disorder is much more complex. While such analysis is beyond the scope of this chapter, it does warrant further attention.

Collecting

Many proponents of insect collecting have argued that much of the challenges vis-à-vis collecting lies with opponents of insect collections and agency protocols (i.e., permits), while the real issue of anthropogenic changes continues unabated (Kahawara and Pyle, this volume; Slone *et al.* 1997). While there is certainly some truth to this argument, the reality is that insect festivals and citizen science attracts participants from various socio-cultural and educational backgrounds (Hvenegaard *et al.*, this volume; Johansen and Auger, this volume). Therefore, activities which were once seen as the mainstay of entomological activities (insect collections), justified on both scientific and recreational grounds, are no longer accepted at face value and are often challenged and/or criticized (Lemelin 2012; Kiff 2005). To state that opponents of collecting are largely uninformed is to dismiss contemporary realities in North America and Europe where consumptive recreational activities like hunting and trapping are in decline, while others like egg-collecting or skin collections are largely prohibited without permits (Kiff 2005; Thomas *et al.* 2001). In my own studies of dragon-hunters (see Lemelin 2009), participants drew a distinction between individual collection and collecting specimens for scientific purposes. In all cases, people supported the latter and opposed the former; arguing that specimen collection for science was understood to be required, however, personal insect collections were often perceived as a deliberate intent to harm animals, and thus unnecessary. If we are serious about involving new participants in these activities, then the issue of intent to harm will require further attention, particularly if it potentially acts as a deterrent to the future involvement of these individuals in such activities.

Hollerback's (1996) article examined public perceptions of cruelty and the growth of entomology in England in the early nineteenth century; from a more contemporary perspective Pearson (this volume) suggests that the history of collecting versus non-collecting can quickly deteriorate into a pro- and anti-collecting schism. To paraphrase Pearson, not only do new initiates need to be shown that there is a time and place for collecting, but the old guard needs to change their preconceptions that a specimen is needed to document every event. This change in mentality, Pearson suggests, will require that in certain instances catch-and-release philosophies be implemented and that suitable technological replacements like digital images be incorporated into socio-entomological studies. The promotion of such new approaches combined with proper communication will help reduce tensions, and hopefully encourage more individuals to participate in non-consumptive insect leisure activities.

Science

The chapter on citizen science (see Johansen and Auger, this volume) suggests that scientific inquiry is largely positivistic or post-positivistic. Perkins (1979, 1982) and Castonguay (2004) suggest that a belief in the utilitarian paradigm and scientific method in entomology continues to provide the philosophical rationale to remain objective and professional. While there is a need for this type of research, such narrow definitions of inquiry have been criticized and challenged by various researchers (Bocking 2004; Fisher 2003; Harding 2010). It has been my observation that such approaches often fixate on nomenclature and taxonomy and discourage other ways of learning or knowing.

By believing in 'true and real science' and failing to acknowledge the various socio-political forces at play during research, we limit the numerous opportunities to incorporate local knowledge from fly-fishers and traditional knowledge by First Nation and Inuit communities in Canada and elsewhere (see Franklin and Ernst, Vinke *et al.*, this volume). To limit ourselves to one approach creates a great disservice to other modes of inquiry like qualitative research, complexity theory and fuzzy logic, and the potential contributions of various knowledge systems to insect education and conservation.

Experiential education and the negative sublime

Through story-telling, experiential activities and technology, the beauty and ferocity of insects have been captured in books such as *The Very Hungry Caterpillar* by Eric Carle (1981) and *How Mildred Became Famous* by James Huebing-Reitinger (2008), documentaries (*Life in the Undergrowth* 2006, produced by the BBC, *Insectia* 2005, produced by George Brossard), movies (see Richards 2011), paintings and photographs (see the recent grand prize winner for the nature category image entitled 'Splashing' National Geographic 2011). These mediums have in some cases fostered an awareness of insects, and in other cases generated an interest in these animals.

As the Blackawton *et al.* (2011) study illustrates, the chidren's drawing competition held as part of National Insect Week at the University Museum of Zoology, Cambridge (Snaddon and Turner 2007), both conducted in the UK, and the annual arthropod events held at Maryland College Park in the United States demonstrate, hands-on experiences represent an excellent opportunity for participants to engage and discover the wonderful world of insects. For these participants, experiential education provides a novel, memorable and exciting experience where insects can simultaneously thrill, chill and enthral. When invited to approach and touch the insects, students' reactions range from tentative to enthusiastic, but there is always tremendous curiosity and a steady stream of questions (See Vinke *et al.*, this volume; see Figure 20.2).

Various educational and outreach strategies like the native pollinator programmes at the University of Illinois and the St Louis Zoo and the All Taxa Biodiversity Inventory at the Boston Harbor Islands national park area, have also promoted the appeal of the negative sublime (See Rykken and Farrell, this volume). These examples illustrate that insects are particularly useful for experiential and applied learning; from a practical standpoint, they are one of the few types of animals that can be brought into a classroom; they are also readily encountered and captured outdoors with minimal equipment, in all types of urban, naturalized and undisturbed environments.

Figure 20.2   Youth interacting with insects in a garden.

Technology

In a radio interview with the Canadian Broadcast Corporation, Kevin Schawinski, the co-founder of GalaxyZoo and Zooniverse, discussed the importance of citizen science and information accessibility through an international medium like the web (see Mitchell, this volume). Originally devised as a project in astronomy, GalaxyZoo evolved into Zooniverse (CBC 2012), home to the internet's largest, most popular and most successful citizen science projects and covering topics such as space, climate, humanities and nature. All of these projects are maintained and developed by the Citizen Science Alliance; as of 21 December 2011, the community consisted of over 500 000 volunteers.

During the same broadcast, John Seely Brown, author of numerous books on radical innovations in learning, suggested that the internet, being one of the most powerful communication tools ever created, combined with new technological tools and interactive programs, can amplify curiosity, foster creativity, and cultivate learning (CBC 2012). These technological opportunities suggest Thomas and Seely Brown (2011) and Mitchell (this volume) have the ability to radically transform how we acquire and share knowledge, increasing collaboration and promoting new ways of learning and discovery (CBC 2012).

OMISSIONS AND RECOMMENDED FUTURE RESEARCH

Missing from this edited work are chapters dealing with other charismatic micro-fauna like ants, lady-bugs, silkmoths and mantises, and such activities as insect combats, gardening in general and biocontrol practices in gardening more specifically, travelling educational exhibitions and insects as pets. The exclusion of these animals or activities was not deliberate, for in some cases this was due to the contributing authors' preferences and space limitations. What these omissions do reveal is that much more work on human–insect encounters is required.

The use of pesticides, besides the discussion on the Pesticide Free Movement in Canada as an urban movement in this chapter, is largely absent. Indeed, the only mention of pesticides is by Yen *et al.* (this volume) who explain that not applying pesticides in the case of harvested and marketed orthopterans for entomophagy can yield more revenue to farmers than the sale of their millet, thereby, providing further incentives to reduce or eliminate the use of pesticides on these crops. The issues surrounding pesticide uses (Alaux *et al.* 2010; Benjamin and McCallum 2009; Cox-Foster and vanEngelsorp 2009; Schacker 2008; Tennekes 2010) and the role or lack thereof of entomologists and entomological societies in these issues (see Perkins 1979; Russell 2001) also requires more attention.

APPRECIATIVE INQUIRY, LEISURE, TOURISM AND INSECTS

The purpose of this book was to address the challenges by providing examples where innovations and pro-active management approaches have been adapted and incorporated in leisure and tourism initiatives involving insects. I also wanted to illustrate how individuals suffering from arachnophobia (see Pyle, this volume) and the fear of insects (see Lemelin, this volume) overcame these afflictions and have gone on to publish books and articles, and even advocate for insects. Last year, my wife and son invited several monarch caterpillars into our home and together we successfully raised and released 14 monarch butterflies back into our garden (Figure 20.3), with the hope that future generations will find a haven in our backyard and throughout North America.

The corollary of Appreciative Inquiry, a positive research methodology, suggests that by focusing on the negative one finds the negative, but focusing on the positive can yield important information to make positive changes (Cooperrider and Whitney 2005; Koster and

Figure 20.3   Raising monarch butterflies for live releases.

Lemelin 2009). Nowhere is this more evident than in research pertaining to insects where negative questions regarding these animals beget negative answers. Instead of focusing on the negative, I decided to focus on the positives, to initiate, in the words of Cooperrider and Whitney (2005) a generative spiral, where:

- instead of blaming and condemning, successes are celebrated and opportunities are examined;
- instead of condemning urbanization, opportunities for conservation and citizen science provided through urbanization are examined;
- instead of critiquing anti-collecting sentiments, opportunities to engage these individuals are promoted;
- instead of accepting anthropomorphism and entomophobia at face value, these concepts are challenged and alternatives provided;
- instead of condemning technology, technology should be implemented to engage and inform;
- instead of dismissing leisure and tourism as frivolous activities, its various contributions to education, awareness, conservation, and well-being should be promoted.

According to Louv (2011), recognizing and respecting the role of insects (and other animals) in ecosystems is fundamental to ending nature-deficit disorder, restoring humanity in the natural world, and initiating reconciliation ecology. As this chapter suggests, leisure and tourism provide crucial opportunities to educate and engage all sectors of society regarding the importance of insects. Therefore as important as experiential education, festivals, and citizen science projects are, all of these events are structured experiences, and we should not overlook the benefits of unstructured time where exploration of backyards, gardens, ponds and parks, complemented in some cases with interactive technologies provided through tablets and smart phones, can help us re-create, re-connect and re-establish our links with nature and animals. Technological sceptics should be comforted with the fact that while guides may be replaced by tablets and smart phones, nets and various storage devices will always be required to capture, store, document and eventually release these animals. The curiosity and openness of children to insects can be cultivated through leisure and tourism opportunities, creating a positive cycle of change whereby insects are recognized for their diversity, adaptability and marvels, and become worthy of our respect and our tolerance.

REFERENCES

Alaux, C., Brunet, J. L., Dussaubat, C. *et al.* (2010) Interactions between *Nosema* microspores and a neonicotinoid weaken honeybees (*Apis mellifera*). *Environmental Microbiology*, **12**, 774–782.

Baldwin, R. W., Koehler, P. G., Pereira, R. M. and Oi, F. M. (2008) Public perceptions of pest problems. *American Entomologist*, **54**(2), 73–79.

Benjamin, A., & McCallum. (2009). *A World Without Bees*. Pegasus Books.

Blackawton, P.S., Airzee, S., Allen, A. *et al.* (2011) Blackawton bees. *Biology Letters*, **7**, 168–172.

Bocking, S. (2004) *Nature's Experts: Science, Politics, and The Environment*. Piscataway, NJ: Rutgers University Press.

British Broadcasting Corporation (BBC) (2006) *Life in the Undergrowth*. Produced by the BBC, London.

Brossard, G. (2005) *Insectia*. Produced by Imavision.

Brown, G. G., Feller, C., Blanchard, E., Deleporte, P. and Chernyanskii, S. S. (2003) With Darwin, earthworms turn intelligent and become human friends. *Pedobiologia*, **47**, 924–933.

Canadian Broadcasting Corporation (CBC) (2012) *ReCivilization: Episode Two: Open-Source Knowledge*. Available at: http://www.cbc.ca/video/news/audioplayer.html?clipid=2190495701, accessed 10 February 2012.

Carle, E. (1981) *The Very Hungry Caterpillar*. New York: Philomel Books.

Castonguay, S. (2004) *Protection des cultures, construction de la nature. Agriculture, foresterie et entomologie au Canada 1884–1995*. Sillery, Canada: Septentrion.

Cheesman, O. D. and Key, R. S. (2007) The extinction of experience: a threat to insect conservation? In *Insect Conservation Biology*, ed. A. J. A. Stewart, T. R. New, O. T. Lewis. Wallingford, UK: CABI, pp. 322–325.

Clark, J. A. and May, R. M. (2002) Taxonomic bias in conservation research. *Science*, **297**, 191–192.

Cooperrider, D. L. and Whitney, D. (2005) *Appreciative Inquiry: A Positive Revolution in Change*. San Francisco, CA: Berrett-Koehler.

Cormier, C. M., Forbes, T. A., Jones, T. A., Morrison, R. D. and McCorquodale, D. B. (2000) Alien invasion: the status of non–native lady beetles (Coleoptera: Coccinellidae) in industrial Cape Breton, Nova Scotia. *Northeastern Naturalist*, **7**, 241–247.

Cox-Foster, D. and vanEngelsorp, D. (2009) Saving the honeybee. *Scientific American*, **300**(4), 40–45.

Dutcher, D. D., Finley, J. C., Luloff, A. E. and Johnson, J. B. (2007) Connectivity with nature as a measure of environmental values. *Environment and Behavior*, **39**, 474–493.

Evans, A.V. (2008) *What's Bugging You? A Fond Look at the Animals we Love to Hate*. Charlottesville, VA: University of Virginia Press.

Fisher, F. (2003) *Citizens, Experts, and the Environment: The Politics of Local Knowledge*. Durham, NC: Duke University Press.

Harding, S. (2010) *Whose Science? Whose Knowledge? Thinking from Women's Lives*. Ithaca, NY: Cornell University Press.

Hardy, T. (1988) Entomophobia: the case for Miss Muffet. *Bulletin of the Entomological Society of America*, **34**, 64–69.

Hollerback, A. L. (1996) Of sangfroid and sphinx moths: cruelty, public relations, and the growth of entomology in England, 1800–1840. *Osiris*, **11**, 201–220.

Huebing-Reitinger, J. (2008) How Mildred became famous. Project InSECT LLC.

Jones, R. (2010) *Extreme Insects*. London: HarperCollins Publisher.

Kellert, S. R. (1993) Values and perceptions of invertebrates. *Conservation Biology*, **7**, 845–855.

Kerley, G. I. H., Geach, B. G. S. and Vial, C. (2003) Jumbo or bust: do tourists' perceptions lead to an under-appreciation of biodiversity? *South African Journal of Wildlife Research*, **33**(1), 13–21.

Kiff, L. F. (2005). History, present status, and future prospects of avian eggshell collections in North America. *The Auk*, **122**(3), 994–999.

Knegtering, E., Hendrickx, L., Windt, H. J. and Shoot Uiterkamp, A. J. M. (2002) Effects of species' characteristics on nongovernmental organizations' attitude toward species conservation policy. *Environment and Behavior*, **34**, 378–400.

Koster, R. and Lemelin, R. H. (2009). Appreciative inquiry and rural tourism: a case study from Canada. *Tourism Geographies*, **11**(2), 259–270.

Leather, S. R. (2009) Taxonomic chauvinism threatens entomology. *Biologist*, **56**, 10–13.

Lemelin, R. H. (2007) Finding beauty in the dragon: the role of dragonflies in recreation, tourism, and conservation. *Journal of Ecotourism*, **6**(2), 139–145.

Lemelin, R. H. (2009) Goodwill hunting? Dragon hunters, dragonflies and leisure. *Current Issues in Tourism*,**12**(3), 235–253.

Lemelin, R. H. (2012) To bee or not to bee: whether 'tis nobler to revere or to revile those six-legged creatures during one's leisure. *Leisure Studies*, doi:10 .1080/02614367.2011.626064.

Lorimer, J. (2007) Nonhuman charisma. *Environment and Planning Development: Society and Space*, **25**(5), 911–932.

Louv, R. (2008). *Last Child in the Woods: Saving our Children From Nature-deficit Disorder*. Chapel Hill: Algonquin Paperbacks.

Louv, R. (2011) *The Nature Principle*. Chapel Hill, NC: Algonquin Books.

National Geographic (2011) National Geographic announces winners of global photography contest. Available at: http://photoblog.msnbc.msn.com/_news/2011/12/19/9566460-national-geographic-announces-winners-of-global-photography-contest#.Tu_gz_lQtZo.twitter, accessed 19 February 2012.

Norton, B. G. (2000). Biodiversity and environmental values: in search of a universal earth ethic. *Biodiversity and Conservation*, **9**(8), 1029–1044.

Perkins, J. H. (1979). The quest for innovation in agricultural entomology, 1945–1978. In *Pest Control: Cultural and Environmental Aspects*, ed. D. Pimentel and J. H. Perkins. AAAS Selected Symposia Series. Boulder, CO: Westview Press, pp. 23–71.

Perkins, J. H. (1982) *Insects, Experts, and the Insecticide Crisis: The Quest for New Pest Management Strategies*. New York: Plenum Press

Pralle, S. B. (2006a) Timing and sequence in agenda-setting and policy change: a comparative study of lawn care pesticide politics in Canada and the US. *Journal of European Public Policy*, **13**(7), 987–1005.

Pralle, S. (2006b) The 'mouse that roared': agenda setting in Canadian pesticides politics. *Policy Studies Journal*, **34**(2), 171–194.

Richards, L. (2011) Insect film invertebrate wins award in France. *Evening Chronicle*. Available at: http://www.chroniclelive.co.uk/north-east-news/evening-chronicle-news/2011/04/08/wylam-film-maker-crawls-over-the-competition-72703-28484555/, accessed 1 June 2011.

Royal Entomological Society (2008) National Insect Week: the good, the bad, the beautiful and the ugly. The survey results. Available at: http://www.nationalinsectweek.co.uk/gbbu.php.

Russell, E. (2001) *War and Nature: Fighting Insects with Chemicals from World War I to Silent Spring*. Cambridge: Cambridge University Press.

Samways, J. M. (2005) *Insect Diversity Conservation*. Cambridge: Cambridge University Press.

Sandberg, A. L. and Foster, J. (2005) Challenging lawn and order: environmental discourse and lawn care reform in Canada. *Environmental Politics*, **14**(4), 478–494.

Schacker, M. (2008) *A Spring Without Bees: How Colony Collapse Disorder has Endangered our Food Supply*. Guilford, CT: The Lyons Press.

Seidl, D. E. and Klepeis, P. (2011). Human dimensions of earthworm invasion in the Adirondack State Park. *Human Ecology*, **39**, 641–655.

Shepherd, M., Vaughn, M. and Hoffman Black, S. (2008) *Pollinator-friendly Parks. How to Enhance Parks, Gardens and other Greenspaces for Native Pollinators*. Seattle, WA: The Xerces Society for Invertebrate Conservation.

Simaika, J. P. and Samways, M. J. (2010) Biophilia as a universal ethic for conserving biodiversity. *Conservation Biology*, **24**(3), 903–906.

Slimak, M. W. and Dietz, T. (2006) Personal values, beliefs, and ecological risk perception. *Risk Analysis*, **26**(6), 1689–1705.

Slone, T. H., Orsak, L. J. and Malver, O. (1997) A comparison of price, rarity and cost of butterfly specimens: implications for the insect trade and for habitat conservation.' *Ecological Economics*, **21**(1), 77–85.

Snaddon, J. L. and Turner, E. C. (2007) A child's eye view of the insect world: perceptions of insect diversity. *Environmental Conservation*, **34**(1), 33–35.

Tallamy, D. W. (2009) *Bringing Nature Home: How Native Plants Sustain Wildlife in our Gardens*. Portland, OR: Timber Press.

Tennekes, H. (2010) *The Systemic Insecticides: A Disaster in the Making.* Zutphen, The Netherlands: Weevers Walburg Communicatie.

Thomas, D. and Seely Brown, J. (2011) *A New Culture of Learning.* Durham, NC: Createspace.

Thomas, M., Elliott, G. and Gregory, R. (2001) The impact of egg collecting on scarce breeding birds 1982–1999. *RSPB Conservation Review*, **13**, 39–44.

Uhl, P. (1998) Conservation biology in our front yard. *Conservation Biology*, **12**(6), 1175–1177.

Van Den Born, R. J. G., Lenders, R. H. J., de Groot, W. T. and Huijsman, E. (2001) The new biophilia: an exploration of visions of nature in Western countries. *Environmental Conservation*, **28**, 65–75.

Waring, P. (2004) Entomologists: born or made? *Entomologist's Record*, **116**, 122–124.